AF684663

Lecture Notes in Geoinformation and Cartography

Series Editors: William Cartwright, Georg Gartner, Liqiu Meng, Michael P. Peterson

Jonathan Li • Sisi Zlatanova •
Andrea Fabbri (Eds.)

Geomatics Solutions for Disaster Management

With 186 Figures

Springer

Editors:

Associate Prof. Jonathan Li
Department of Geography
University of Waterloo
200 University Avenue West
Waterloo, Ontario, N2L 3G1
Canada
Email: junli@uwaterloo.ca

Assistant Prof. Sisi Zlatanova
OTB Research Institute for Housing,
Urban and Mobility Studies
Delft University of Technology
Jaffalaan 9, 2628 BX Delft
The Netherlands
Email: s.zlatanova@tudelft.nl

Prof. Andrea G. Fabbri
Spatial Information Laboratory
SPINlab
Institute of Environmental Studies
(IVM)
Free University
De Boelelaan 1105
1081 HV Amsterdam
The Netherlands
Email: andrea.fabbri@ivm.vu.nl

ISSN 1863-2246
ISBN 10 3-540-72106-1 **Springer Berlin Heidelberg New York**
ISBN 13 978-3-540-72106-2 **Springer Berlin Heidelberg New York**

Library of Congress Control Number: 2007924820

This work is subject to copyright. All rights are reserved, whether the whole or part of the material is concerned, specifically the rights of translation, reprinting, reuse of illustrations, recitation, broadcasting, reproduction on microfilm or in any other way, and storage in data banks. Duplication of this publication or parts thereof is permitted only under the provisions of the German Copyright Law of September 9, 1965, in its current version, and permission for use must always be obtained from Springer-Verlag. Violations are liable to prosecution under the German Copyright Law.

Springer is a part of Springer Science+Business Media
springeronline.com
© Springer-Verlag Berlin Heidelberg 2007

The use of general descriptive names, registered names, trademarks, etc. in this publication does not imply, even in the absence of a specific statement, that such names are exempt from the relevant protective laws and regulations and therefore free for general use.

Cover design: deblik, Berlin
Production: A. Oelschläger
Typesetting: Camera-ready by the Editors
Printed on acid-free paper 30/2132/AO 54321

Preface

During the last few years, disasters stemming from natural and technological sources have affected many societies with increasing frequency and violence. The number of victims worldwide runs in to hundred thousands or millions, in addition to enormous socio-economic losses. Hazard prediction and risk avoidance/mitigation are in critical need of research improvements and developments to curb damages and protect the lives in both the developing and the developed world.

Anticipating the obvious and growing importance of geo-information for disaster and risk management we have initiated a worldwide discussion concerning the collection, management, analysis, sharing and visualization of geo-information by organizing the 1st International Symposium on Geoinformation for Disaster Management (Gi4DM2005) in Delft, The Netherlands, March 21-23, 2005 and the 2nd Symposium (Gi4DM2006) in Goa, India, September 25-26, 2006. The fundamental goal of the symposia was to begin a dialogue on disaster management problems in their entirety by considering geospatial technology applicable to disaster management, user requirements for spatial data and standards.

This conference (Gi4DM2007) concentrates on geomatics technologies (satellite positioning, remote sensing, GIS, geodetic and hydrographic surveying). Lessons learned in the last few years have clearly indicated that the prediction and efficient monitoring of disasters represent two critical factors in the decision-making process, and that space-based technologies have great potential in meeting such needs, in a fast, near-real-time fashion.

Global navigation satellites and Earth observation satellites have already demonstrated their flexibility in providing data for a broad range of applications: weather forecasting, vehicle tracking, disaster alerting, forest fire and flood monitoring, oil spill detection, desertification spread monitoring, and crop and forestry damage assessment. Monitoring and management of recent natural disasters have also benefited from satellite imagery, such as the Indian Ocean tsunamis in 2004, floods (Austria, Romania, Switzerland, and Germany in 2005), hurricanes (USA in 2005), forest fires (Portugal, France in 2005), earthquakes (Pakistan in 2005, Indonesia in 2006), etc. Charters and international organizations have already launched various initiatives on the extended utilization of satellite positioning and remote sensing technologies in disaster monitoring and management.

Effective utilization of satellite positioning and remote sensing in disaster monitoring and management requires research and development in numerous areas: data collection, access and delivery, information extraction and analysis, management and their integration with other data sources (airborne and terrestrial imagery, GIS data, etc.), data standardization, organizational and legal aspects of sharing remote sensing information.

Goal and Objectives

The fundamental goal of Gi4DM 2007 is to use the latest state-of-the-art space-based geomatics technologies to understand dynamic Earth processes and geo-hazards. This will be accomplished through the use of geo-information for risk prevention, reduction, mitigation, monitoring and response to disasters, including:

- Geomatics technologies and advances
- Geospatial information
- Geodesy and Earth processes

The objectives of the Gi4DM2007 are

- To establish the role of satellite positioning and remote sensing in disaster monitoring and prediction
- To examine existing tools, software, and geospatial data sources, organizational structures and methods for work in crisis situations.
- To outline limitations of their current use, discovery, integration and exchange of remote sensing data and geo-information.
- To explore the future of risk prevention and disaster management and make suggestions for future research directions.

Conference Topics

This conference will address the following aspects:

- State-of-the-art geomatics technologies for risk prevention, reduction, monitoring and response to disasters.
- Methods and software tools for real-time data collection, processing and information extraction for disaster monitoring and damage assessment.
- Web-based information services based on satellite imagery and GIS.

- Spatial decision support systems/tools for decision-making.
- Methods and software tools for data management, data integration and data analyses with emphasis on multi-source data integration.
- Dynamic Earth processes and geo-hazards, climate change and meteorological hazards, anthropogenic activities and environmental hazards.
- Legal aspects, international and national activities in risk prevention and disaster management.
- Methods and systems for risk and disaster mapping, management of risk maps and multi-user sharing.
- Link between risk prevention and crisis response.
- A 2030 futures panel will develop scenarios for disaster management and implications of converging technologies (bio, nano and geo-info) and adaptive management systems.

Paper Selection Process

Nearly 140 abstracts were submitted to the Joint CIG/ISPRS Conference on Geomatics for Disaster & Risk Management, May 23-25, 2007, Toronto, Canada. After the reviewing process by the editors, the authors of a total of 46 abstracts were invited to submit their complete papers. After a peer view process, 27 full papers are finally included in this volume

Acknowledgements

The editors of this volume would like to thank many people who contributed to it. The authors and reviewers gave generously of their time to produce a state-of-the-art analysis and evaluation of the applications of remote sensing and geospatial information technologies in risk and disaster management. All peer reviewers, Rifaat Abdalla (York University, Canada), Piero Boccardo (Politecnico di Torino, Italy), Eric Chang (Ryerson University, Canada), Michael A. Chapman (Ryerson University, Canada), Zhi Chen (Concordia University, Canada), Suzana Dragicevic (Simon Fraser University, Canada), Andrea Fabbri (Free University Amsterdam, The Netherlands), Karean Fabbri (ICT for Environment, European Commission), Verda Kocabas (Simon Fraser University), Jonathan Li (University of Waterloo, Canada), Songnian Li (Ryerson University, Canada), Yu Li, (University of Waterloo, Canada), Steve H.L. Liang (York University, Canada), Jie Shang (Purdue University, USA), and Sisi Zlatanova (Delft University, The Netherlands) who helped in evaluating the many manuscripts with assigned a short time interval. The assistance from Xiangyu Si, a PhD student at the Department of Geography, University of Waterloo is highly appreciated in handling the communication with authors and reviewers during the peer review and revision process as well as some editing work. Our special thanks are for the contributions of the entire CIG headquarter staff, and especially Lucie Lebrun-Ginn, for her efforts in handling the collection of abstracts. Last but not least, we wish to acknowledge the supporting organizations and institutions ISPRS, UNOOSA, ICA, FIG, JBGIS, EC 'ICT for the environment' and OGC.

Jonathan Li, Sisi Zlatanova and Andrea Fabbri
May, 2007

Table of Contents

Author Index

Programme Commitee

Co-Conference Chairs
Ahmed El-Rabbany, Ryerson University, Canada (Chair, CIG Toronto Branch)
Jonathan Li, University of Waterloo, Canada (Co-Chair, ISPRS WG IV/8)

Chair, Technical Program Committee
Songnian Li, Canada (Chair, ISPRS WG IV/5)

Co-Chairs
Sisi Zlatanova, Delft University of Technology, The Netherlands (Chair, ISPRS WG IV/8)
David Stevens, UN (Office for Outer Space Affairs)
Karen P. Fabbri, EC (ICT for Environment)

Members
Costas Armenakis, Geomatics Canada
Sunil Bisnath, York University, Canada
Piero Boccardo, Politecnico di Torino (Chair, ISPRS WG VIII/2), Italy
Alexander Braun, University of Calgary, Canada
Katarzyna Dabrowski-Zielinska, Institute of Geodesy & Cartography (Co-Chair, ISPRS WG VII/6), Poland
Suzana Dragicevic, Simon Fraser University (Secretary, ISPRS WG IV/5), Canada
Andrea G. Fabbri, Free University Amsterdam (Secretary, ISPRS IV/8), Netherlands
Jianya Gong, Wuhan University (Chair, ISPRS WG VII/5), China
Alain Grignon, Geological Survey of Canada
Ayman Habib, University of Calgary, Canada
Bo Huang, Chinese University of Hong Kong, China
Guangyun Li, University of Information Engineering, China
Darka Mioc, University of New Brunswick, Canada
Jacynthe Pouliot, Laval University, Canada
Raad Saleh, University of Wisconsin at Madison (Co-Chair, ISPRS WG I/3), USA

Jie Shan, Purdue University (Co-Chair, ISPRS WG VII/6), USA
Michael Sideris, University of Calgary, Canada
Julian Smit, University of Cape Town, South Africa
Vincent Tao, MicroSoft Virtual Earth (Chair, ISPRS WG I/3), USA
Jianguo Wang, York University, Canada
Qing Zhang, Changan University, China
Bert Veenendaal, Curtin University of Technology (Co-Chair, ISPRS WG IV/5), Australia

Chair, Local Organizing Committee
Georgia Fotopoulos, University of Toronto, Canada (IAG)

Co-Chair
Maureen Mountjoy, Canada (AOLS)

An Online Colour 2D and 3D Image System for Disaster Management

Yun Zhang, Pingping Xie and Hui Li

Department of Geodesy and Geomatics Engineering, University of New Brunswick, Fredericton, New Brunswick, Canada E3B 5A3
yunzhang@unb.ca

Abstract

This paper presents a new automatic system for fast generation of multi-scale colour 2D and 3D satellite images and for online dynamic visualization of the 2D and 3D information of the areas of interest. Medium-resolution satellite images such as Landsat 7 and high-resolution satellite images such as Ikonos or QuickBird are the main data sources for the multi-scale 2D and 3D images. Since Landsat imagery has a global coverage and the Ikonos and QuickBird images can be acquired quickly for the areas of interest, the generation and online visualization of global colour 2D and 3D satellite images at different scales is possible. Ground objects from mountain ranges, such as Rocky Mountains, to individual family houses and trees can be dynamically visualized and analyzed in 2D and 3D through the Internet. The system presents a great potential for fast and effective visualizing, monitoring, and analyzing disaster situations in 2D and 3D within a short time period, which can provide decision makers with important information for emergency response and disaster management. The concept of the 3D satellite image generation and online dynamic visualization are presented in this paper. Some examples on the potential of using online 3D for disaster management are given.

1 Introduction

Fast data acquisition, visualization and analysis are essential factors for disaster management. Online mapping technologies have proven to be effective tools for fast communication, visualization and analysis of disaster information among a broad community, which is important for informed decision making in an emergency situation. For many natural disasters, such as flood, land slide and forest fire, 3D information plays a crucial role for an effective management or response.

Currently, online mapping technologies can effectively transfer and visualize 2D images of the disaster areas, making the analysis of disaster situations faster and easier than using traditional hard copy images omaps. However, current online 3D mapping technologies are still mainly based on the generation of 3D models, which is time consuming and, hence, not suitable for a fast response. For example, Google Earth (http://earth.google.com/) and Microsoft Virtual Earth (http://maps.live.com/) both utilize the technology of draping satellite or aerial images on top of digital elevation models (DEMs) for 3D visualization of mountains or hills, and draping images on manually generated 3D building models for 3D visualization of urban areas. The technology is capable of visualizing small scale 3D information in a cost effective fashion. But, it is not possible to generate detailed 3D models to represent the real 3D environment of disaster areas within a short time period and at an affordable cost. Therefore, it is impossible to provide reliable, large scale 3D information for detailed 3D analysis of the disaster areas, when the current 3D online technology is applied.

To date, a global coverage of Landsat images and DEMs has been made available at affordable costs, even free of charge for some areas. High resolution satellite stereo images from QuickBird or Ikonos can be acquired on demand for the areas of interest. Based on these major data sources (Landsat, global DEMs, and QuichBird or Ikonos stereo), it is theoretically possible to build a multi-scale global colour 2D and 3D image database. Landsat images and DEMs can be utilized to produce large coverage, low resolution 3D images, while stereo images from QuickBird or Ikonos can be used to produce high resolution 3D images for detailed 3D visualization and analysis of emergency areas.

If a new technology can be developed to automatically generate natural colour 3D satellite images at different scales and dynamically visualize the colour 3D images online, the impact of satellite remote sensing on disaster monitoring and management will be significantly increased. Such a technology has also a great potential to upgrade or complement the state-of-the-art 2D online mapping technologies and 3D model based technologies of Goolge Earth and Microsoft Virtual Earth.

This paper presents a new automatic system for fast generation of multi-scale colour 3D satellite images and for online dynamic visualization of the 2D and 3D information of the areas of interest. Medium-resolution satellite images such as Landsat 7 and high-resolution satellite images such as Ikonos or QuickBird are the main data sources for the multi-scale 2D and 3D images. Through the system, objects on the Earth's surface from mountains to individual family houses and trees can be dynamically visualized and analyzed in 2D and 3D on the Internet. It presents a great potential for fast and effective visualizing, monitoring, and analyzing disaster situations in 2D and 3D within a short time period, providing important information for emergency response and disaster management.

The concept of the 3D satellite image generation and the online dynamic visualization are presented below. Some examples on the potential of using the online 3D for disaster management are given.

2 Proposed 3D System

Based on the general concept of Zhang (2002), an automatic system has been developed to automatically generate colour 3D images and dynamically visualize natural colour 3D images on the Web.

Figure 1 illustrates the major components of the system for the generation of medium resolution Landsat colour 3D images and high resolution colour 3D images using QuickBird or Ikonos stereo pairs, as well as for the online visualization of the 3D images. Because Landsat can not collect stereo images, the medium resolution 3D images are reconstructed using monoscopic Landsat images and the corresponding DEMs, while the high resolution 3D images are constructed using stereoscopic QuickBird or Ikonos images.

From Fig. 1 it can be seen that the system consists of seven major components:

1. automatic image fusion to
   - produce 15 m pan-sharpened medium resolution Landsat natural colour images, and
   - produce 0.7 m pan-sharpened high resolution QuickBird (or 1 m Ikonos) natural colour image pairs;
2. automatic reconstruction of medium resolution stereo images using Landsat colour images and DEMs (because Landsat can not collect stereo images);

3. automatic colour enhancement for better colour 3D visualization;
4. automatic image matching to find corresponding matching points in the stereo pairs;
5. automatic epipolar image generation and colour 3D image generation;
6. fast data transfer through the Internet for effective visualization of large volume of 3D images; and
7. automatic parallax refinement for effective colour 3D viewing on the client end.

The general concept and results of the individual components are described in the following sections.

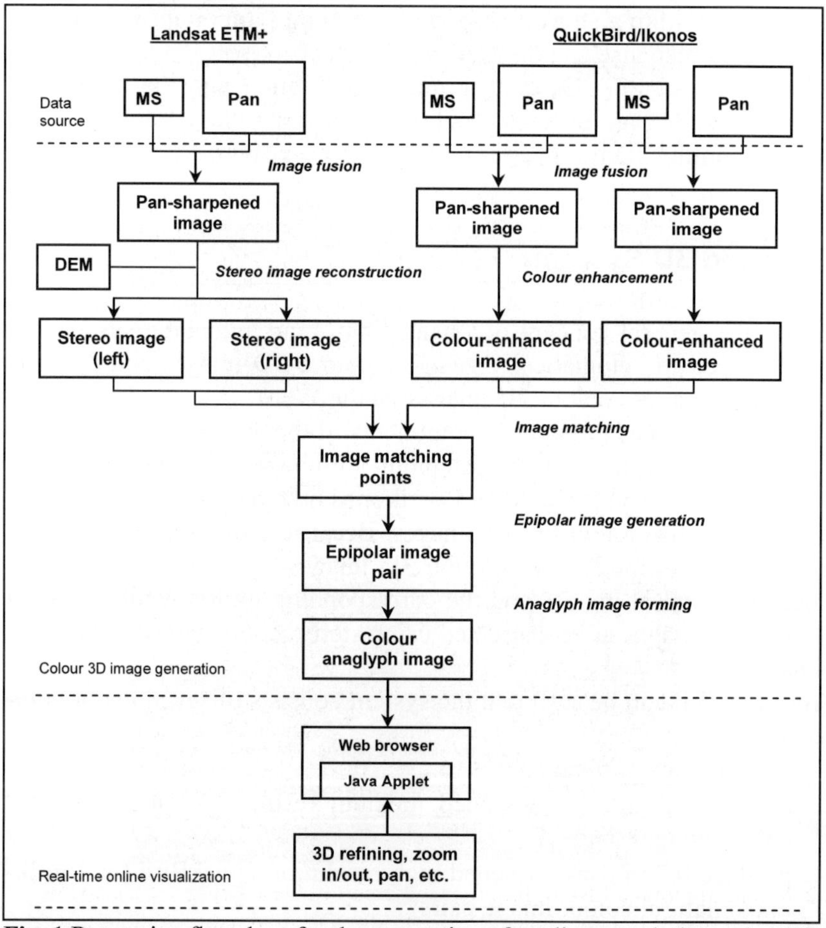

Fig. 1 Processing flowchart for the generation of medium resolution colour 3D images from monoscopic Landsat images and high resolution colour 3D images from QuickBird/Ikonos image pairs, and the online visualization

2.1 Automatic image fusion

A new automatic image fusion approach has been developed (Zhang 2002, 2004) and integrated into the automatic system for colour 3D image generation.

Fig. 2 Automatic image fusion, Ikonos, University of New Brunswick, Canada. Top: Original panchromatic image (1 m); Middle: Original natural colour image (4 m); Bottom: Fused natural colour image (1 m).

The fusion technique can automatically fuse panchromatic and multispectral images of Landsat, QuickBird, Ikonos or any other sensors through a one-step process, without losing spatial detail or altering colour information. Fig. 2 shows an example of the images before and after the image fusion using the fusion technique developed by Zhang (2002).

2.2 Automatic stereo image reconstruction

To produce medium resolution colour 3D satellite images for global 3D visualization, Landsat imagery is an ideal data course because of its high image quality, global availability, and low cost. However, Landsat series does not collect stereo images, so that it is not possible to generate colour 3D images directly using Landsat imagery. An alternative solution was found and an automatic software tool was developed which can reconstruct Lansdat colour 3D images using available Landsat monoscopic images and corresponding DEMs. Fig. 3 shows a Landsat image before and after 3D reconstruction.

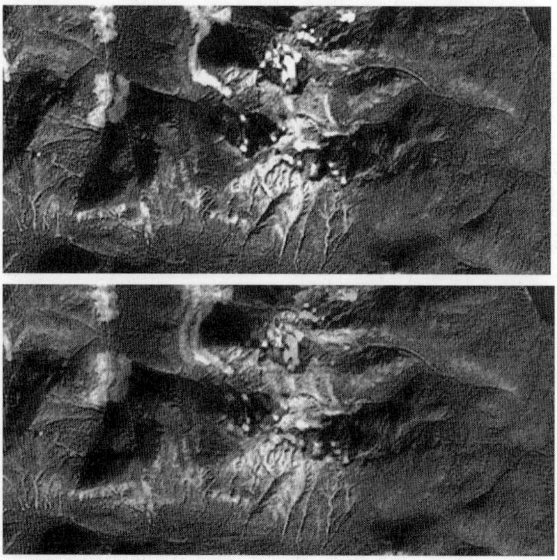

Fig. 3 Reconstruction of Landsat stereo imagery. Top: 2D colour imagery collected by Landsat; Bottom: Reconstructed colour 3D imagery from the 2D image above and a corresponding DEM.

2.3 Automatic colour enhancement

Compared to colour aerial photos, the colour of the natural colour QuickBird or Ikonos images usually appears somewhat unnatural or distorted (Fig. 4, top and middle). This unnatural colour also affects the 3D effect when a colour 3D image is generated. A colour enhancement algorithm has, therefore, been developed and integrated into the system for automatic colour 3D image generation to improve the colour 3D visualization effect. The bottom image of Fig. 4 shows an example of a colour enhanced Ikonos image.

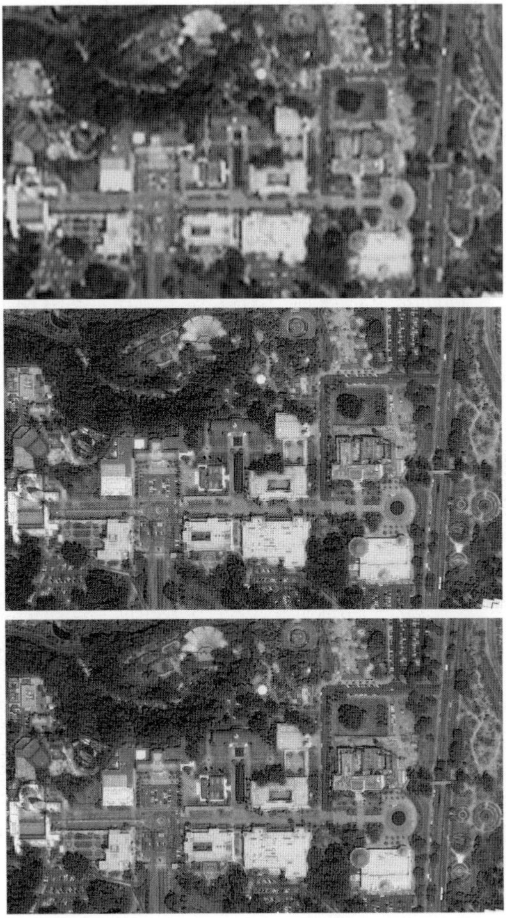

Fig. 4 Colour enhancement of Ikonos natural colour image. Top: original Ikonos colour image; Middle: pan-sharpened Ikonos colour image; Bottom: pan-sharpened and colour enhanced Ikonos image.

2.4 Automatic image matching

The purpose of image matching is to find corresponding points in the stereo image pairs to automatically form colour 3D images.

An automatic image matching technique has been developed which integrates both feature-based and area-based matching approaches. The overview of the matching algorithm is illustrated in Fig. 5. In the matching process, an epipolar constraint is applied to eliminate the outliers. An image tiling technique is also employed to allow an effective process of big images piece by piece.

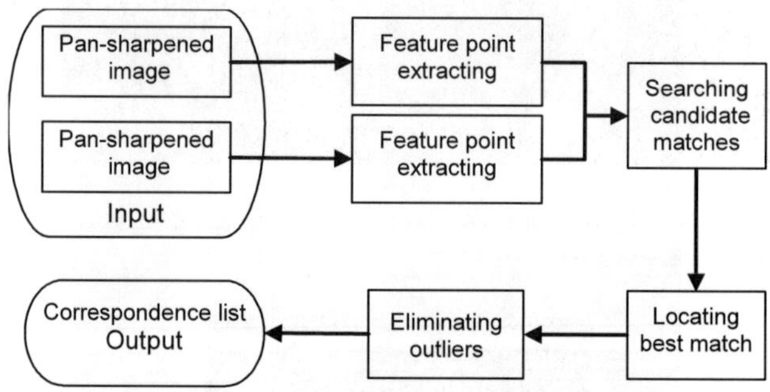

Fig. 5 Overview of the image matching process involved in the colour 3D system

2.5 Epipolar image resampling and colour 3D image generation

According to corresponding matching points found from the image matching step and the epipolar geometry of the stereo pair from QuickBird or Ikonos, an epipolar image resampling was applied as follows:

1. Carrying out the affine transformation from the right image to the left image.
2. Rotating both the left and the transformed right images by the angles of the epipolar line.

Fig. 6 shows a pair of Ikonos images before and after epipolar resampling. After spipolar images are generated, colour 3D (or anaglyph) images can then be generaten by superimposing one colour band from one image with two colour bands from another image of the epipolar stereopair. The criterion of superimposing in the horizontal direction is to compel the smallest horizontal disparity of the matching points to zero. Fig. 7 shows the colour 3D image automatically generated using the epipolar image pair of Fig. 6.

Fig. 6 Epipolar geometry resampling. Top: Ikonos image pair before epipolar resampling; Bottom: epipolar images.

Fig. 7 Colour 3D image generated using epipolar images

2.6 Fast data transfer

Most Internet users access to the Internet through relatively low speed modems. The problem of bandwidth constraint exists for viewing large volume colour image online. To overcome this problem, a technique was developed to divides high-resolution images into bite-size tiles and deliver "pixels on demand". Therefore, the data transmission is done only when there is a downloading demand, which allows high resolution images being transferred over all bandwidth environments.

2.7 Automatic 3D refinement

The depth perception of a stereo image in binocular vision is caused by horizontal parallaxes. However, a larger horizontal parallax will cause difficulties in depth perception and stereo fusion problems. Therefore, the overall horizontal parallaxes should be maintained in a limited range for an acceptable stereo fusion (Carr 1993). Since satellite images cover a larger area, some areas may have larger horizontal parallaxes, as illustrated in Fig. 8 (left), causing stereo fusion problems. A tool for automatic 3D refinement is therefore necessary to reduce the overall horizontal parallaxes to maintain an acceptable parallax range for effective 3D viewing (Fig. 8, right).

Fig. 8 Influence of overall horizontal parallaxes for effective colour 3D viewing. Left: larger overall horizontal parallaxes may cause stereo fusion problems; Right: small overall horizontal parallaxes are more comfortable for colour 3D viewing.

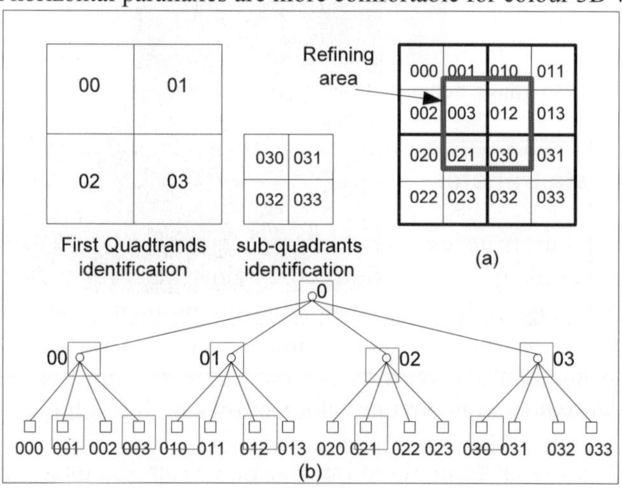

Fig. 9 Quadtree method for parallax refining. (a) decomposition of an image, and (b) quadtree representation of the image.

A quadtree-based approach was developed to allow 3D refinement being performed in real time. To refine 3D image effetely, only those quadrants crossing the refining area (Fig. 9) are searched.

3 Potential Applications for Disaster Management

Using the new system developed in this study, satellite (also airborne) images of disaster areas can be rapidly posted online, providing colour 2D and 3D information for dynamic visualization and analysis. Professionals and decision makers can join the analysis of the disaster situations remotely through the Internet, which can significantly increase the response speed. Because colour 3D information of the real situation can be visualized timely by the system, more useful information can be obtained through the system than through other systems, such as Google Earth and Microsoft Virtual Earth.

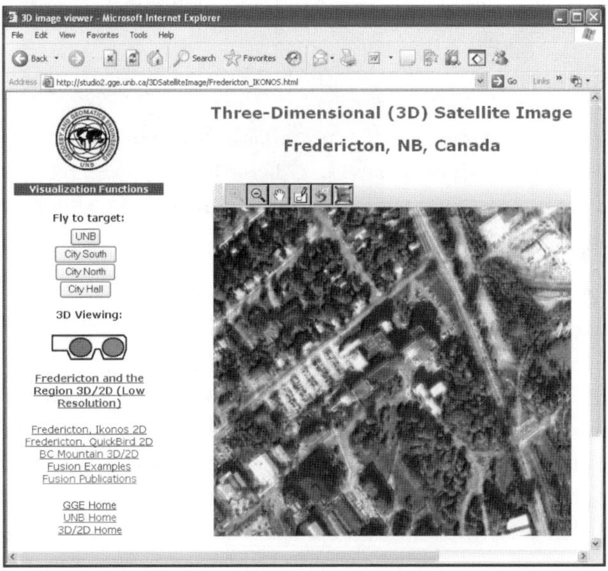

Fig. 10 Window view of a large scale colour 3D image of the Engineering Building of the University of New Brunswick, Canada

Fig. 10 shows an example of large scale colour 3D images generated using Ikonos images and dynamically visualized online through the new system. Individual houses and trees can be viewed in colour 3D. The same scene can also be visualized in colour 2D by selecting the 2D option. Fig.

11 shows a window view of a small scale colour 3D image generated using a Landsat-7 ETM+ image. The 3D and 2D colour images can also be visualized in full screen by selecting the full screen option, at different scales by zooming in or out, and can move or fly from one location to another by panning or clicking the flying-to-target button.

A website example of the 2D and 3D visualization can be found at: http://studio2.gge.unb.ca/3dSatelliteImage/.

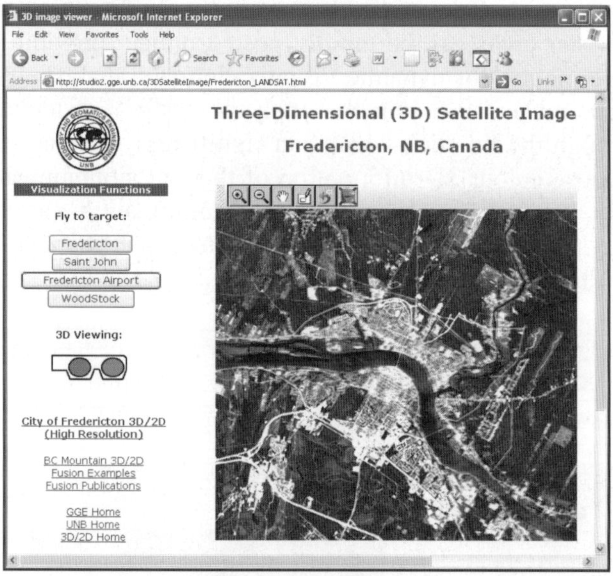

Fig. 11 Window view of a small scale colour 3D image of the city of Fredericton, New Brunswick, Canada

The difference between 2D visualization and 3D visualization can be seen from the two images in Fig. 12, showing a hydropower plant and the surrounding area. The left image is a 2D colour image visualized by the new system which is the same as those provided by Google Maps. In the 2D image (Fig. 12, left) ground objects around the plant can be seen. But, no height information can be observed. However, when the right image of Fig.12 is viewed using a pair of 3D classes, not only the ground objects but also the height information of individual objects can be seen. This height information is particularly important for the analysis and prediction of potential flooding areas for the case if a disaster happened to the dam of the power plant. Fig. 12 is just one of examples which shows the importance of 3D information for disaster analysis and management.

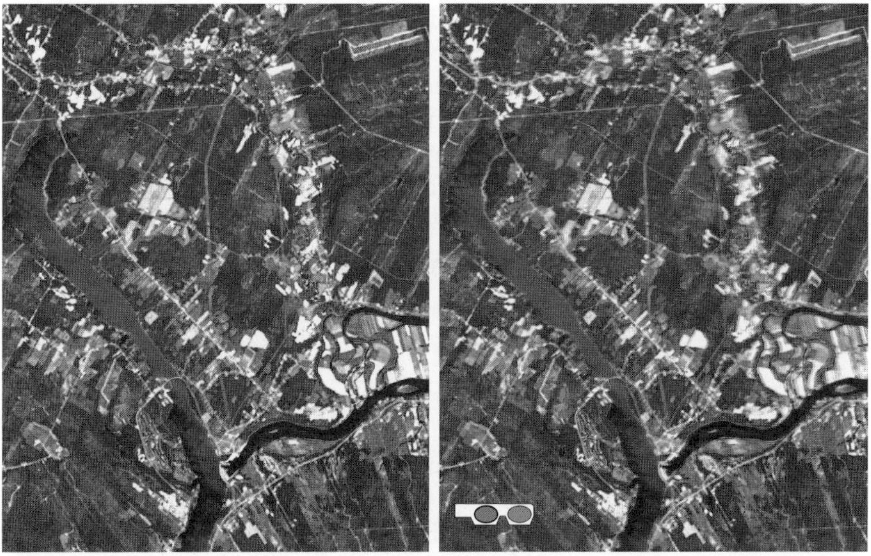

Fig. 12 The difference between 2D and 3D visualizations (Landsat-7 EMT+). Left: 2D visualization of a hydropower plant and the surrounding (all the objects in the area appear on the same plane); Right: 3D visualization of the hydropower plant and the surrounding (individual objects show individual height information).

Fig. 13 Forest fire in a 2D satellite image (from which the information on the accessibility to the fire areas is not easy to be obtained) (courtesy of DigitalGlobe)

The importance of quickly accessing 3D information of a disaster area can also be illustrated by another example—forest fire—in Fig. 13. For such a disaster, a 2D satellite image can though show the damaged areas of the forest fire (Fig. 13), a colour 3D image would provide more useful information for the disaster monitoring and management. For example, 3D information can provide crucial information on the accessibility of the area, which is important for an emergency rescue while the forest fire and for an effective rehabilitation after the disaster.

4 Conclusions

This paper presents a new system for producing colour 3D satellite images and for online visualization of colour 2D and 3D images. To utilize the spectral and spatial information of the available multispectral and panchromatic images of the Landsat-7 ETM+ and QuickBird/Ikonos, and to explore an effective and affordable way for detailed colour 2D and 3D visualization, an image fusion process is integrated into the system. To achieve an effective colour 3D visualization, a colour enhancement technique is employed to make the colour of QuickBird and Ikonos images more natural.

The results of this study have demonstrated that it is possible to develop a system for online colour 2D and 3D visualization, using Landsat images and DEMs for low resolution and large coverage visualization, and using QuickBird or Ikonos for detailed visualization of specific areas. Different objects on the Earth's surface, from mountain ranges to urban family houses and trees, can be dynamically visualized and analyzed through the Internet in colour 2D and 3D. As the 2D and 3D images can be quickly generated and posted online (within a day and so after receiving the raw image data), the system is especially useful for disaster monitoring and management for informed and timely responses.

Acknowledgements

The research was supported by the I2I program of the Natural Science and Engineering Research Council (NSERC), Canada. The authors thank Mr. Rob Lunn, GIS supervisor of the City of Fredericton, NB, Canada for providing the raw Ikonos image data. The Landsat image was obtained from GeoBase (http://geobase.ca/).

References

Carr DB (1993) Production of stereoscopic displays for data analysis, *Statistical Computing and Statistical Graphics Newsletter*.

Zhang Y (2002) Natural colour urban 3D modeling: A stereoscopic approach with IKONOS multispectrsal and panchromatic data. *International Archives of Photogrammetry and Remote Sensing,* Vol. 34, Part 4.

Zhang Y (2002) A new automatic approach for effectively fusing Landsat 7 as well as IKONOS images. *Proceedings of IEEE/IGARSS'02*, Toronto, Canada, June 24-28, 2002.

Zhang Y (2004) Understanding image fusion. *Photogrammetric Engineering & Remote Sensing*, 70 (6): 657-661.

On the Application of Nighttime Sensors for Rapid Detection of Areas Impacted by Disasters

A. de la Cruz[1] G. Laneve[2] D. Cerra[1] M. Mielewczyk[1]M.J. Garcia[1] G. Santilli[2] E. Cadau[2] and G. Joyanes[1]

[1]EUSC, Apdo de Correos, 511. E28850 Torrejon de Ardoz, Madrid, Spain
[2]CRPSM – Università di Roma "La Sapienza", Via Salaria, 851, 00138 Roma, Italy, laneve@psm.uniroma1.it

Abstract

Today a few sensors operating at night are available in the visible/near infrared part of the spectrum, e.g., the U.S. Defense Meteorological Satellite Program Operational Linescan System (DMSP/OLS). However, in the case of DMSP/OLS, the availability of a series of satellites arranged in a constellation and the width of the sensor's swath allows Earth coverage twice nightly. This can result useful in the aftermath of a natural disaster such as earthquake, when first responders providing relief action need to know the location and the extent of the areas of damages, the potential amount of population involved and the place where survivors are concentrated. Naturally, after this prompt detection of the areas affected by the event, the corresponding very high spatial resolution satellite images can be acquired to obtain an accurate overview of the actual damages. In fact, the availability of a preliminary fast estimate of the areas mainly impacted can support a suitable selection of the very high spatial resolution (VHSR) satellite images acquisition time because these sensors are characterized by a very small frame size that makes unpractical a blind acquisition of the whole region possibly impacted. This way to proceed is also compatible with the longer time usually needed to obtain a VHSR image of a given area of interest, due to the orbital and observation geometry constraints.

Even if it is high, the OLS sensor's sensitivity could be insufficient to detect settlements with reduced artificial lights, as is often the case in the selected regions of interest. Moreover, in many cases, as for informal settlements following the occurrence of natural or man-made disasters, only the presence of bone-fires could reveal the presence of a human community. As a consequence, it would be necessary to observe the affected areas using wavelengths in the Middle-Wave Infrared region of the spectrum (~ 4 μm), which is presently not feasible due to the limited sensitivity of available sensors. Nevertheless a couple of examples of the results obtainable using night-time images in these scenarios are provided.

To overcome the difficulties described above, this study focuses on a design analysis of a new night-time sensor. The study is based on accurate simulations of the expected radiance scenario reaching the sensor. This, in fact, is required to assess the characteristics of a new sensor capable of detecting the desired target sources (lights/bone-fires).

1 Introduction

High resolution satellite imagery can provide a good insight into the magnitude of a disaster and a detailed assessment of the damage. To meet these objectives, HR imagery has to be collected immediately after the disaster and precisely in the areas that have been damaged by the earthquake. In the general confusion that follows an earthquake with the lack of reliable data, this information either is not known or is incomplete. Presently, space based remote sensing systems result unsuitable to provide useful information when disastrous events require simultaneously high temporal and spatial resolutions. Mainly this is due to the fact that, even if many high resolution satellites are available, they are not organized in a constellation and the image acquisitions, for observational reasons, are concentrated around a convenient local time. Further, due to the technological limits of the transmission systems a very high resolution is usually coupled with a reduced sensor swath (typically $\approx$10x10 km^2). This means that the observation can be carried out when the area to be imaged is known. On the other side low-resolution satellite could provide, in principle, some information with the required promptness. In the opinion of the authors this is the case of sensors able to acquire night-time images of the Earth. In fact, they can have sensitivity high enough to provide information useful in the aftermath of natural disasters such as earthquakes, when first responders providing relief action need to know the location and the extent of the damaged areas, the potential amount of population involved and the place where survivors are concentrated. Naturally, after this prompt detection of the areas affected by the event, the corresponding very high spatial resolution satellite images (VHSR) can be acquired to obtain an accurate overview of the actual damages. In fact, the availability of a preliminary fast estimate of the mainly impacted areas can support a suitable selection of the VHSR images acquisition time because these sensors are characterized by a very small frame size that makes unpractical a blind acquisition of the whole region possibly impacted. This way to proceed is also compatible with the longer time interval usually needed to obtain a VHSR image of a given area of interest, due to the orbital and observation geometry constraints.

Presently, few sensors operate at night in the visible/near infrared part of the spectrum (DMSP/OLS, SAC-C/HSTC). However, in the case of DMSP/OLS, the availability of a series of satellites organized in a constellation and the width of the sensor's swath allows earth coverage twice nightly.

The nightlights data of the Operational Linescan System (OLS) aboard the constellation of Defence Meteorological Satellite Program (DMSP) satellites can provide the rapid overview with the geospatial location of damage areas needed for disaster management and guiding the acquisition of HR imagery. The areas of damage in urban areas are indicated by the reduction of "stable" lights after the earthquake due to power failure as well as building damage and the reduction of human activity.

As an important follow up, there is a related increase of the so called "ephemeral" lights in the vicinity of the damaged areas. These are caused by the affected population that relocates in open areas lighting bonfires to cook and to keep warm. The occurrence of aftershocks prevents them from returning to their homes if they have not collapsed. The geospatial location of these areas of ephemeral lights is important information for first responders to provide assistance to the affected population in the areas they are relocated to. Therefore, a methodology is needed to process nightlights imagery in the most suitable way to highlight the areas of both damage and relocation. Furthermore, the high frequency of observations of nightlights data, provided by a constellation of satellites, makes it very suitable for emergencies during natural or manmade disasters.

In the mainframe Network of Excellence (NoE) GMOSS (Global Monitoring for Stability and Security) the European Union Satellite Centre (EUSC) and the University of Rome (Centro di Ricerca Progetto San Marco, CRPSM) are studying the performances and limits of applicability of night-time satellite images for the detection and occupancy monitoring of human settlements and for a prompt assessment of damages caused by disasters of natural or anthropic origin. In particular, their paper concerns two application examples: Kashmir, in consideration of the recent (Oct. 2005) earthquake event and Lebanon, after the summer 2006 Israel-Lebanon crisis.

The results have been validated by using high-resolution images and are supported by extensive collateral data.

The experience accumulated by means of the applications herein described has led to a methodology to obtain rapid results from the processing of nightlight data. This methodology could be useful in GMES initiatives such as RESPONSE.

A first application of satellite night-time images, for humanitarian purposes, was carried out in the African region of the Great Lakes using data from 1994, 1995 and 1996, and concerned the suitability of these data for monitoring the refugee influx in the countries bordering the Great Lakes during that period. The nightlight data was compared with collateral information and maps kindly provided by UNHCR. The results can be summarized as follows:

- during the Rwandan refugee crisis in 1994, a good correlation existed between the location of refugee camps and occurrence of ephemeral lights produced by refugees lighting bonfires.
- Nightlight data also suggested that the high refugee activity persisted in the Goma area during 1995 as well as the location of refugee camps in the region during that year.
- During the Great Lakes refugee crisis in 1996 a good correlation could be established between the ephemeral nightlights and the location of refugee camps provided by UNHCR.

In certain locations (refugee camps near Lake Vert), the lack of nightlights, from a certain date onwards, indicated that former refugee camps could have been abandoned. In other areas (north of Goma and south of Uvira) nightlight data suggested a possible presence of refugee camps not present in the UNHCR maps. Furthermore, in other areas, UNHCR admitted that the exact location of refugees was unknown. Nightlights could have provided input to clarify this uncertainty.

2 Data and Methods

The OLS is a scan radiometer with two spectral bands (VIS and TIR) with an extreme high capability to detect faint lights that is four orders of magnitude more sensitive than any other NOAA and Landsat sensors. The OLS sensor flies on board of a constellation of sun-synchronous polar orbiting satellites (F 10, F 11, F 12, F 13, F 14, F15 and F 16) at mid altitude (830 km). At present two of these satellites (F 15 and F 16) cover the earth twice daily. The archiving of nightlights imagery has been recorded from 1972 until present and data availability has been planned up to 2010. Spatial resolution is available at 0.56 km (fine resolution) and 2.7 km (average fine resolution). A methodology has been developed by NOAA to obtain sub-pixel spatial resolution from nightlights data that uses a geo-location algorithm that projects the central point of each pixel onto the earth's surface. Satellite altitude, orbit and a DEM are also considered (Elvidge et al. 1999).

The relative lack of operational applications of the nightlights data is probably due to the low spatial resolution. Potential users are not aware of its extremely high sensitivity to the light. This factor plays an important role in the detection of very faint light sources that could be very useful for providing a rapid overview of damage and population dynamics after natural or man-made disasters.

The analysis carried out in this paper are based on time series of DMSP/OLS night-time images, covering the Kashmir area (hit in October 2005 by an earthquake) and the Middle East (summer Lebanon-Israel crisis). These images are characterized by a spatial resolution of 1.0 km. According to (Cinzano et al. 2000; Eldvige et al. 1999) this sensor is able to detect a radiance level as low as 1.54 10^{-9} [W/cm^2/sr/μm] and up to 3.17 10^{-7}. In terms of luminance this means that the sensor is able to detect radiation corresponding to 3.0 mcd/m^2. In other words, two unshielded 250 W lamps with 125 lm/W efficiency, placed every square kilometre, are sufficient to produce the required luminance (Cinzano et al. 2000). These numbers give an idea of the sensitivity of the sensor. From the point of view of our problem of estimating natural disasters impact on the population, the digital number (DN) read from the image has to be converted to radiance and then to inhabitants for radiance unit. This last quantity is strongly variable with the development conditions of the area of interest. From the point of view of the DN – radiance relationship a logarithmic scale has been obtained (Cinzano et al. 2000): $L = (DN)^{3/2} \cdot 10^{-10}$ [W/cm^2/sr/μm].

In order to estimate in an early way the population impacted by a catastrophic event we can follow, in principle, two approaches:

- In one case, if a global population density map (like the one provided by LandScan) is available, the amount of people affected by an event can be computed simply multiplying the pixels, where a strong decrease of the detected light is measured, for the corresponding population density.
- In the second case, we can use the relationship between population and digital number obtained, using OLS data, for the 60 most populous countries of the Earth (Sutton et al. 2001).

Sutton et al. (2001) give also an interesting view of the minimum detectable population that is not strictly related to the economic development of the country. In fact, it results in about 28000 people in Netherlands and in 169 people in Egypt. For an explanation of this result the reader is addressed to the above quoted paper. From our point of view it is interesting to observe that for the area of interest the minimum detectable population results of the order of (554+386)/2 persons/km^2. This number comes from the minimum detectable population given for the two countries Pakistan (386) and India (554) to which Kashmir belongs.

To be useful to our purposes, night-time images require an accurate preprocessing, because several sources of misinterpretation are present, such as clouds, forest fires, etc. In particular, the following factors have been addressed:

- Image geo-referencing. Sometimes the nightlight imagery may have some offset due to gyroscope problems. A suitable procedure for co-registering the images has been implemented.
- Cloud effects. Nightlight data require cloud-free conditions. The TIR channel should be used to estimate cloud coverage.
- Blooming effect.
- Ephemeral lights. The presence of temporary sources of lights (e.g. wild fires) can alter the results. For instance, in the case of the problem of detecting new informal settlements, it may result not easy to discriminate between wild fires and bone-fires.
- Change detection. To perform a more robust analysis and reduce the uncertainty of results, improved change detection algorithms should be applied. A work flow is proposed in Fig. 1.

It is worthwhile to note that, in principle, analysing a time series of night-time images on an area affected by a disaster, we expect to observe:

- a reduction of lights immediately after the event,
- a displacement of lights some time after the event, as new settlements are established.

Fig. 1 shows a flow chart of a general methodology proposed for processing OLS night-time images aiming to obtain, in a rapid way, information suitable to support humanitarian activities.

This general procedure has been applied to two events:

- the Kashmir earthquake of October 2005,
- the Israel-Lebanon crisis of July/August 2006.

Emphasis has been placed on the development of a simple and rapid way of obtaining results that could be used to support first responders to know exactly where the areas of damage are located and where to assist the affected population. This information is important to allocate the limited human and physical resources quickly and effectively after a natural disaster such as an earthquake.

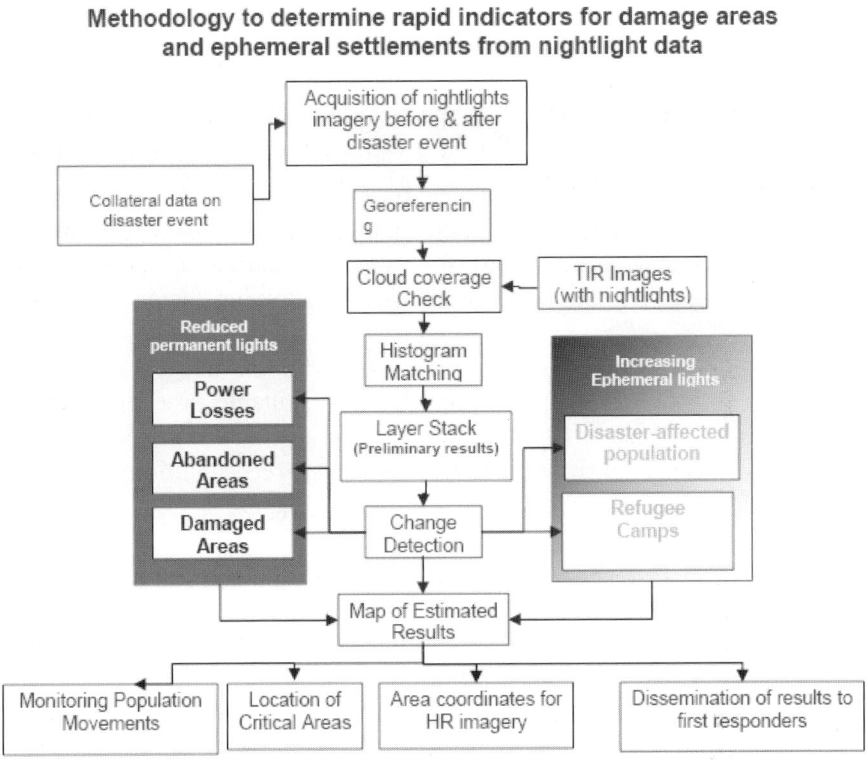

Fig 1 Flowchart of the proposed methodology.

3 Results

3.1 The Kashmir earthquake

Night-lights data, before and after the Kashmir earthquake, were requested by EUSC from NOAA. The following data were provided:

- F 15 satellite: 04th, 05th, 06th, 07th, 08th (day of the earthquake), 09th and 10th October 2005
- F 16 satellite: 04th, 05th, 06th, 07th, 08th (day of the earthquake), 09th and 10th October 2005

These data were made available by EUSC to the CRPSM.

A preliminary observation suggested the reduction of night-lights in the areas worst affected by the earthquake of Muzaffarabad, Balakot and Rawalakot and the increase of ephemeral lights in the nearby areas originated by the concentration of survivors´ bonfires (Fig. 2).

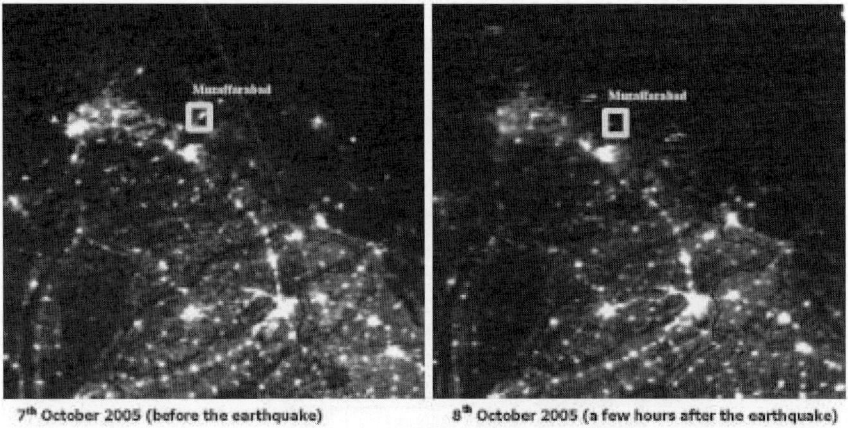

Fig. 2 Nightlights before and after the Kashmir earthquake showing a considerable reduction of permanent lights (suggesting damage in Muzaffarabad and surrounding areas) after the earthquake.

The layer stack of the VIS bands before and after the earthquake (4th and 9th October 2005) produces a general overview image that displays, in a comprehensive way, the three critical areas (red circles) in Kashmir and India (Fig. 3) where the damage was more intense (reduction of permanent lights). In association with these three critical areas and in the vicinity of each of them, there are large areas (in light blue) of new ephemeral settlements (increasing ephemeral lights) indicating that earthquake survivors have moved from the damaged areas to other urban areas where damage has been minimal or into open areas. This interpretation is supported by collateral data (NDTV.com).

To perform a more robust analysis and reduce the uncertainty of results, improved change detection algorithms should be applied. On the other hand, the use of averaged images helps to smooth out noises. Furthermore, a comparison with the local standard deviation of each image pixel is calculated: a change index C(x) is assigned to each pixel P by dividing its absolute variation by its standard deviation value. This pixel P enters the map of changes if C(P) > 2. The value chosen as a threshold guarantees the geolocation of significant areas of change, simultaneously rejecting most of the false alarms. The areas of damage are displayed in red (decreasing permanent nightlights) and the areas of displaced population are displayed in light blue (increasing ephemeral nightlights); at this stage the results are available to first responders in a natural or man-made disaster.

These critical areas, where most of the earthquake damage is concentrated, are the following:

- The area of the town of Muzaffarabad and its vicinity (near the epicentre). The permanent nightlights of Muzzaffarabad and those of several towns located to the SE of it mostly disappear after the earthquake. On the contrary, the ephemeral lights increase in Manshira, located west of Muzaffarabad and in open areas to the north of this town. This could indicate that the affected population from Muzaffarabad has moved to the area of Manshira (Fig. 3).
- The area South of Islamabad. Fig. 3 suggests areas of damages at approximately 107 km south of Islamabad where the permanent lights have been reduced (towns of Bhera and Miami). These results are in line with the collateral data describing the effects of the earthquake in the region of Islamabad.
- The Haryana region in India, located NE of Delhi, suffered strongly the effects of the earthquake. Several cities like Sirsa experienced power cuts due to damage in buildings, electricity networks and power stations (source:NDTV.com). The permanent lights in the region show a reduction after the earthquake indicating widespread damage. However, there were towns like Fatehabad and Hisar mostly unaffected by the earthquake that display an increase of ephemeral lights suggesting that an influx of the population moved from damaged areas like Sirsa. On this basis, the vectors reflecting the movements of the population affected can be drawn as shown in Fig. 4.

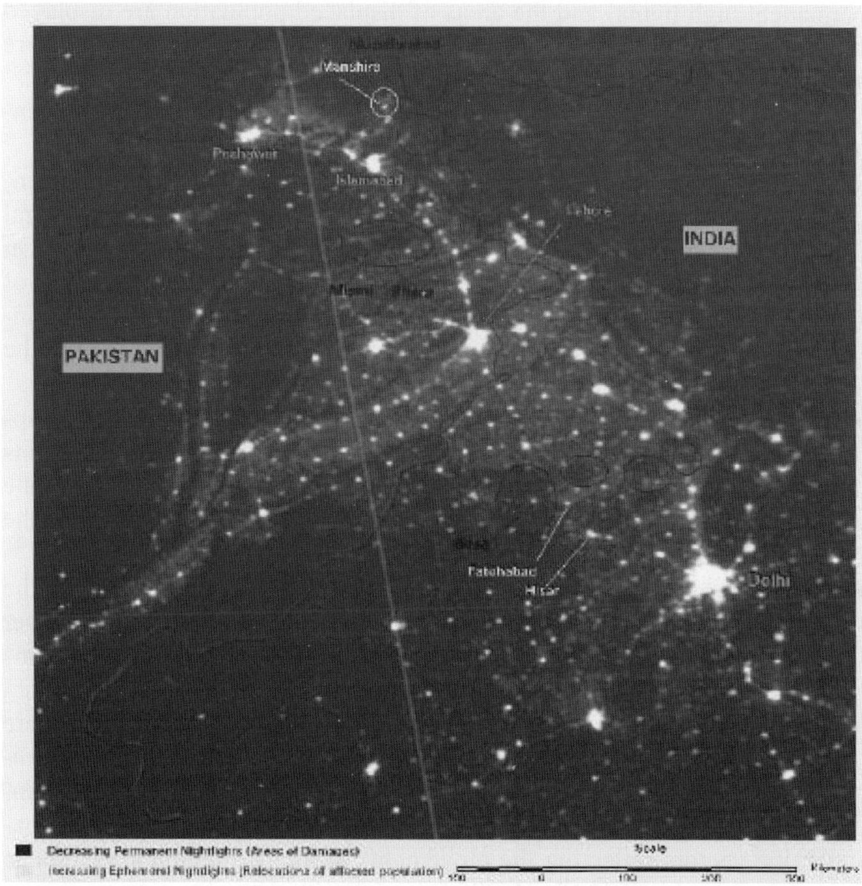

Fig. 3 Layer stack (4-9 October 2005). Comprehensive results displaying areas of damage (in red) and displacement of the affected population (in light blue) after Kashmir earthquake.

At CRPSM a similar change detection technique was applied to the averaged radiances obtained using three images preceding the earthquake and three images after the event.

Figs. 5 shows the light differences map computed, for the part of the

Fig. 4 Detail of the population dynamics in the Haryana region as deducible by using Fig. 3.

$$\Delta L = \frac{L_{af} - L_{pr}}{L_{\max}} \tag{1}$$

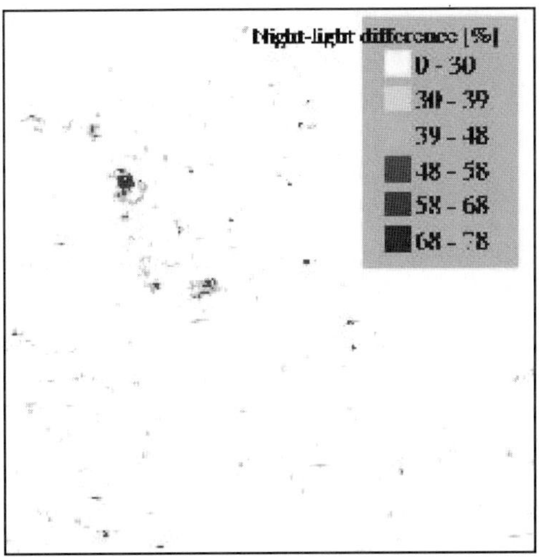

Fig. 5 Map of the difference (in percentage) between night-lights detected by DMSP/OLS sensor before and after the October 2005 earthquake.

where L_{pr} and L_{af} represent the mean Digital Number corresponding to the pre- and post-event average, respectively. Averages have been computed taking into account the clouds effects using images acquired simultaneously in a thermal band (the TIR detector is sensitive to radiation from 10.0 to 13.4 μm). At a first stage of the analysis, pixels characterized by a relative difference lower 30% as well these with a positive change (corresponding to an increase in the intensity of the light detected by the sensor) are excluded among the areas affected by significant damages. This can help, in general, to take into account the effect of presence of the moon. Actually, in our case, the days before the earthquake correspond to new moon conditions whereas nights of the days 9th and 10th are illuminated by a moon at 40%. This could explain the presence of diffuse lights observable in the after-event images with respect to the pre- event case. Fig. 5 shows the map of the differences between night-lights available after the earthquake with respect to the pre-event distribution. This figure has been obtained assuming as affected by a real change these pixels where a ΔL (see Eq. 1) negative and lower than 30% takes place. The value of 30% arises from a statistical analysis of the distribution of the differences and corresponds to the limit of 2σ, where σ is the standard deviation of the distribution. Fig. 6 shows a sample of the Landscan population density map, as people per km^2, for the area of interest. In order to proceed to assess the possible impact of the catastrophic event in the region the human settlements must be identified. In particular, pixels affected by a significant change and corresponding to the same town/city must be recognized. This problem can be managed by using Mathematical Morphology. Fig. 7 shows the image obtained applying such mathematical morphology techniques. The objects recognized as areas of changes are shown. On the image are also reported the cities corresponding to the detected objects. At this point the potential amount of population involved in the event can be computed using the values given by the Landscan map in correspondence with the detected objects. Fig. 8 reports the estimate population involved for the 3 cities indicated in Fig. 7. These values have to be compared with the numbers given by humanitarian organization given in Table 1. It follows that quite a good estimate can be obtained. According to our results the population affected by the event is in the order of 120000, 27000 and 7000, respectively.

Table 1. Population affected by earthquake according to humanitarian agencies reports.

City	Population	Victims
Muzaffarabad	200,000	11,000
Rawalakot	60,000	10,000
Balakot	25,000	17,000

It is worthwhile to recall here that a possible estimate problem can arise from the so-called blooming effect present on the DMSP/OLS images. It consists in an overestimation of the actual size of the human settlements due to the OLS capabilities to detect sub-pixel light source, to geo-location errors and surface effects (Sutton et al. 2001). This phenomenon, in our case, introduces only a small error because when the apparent lights are affecting a not inhabited pixel the corresponding contribution to the amount of population possibly affected by the disaster is negligible. Then, the technique herein described seems promising.

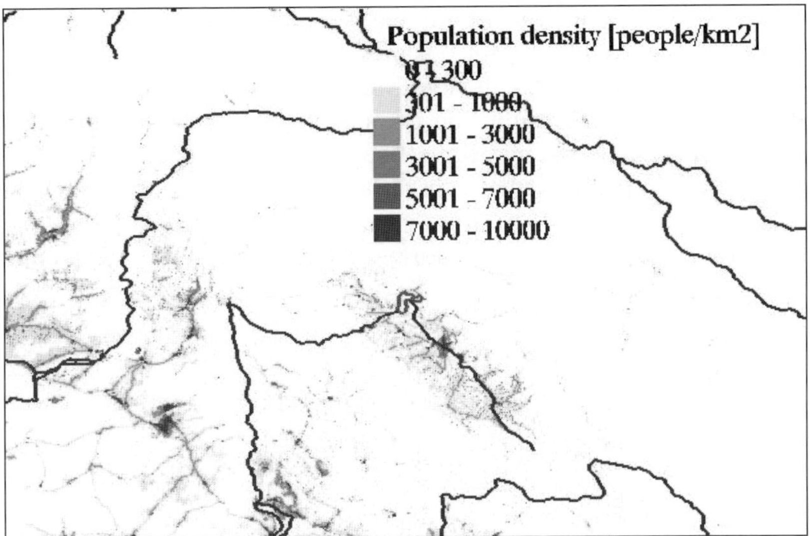

Fig. 6 Sample of the Landscan population density [people/km^2] map of the area of interest.

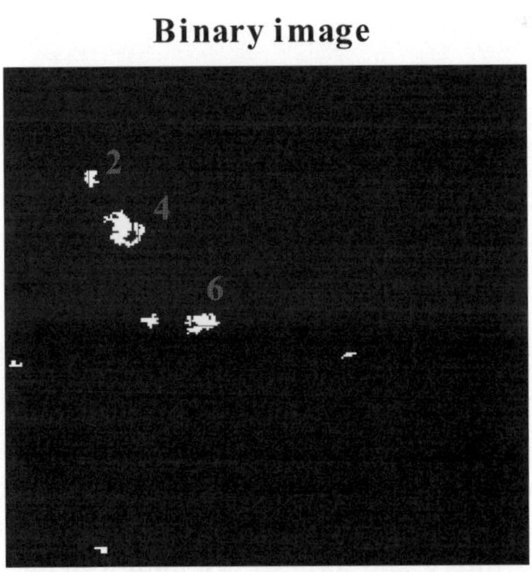

Fig. 7 Objects corresponding to areas affected by illumination changes resulting by the application of morphological techniques. Three of the seven detected objects correspond to towns affected by the earthquake.

3.2 The Israel-Lebanon crisis

The night-lights images of the Israel-Lebanon region were processed as a further test of the suitability of the night-time data to obtain a rapid indication of damage areas and population dynamics during a security crisis. Fig. 9 shows the results obtained by EUSC analyzing the nightlight data before and after the summer 2006 crisis had started (mid July). These results have been obtaincd following the methodology shown in Fig. 1. Damaged areas (in red), as a result of the reduction of permanent lights in urban areas, caused by the bombing and by people fleeing their homes in the areas of risk, are displayed. The crisis has caused large population movements that have originated new relocation areas such as the ones displayed by the nightlight results on the eastern outskirts of Damascus (Syria) and around

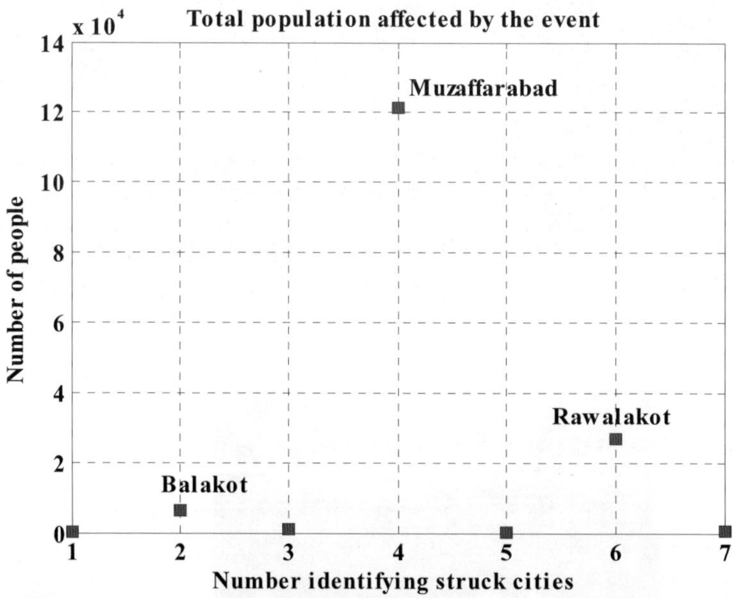

Fig. 8 Number of people affected by the event estimated using the object affected by illumination changes and the Landscan population density map.

the urban areas of Yabrud and An Nabk to the N.E of Damascus.

In addition, the nightlight results have been compared with the sites of confirmed attacks by artillery and aircraft bombing (green crosses) and a fair correlation has been obtained with the areas of reduced permanent lights (Figs. 9 and12).

Comparison with collateral data indicates that these hotspots have not been caused by other sources (e.g. forest fires). Furthermore, these results also correlate well with the Quickbird imagery collected after the bombing

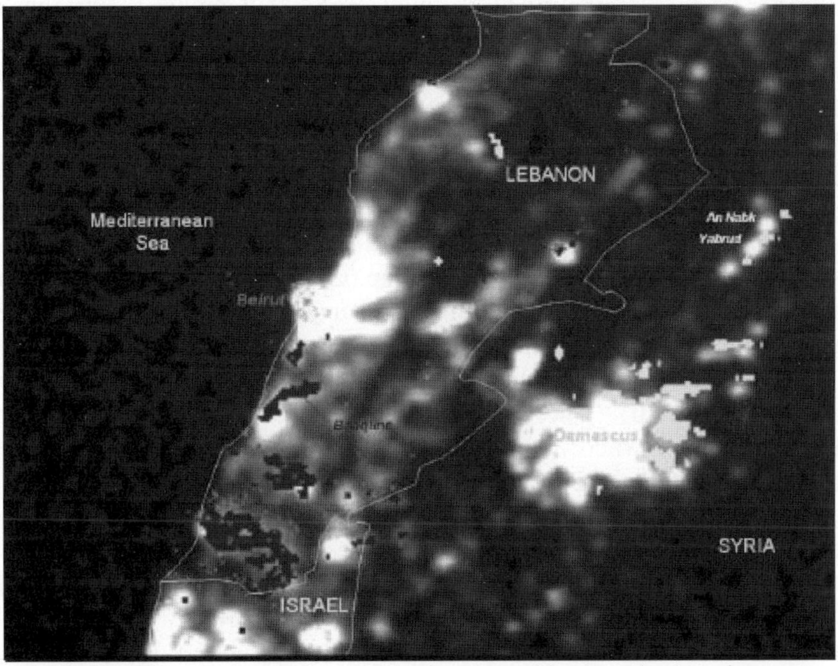

Fig. 9 Correlation of damaged areas (in red) suggested by the reduction of permanent lights with confirmed attacks (green crosses). Refugees from Lebanon have concentrated on the eastern outskirts of Damascus, Syria (large increase of ephemeral lights – in blue).

of the fuel depots at the Beirut airport as well as in other locations (work in progress).

Also in this case the time series of OLS images were provided to CRPSM by EUSC. Six images, three pre- and three post-event were analyzed (see Table 2).

Table 2. List of images used for the analysis herein described.

Sensor	Acquisition date	Spatial res. [km]	Image size [pixels]	Pixel shift [x,y]
OLS-DMSP F15	25-June-2006	1	1201x1201	X = 7, Y = 3
OLS-DMSP F15	26-June-2006	1	1201x1201	X = 1, Y = 1
OLS-DMSP F15	27-June-2006	1	1201x1201	X = 2, Y = 1
OLS-DMSP F15	21-July-2006	1	1201x1201	X = 3, Y = 2
OLS-DMSP F15	22-July-2006	1	1201x1201	X =-2, Y = 1
OLS-DMSP F15	23-July-2006	1	1201x1201	Ref. image

An automatic co-registration of the images has been obtained by using an algorithm based on the Mutual Information technique and software developed in the Matlab environment. This technique is based on the minimization of the so called Shannon joint entropy (Pluim et al. 2003). From the analysis of the 6 images it results that they can be shifted with respect to one another of several pixels (see Table 2). Fig. 10 shows the behaviour of the Mutual Information parameter as function of the relative position of the images. After the images registration averaged values of night illuminance pre- and post-event were computed and the differences were estimated. The results, for a sector of the images around the area of interest are reported in Fig. 11. This figure shows a RGB combination of the Landscan map (R), the relative difference map (G) and DMSP (OLS) night-time image (B). Some areas of significant change are clearly detected (light green pixels). As mentioned above, these areas have been obtained assuming as affected by a real change the pixels where ΔL (see Eq. 1) is negative and lower than 30%. These areas correspond to the region near the Israel-Lebanon border where mainly the Israeli bombing activity was concentrated and south of Beirut. The different variation in the illuminance around the two sides of the border (a very small number of pixels result changed in the Israeli sector) gives an idea of the difference in the heaviness and efficiency of the military activity, confirmed by Fig. 12. From the analysis it results that 198,000 people were, potentially, affected by the crisis. To well understand the meaning of this number a discussion is required. In particular, it must be recalled that the conflict started on 12 July 2006 and continued up to August 14, 2006. According to the humanitarian

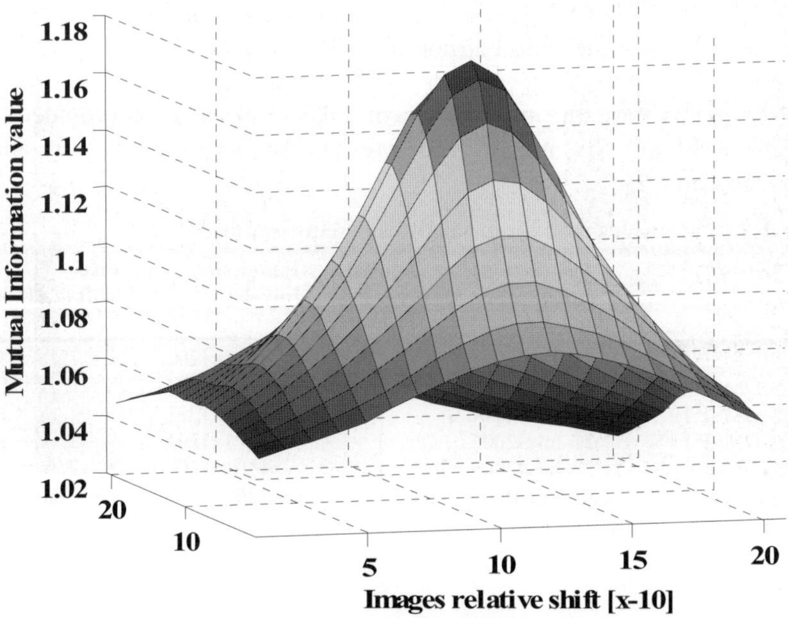

Fig. 10. Behaviour of the Mutual Information index, as function of a shift in x or y of one of the OLS images with respect to the reference image.

agencies the conflict caused the displacement of about 975,000 Lebanese and 300,000 Israelis. As December 1 an estimated number of 200,000 Lebanese remained internally displaced or refugees. Differently from the earthquake case now we know quite well the area affected by the event, then we can restrict our analysis to that. Furthermore, in the earthquake case, where probably only the people living the area stroke by the event left their destroyed or damaged houses, we can expect a higher correspondence between the estimated and the actual number of potentially affected people whereas, in the present case, we have a reduced probability of capturing the number of displaced people owing to the fact that they abandoned their houses even if these were in good shape fearing bombing or a military attack. In fact most of the displaced people were leaving in Beirut where only strategic objectives were hit (airport strip, etc.) with a reduced impact on the nocturnal illuminance. Thus, what we can reasonably estimate is the amount of population residing in the area effectively stroked by war actions where living is not anymore possible because no electric power is guarantee. Possibly those people will correspond to the ones who, still some time after the crisis' end, will stay far from home as refugees or as internally displaced people. To confirm this point it is worthwhile to note that only a couple of pixels results significantly changed in corre-

Fig. 11 RGB representation of the results obtained using an automatic detection of areas with significant night-lights changes.

spondence of the city of Beirut (possibly as a consequence of the Harat Hurayk neighbourhood attach of July 12 but, most of the refugees (displaced people) were from this city. As said above the number of them in December was quite similar to the number of people residing in areas where the night-lights changes were significant (200,000 versus 198,538). This result seems to confirm the utility of using night-time images to promptly estimate the amount of persons potentially affected by a disaster.

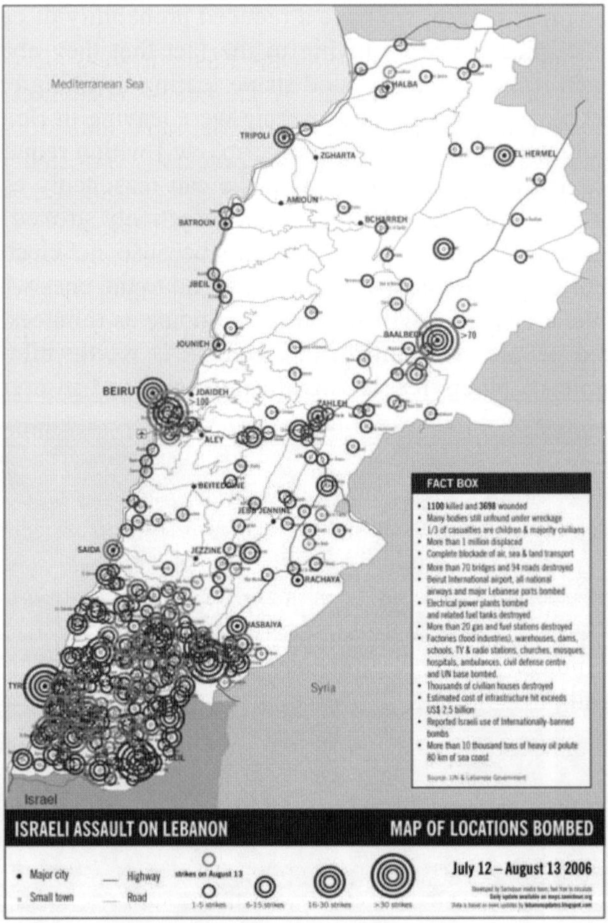

Fig. 12 Distribution of the Israeli assaults on the Lebanese territory.

Let us say something more about the night-light changes in the city of Beirut, of course the following considerations can be applied to almost any "bright" city in the world. As a consequence of its high sensitivity the OLS sensor can easily reach the saturation level in correspondence of cities in developed countries. This fact coupled with the so-called blooming effect (previously introduced) can result in a certain insensitivity to partial changes. In other words, in order to detect relatively small (small from the point of view of the impact on the intensity of the night lights) damages we have to reduce the threshold defining a "real" change. In particular, the area in Beirut affected by the July 12th attack presents a variation of the illuminance of the order of 9%. However, in this application we can better define a suitable threshold because we know "a priori" the area we are interested in analyzing, when looking for the amount of damages there produced.

4 Conclusions

The results obtained from nightlights data over the Kashmir areas affected by the earthquake and over the area affected by the Israel-Lebanon crisis have shown the high potential of these data to provide: rapid indicators to locate areas of damage, the amount of population potentially affected and the population dynamics after a natural disaster and a situation of crisis. This information will save precious time and could be valuable to first responders in their efforts to assist the population affected. In view of this high potential, a general methodology for processing nightlight data, with particular attention to the image co-registration problem and the post-processing analysis, have been proposed. In particular, the results obtained by using a co-registration method based on the Mutual Information technique and a post-processing analysis based on Mathematical Morphology has been shown.

The results seem very promising and future research work will focus on testing the change detection algorithms in different environmental conditions and developing more robust mapping results.

A further, more detailed research report of the Israel-Lebanon crisis will be completed in 2007. This report will provide a monitoring of the Israel-Lebanon crisis with multi-temporal imagery and integration of data from several sources (nightlights, MODIS hotspots and location of confirmed attacks).

Acknowledgments

We wish to acknowledge the continued support of Chris Elvidge of NOAA, National Geophysical Data Centre, providing helpful advice and provision OLS –DMSP nightlight data for the GMOSS project.

References

Amaral S, Camara G, Monteiro AMV, Eldvige C, Quintanilha JA (2001) Assessing nighttime DMSP/OLS data for detection of human settlements in the Brazilian Amazonia, http://www.dpi.inpe.br/gilberto/papers/silvana_sbsr2001.pdf (last date accessed: February 26, 2007)

Cinzano P, Falchi F, Eldvige CD, Baugh KE (2000) The artificial night sky brightness mapped from DMSP satellite Operational Linescan System measurements, Mon. Not. R. Astron. Soc. 318, 641-657, (http://www.inquinamentoluminoso.it/cinzano/download/mnras_paper.pdf, last date accessed February 26, 2007)

Elvidge CD, Baugh KE, Kihn EA, Kroehl HW, Davis ER (1999) Mapping city lights with nighttime data from DMSP-OLS, Photogrammetric Engineering & and Remote Sensing, 64(6): 727-734.

De la Cruz A et al. (2005) EUSC presentation: "Indo-Pakistan Integrated Test Case: Nightlight Methodology for Humanitarian Response". GMOSS Graz meeting. December 13 – 14, 2005.

De la Cruz A et al. (2004) EUSC presentation: "Nightlight monitoring of refugee camps during the crises of Rwanda (1994) and the African Great Lakes (1996)".GMOSS Population Monitoring Workshop at JRC.

Pluim JPW, Maintz JBA, Viergever MA (2003) Mutual Information based registration of Medical Images: a survey, IEEE Transactions on Medical Imaging, 22 (8): 986-1004.

Sutton C, Roberts D, Eldvige C, and Baugh K (2001) Census from Heaven: an estimate of the global human population using night-time satellite imagery, International Journal of Remote Sensing, 22 (16): 3061-3076.

A Fuzzy Relational Method For Image-Based Road Extraction For Traffic Emergency Services

Yu Li[1], Jonathan Li[1] and Michael A. Chapman[2]

[1] Geomatics Program, Department of Geography, University of Waterloo, 200 University Avenue West, Waterloo, Ontario, Canada, N2L 3G1, {y62li, junli}@fes.uwaterloo.ca
[2] Geomatics Engineering Program, Department of Civil Engineering, Ryerson University, 350 Victoria Street, Toronto, Ontario, Canada, M5B 3K3, mchapman@ryerson.ca

Abstract

The new generation of high spatial resolution satellite imagery such as IKONOS and QuickBird provide a viable alternative to high resolution aerial imagery. These data have been readily adopted on various applications such as city planning, metropolitan mapping, emergency response and disaster management. In the applications of high resolution remote sensing images, road extraction is one of the most basic and important tasks. Though many image-based road extraction algorithms have been proposed in the past years, most of them extract roads either from grayscale imagery or low resolution satellite imagery. Approaches designed to process low-resolution satellite imagery generally describe roads as curvilinear structures and model roads as relatively homogeneous areas satisfying certain shape and size constraints.

With the increasingly availability of multi-spectral remote sensing images, colour provides another important feature for extracting road networks. The purpose of this study is to develop an efficient algorithm which combines the colour and shape features to extract road networks automatically from high resolution satellite imagery. The proposed method adopts the fuzzy relation based segmentation algorithm and colour similarity measure in the RGB colour space, and follows three steps, (1) calculating colour similarity measurement in the RGB colour space, (2) segmenting colour satellite imagery using fuzzy relation based segmentation algorithm,

(3) extracting central lines of roads by a post-processing procedure. The advantage of this method is to automatically determinate the number of classes in segmentation. In most of situations, it is difficult to specify any desired number of clusters. For example, the situations often happen in the segmentations of remote sensing images, because the ground truth is always not available for the scenes covered by those images. The proposed method is examined by extracting road networks from QuickBird and IKONOS imagery. The results show that the proposed method for road extraction is very effective. In order to illustrate the accuracy, the extracted road centerlines are overlaid on the original images.

1. Introduction

The natural and human-caused disasters such as earthquakes, volcanic eruptions, floods, droughts, fire, oil spills, and industrial pollutions have been causing extensive loss of life, damage to property, and harm to the environment. Recently, disaster studies have been paid much more attentions by governments, associations and originations around the world. In order to scientifically deal with increasing disasters, the concept of disaster management has been presented. The strategy to disaster management emphasizes issues related to disaster analysis, prediction, prevention, assessment, response and post-disaster rehabilitation and reconstruction. From informatics point of view, disaster management depends on intelligent systems which are capable of exactly and automatically extracting information from complex, uncertain, sometimes even contradictory information sources from different kinds of sensors such as visible-infrared, radar, LIDAR and hyper-spectrum sensors and managing the extracted information to help users to make decisions. To this end, the major focuses of the research is directed towards to develop efficient road extraction techniques.

Recently, new generation of high resolution satellite imagery, such as IKONOS and QuickBird, has provided a viable alternative to aerial photography. Unfortunately, these data have not been readily adopted on civil engineering application such as city planning, metropolitan mapping and emergency responding. In the applications of high resolution remote sensing images, road extraction is the most basic and important task. Though many road extraction algorithms for satellite or aerial images have been proposed in the past years, most of them extract roads either from grayscale and low resolution remote sensing images (Gruen *et al.*, 1995, 1997; Agouris and Stefanidis, 1999; Baltsavias *et al.*, 2001) or focusing on rural areas (Heller *et al.*, 1998; Wang and Trinder, 2000; Wiedemann and Ebner, 2000). The objective of this paper is to present an effective approach to road extraction from high resolution remote sensing images which is suit for complicated urban areas, especially city resident areas. The proposed method utilizes the fuzzy relation based segmentation algorithm and colour similarity measure in the RG colour space which has been developed in our previous works. The proposed method follows three stages, (1) calculating colour similarity measurement in the RGB colour space, (2) segmenting colour satellite imagery using fuzzy relation based segmentation algorithm, (3) extracting central lines of roads by a post-processing procedure. Many high resolution images with different scenes from simple rural road to complicated roads in urban area and residential area have been examined and the testing results show that most centrelines match well their roads, though they do not locate accurately on the centres in some parts of the roads. Those situations occur because of either the existence of cars or shadows on roads or the irregular colour feature for the roads. From the analysis of results, it can be concluded that the proposed method is effective in both visual effect and positional accuracy.

The paper is organized as follows. Section 2 introduces notations for fuzzy relation and colour similarity measure from our previous work. The proposed new method for road extraction is described in Section 3. In section 4, the results on extracted road network from QuickBird and IKONOS colour images by using the proposed method are given. Finally, in section 5, conclusions are drawn.

2. Backgrounds

2.1 Fuzzy Relation

In this section, the basic concepts for fuzzy relation are briefly introduced. Details on fuzzy set and fuzzy logic can be found in Gottwald (1979), Dubois and Prade, (1980), George and Bo (1995).

Let $X = \{x_i, i \in I, I = \{1, 2, \ldots, n\}\}$ be a finite set. A binary relation on X is a subset of the Cartesian product $X \times X$, that is, $R(X, X) \subseteq X \times X$. On the other hand, a relation can also be defined by a characteristic function which assigns a value of 1 to every 2-tuple of X belonging to the relation and 0 to every 2-tuple no belonging to it as follows.

$$R(x_i, x_j) = \begin{cases} 1 & \text{iff } < x_i, x_j > \in R \\ 0 & \text{otherwise} \end{cases} \quad (1)$$

The characteristic function of the relation can be generalized to allow 2-tuples of X to have degrees of membership to define a fuzzy relation on X. Thus, a fuzzy relation on X can be expressed by the characteristic function R: $X \times X \rightarrow [0, 1]$.

Generally, fuzzy relations on X can be distinguished on the basis of three different properties: reflectivity, symmetry, and transitivity. Let $R(X, X)$ be a fuzzy relation on X. $R(X, X)$ is reflexive iff $R(x_i, x_i) = 1$ for all $x_i \in X$; $R(X, X)$ is symmetry iff $R(x_i, x_j) = R(x_j, x_i)$ for all x_i, $x_j \in X$; $R(X, X)$ is max-min transitive if $R(X, X) \geq \max \min\{ R(x_i, x_k), R(x_k, x_j)\}$ for all $<x_i, x_j> \in X \times X$ and $x_k \in X$.

A fuzzy binary relation $R(X, X)$ that is reflexive and symmetric is called proximity relation. If a fuzzy binary relation $R(X, X)$ is reflexive, symmetric and transitive, it is called the similarity relation.

2.2 Color Similarity in the RGB Color Space

The RGB color space can be viewed as a 3-dimensional vector space, in which each color is represented by a discrete vector $\boldsymbol{v} = [v_R, v_G, v_B]$, where v_R, v_G and v_B are the Red, Greed, and Blue components of the color. In order to characterize the relationship among colors, the color similarity measure (μ) in the RGB color space is defined as follows (Li, 2004).

$$\mu(\boldsymbol{v}_i, \boldsymbol{v}_j) = e^{-k_1 d(v_i, v_j)} \cos\left(k_2 \theta(\boldsymbol{v}_i, \boldsymbol{v}_j)\right) \quad (2)$$

where k_1, $k_2 \in [0, \infty)$, $\boldsymbol{v}_i$ and $\boldsymbol{v}_j$ are any two color vector in the RGB color space, and d and θ are distance and angle between $\boldsymbol{v}_i$ and $\boldsymbol{v}_j$ and defined as follows.

$$d(\boldsymbol{v}_i, \boldsymbol{v}_j) = \left((v_{iR} - v_{jR})^2 + (v_{iG} - v_{jG})^2 + (v_{iB} - v_{jB})^2\right)^{1/2} \quad (3)$$

$$\theta(\mathbf{v}_i, \mathbf{v}_j) = \arccos \frac{v_{iR} v_{jR} + v_{iG} v_{jG} + v_{iB} v_{jB}}{\left(v_{iR}^2 + v_{iG}^2 + v_{iB}^2\right)^{1/2} \left(v_{jR}^2 + v_{jG}^2 + v_{jB}^2\right)^{1/2}} \quad (4)$$

3. Fuzzy relation based road extraction method

This section describes the proposed road extraction method which follows two steps. Firstly, by using a fuzzy relation based colour segmentation algorithm, the objects in a colour remote sensing are separated and roads are extracted in terms of the colour features of roads. Then a post-processing procedure is executed on extracted road images, which involves filtering, thinning and overlying.

3.1 Fuzzy Relation Based Colour Segmentation Algorithm

The idea behind the proposed algorithm can be stated as follows. Let R is a similarity relation on set X, each α-cut of the fuzzy relation, R_α, is a crisp equivalence relation that represents the presence of similarity between the elements to the degree α and induces a partition of X. Let $\pi(R_\alpha)$ be the partition corresponding to the equivalence relation R_α. Then each similarity relation R is associated with the set $\Pi(R) = \{\pi(R_\alpha) \mid \alpha \in (0, 1]\}$ of partition of X. The critical issue of using this idea to make a partition is how to create a similarity relation with respect of considered problem.

In this study, a colour image represented in the RGB colour space is viewed as a vector set $V = \{\boldsymbol{v}_1, \boldsymbol{v}_2, \ldots, \boldsymbol{v}_n\}$ where $\boldsymbol{v}_i$, $i = 1, 2, \ldots, n$, is the colour vector corresponding to ith pixel and n is the number of pixels in the image.

It is worth note that the colour similarity measure defined in Equation (2) induces the fuzzy relation on V, that is, $R\ (\boldsymbol{v}_i, \boldsymbol{v}_j) = \mu\ (\boldsymbol{v}_i, \boldsymbol{v}_j)$, $\forall \boldsymbol{v}_i, \boldsymbol{v}_j \in V$. it can be concluded that the fuzzy relation is a proximity relation since it is reflexive and symmetric. For the proximity relation, its transitive closure R_T can be determined by a simple algorithm that consists of the following three steps (George and Bo Yuan, 1995):

- $R' = R \cup (R \circ R)$
- if $R' \neq R$, $R = R'$
- if $R' = R$, $R_T = R$

where the operator $\circ$ is the standard composition operation (Let $O = [o_{ik}]_{n\times n}$, $P = [p_{kj}]_{n\times n}$ and $Q = [q_{ij}]_{n\times n}$ are binary relations such that $Q = O \circ P$, then $q_{ij} = \max\ \{\min\ \{o_{ik}, p_{kj}\}, k = 1, 2, \ldots, n\}$) and the operator $\cup$ is the maximum operator Let $O = [o_{ik}]_{n\times n}$, $P = [p_{kj}]_{n\times n}$ and $Q = [q_{ij}]_{n\times n}$ are binary relations such that $Q = O \cup P$, then $q_{ij} = \max\ \{o_{ik}, p_{kj}\}$). It can be proved that R_T is a fuzzy equivalence relation on V.

After obtaining R_T, the colour image segmentation algorithm can be designed as follows.

***Step* 1:** Computing the level set of R_T,

$$\Lambda = \{\alpha \mid r_{ij} = \alpha, \forall\ r_{ij} \in R_T\} \tag{5}$$

***Step* 2:** For $\alpha \in \Lambda$ from step 1, the α-cut of R_T is a matrix $R_{T\alpha} = [r_{\alpha ij}]_{n\times n}$ given by

$$r_{\alpha ij} = \begin{cases} 1 & r_{ij} \geq \alpha \\ 0 & r_{ij} < \alpha \end{cases} \tag{6}$$

***Step* 3:** By constructing a partition tree for all α-cut matrixes, a hierarchical partition can be obtained. Giving α, a partition of R associated to α can be obtained.

3.2 Post-processing Procedures

The post-processing procedures involve filtering and thinning of roads extracted by segmentation procedure.

Since the complexity of the scenes in remote sensing images, the extracted objects always are contaminated by other objects. For example, the shapes of roads can be affected by cars and shadows (formed by trees and buildings) on the roads. Otherwise, some no-road objects having similar colour feature with roads can also be misclassified as the roads. In order to extract roads correctly, some filtering procedures is necessary. To this end, a series of binary morphological operations are used in this study. For example, binary dilation is employed to eliminate the effects brought by cars and shadows. And binary direct dilation is used to clear off misclassified no-road objects having nonlinear shapes (Li *et al.*, 2002).

In order to obtain the centrelines of roads, the thinning procedure is carried out (Zhang *et al.*, 1984). Finally, road centrelines are overlaid on the original testing images to demonstrate the efficiency of the proposed algorithm.

4. Experiments and results

The proposed road extraction algorithm has been tested on two types of high resolution remote sensing images, including QuickBird and IKONOS images. In this paper, the results of road extraction for two typical resident areas are illustrated. The reason for this choice is that they have complicated scenes and irregular road shapes. The test images have a size of 250 × 250 pixels and 1 *m* resolution form Toronto area, see Figure 1.

(a) (b)

Fig. 1 Testing images: (a) QuickBird, (b) IKONOS.

The pseudo-colour segmented images generated from the tested images are given in Figure 2 which are obtained by $\alpha = 0.9$. it can be observed from the segmented images that the objects in Figure 1 (a) are automatically catalogued four classes where the road network is indicated by yellow, while the objects in Figure 1 (b) are classified five groups where the road network is demonstrated by red.

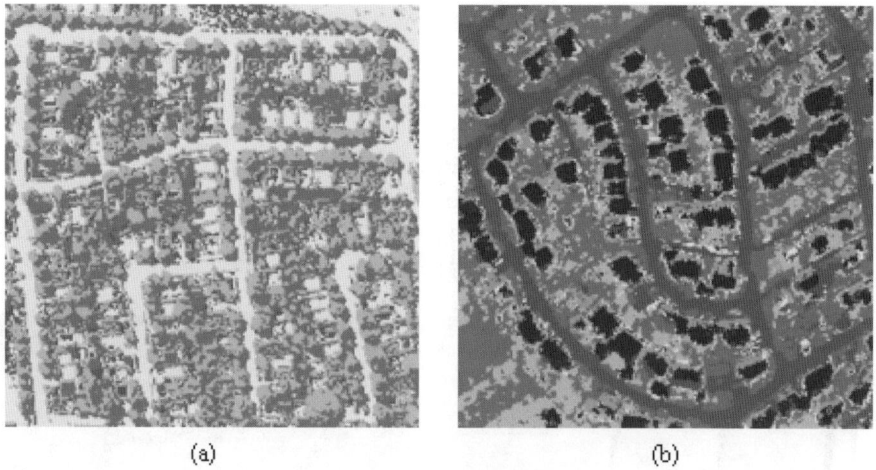

Fig. 2 Segmented images: (a) QuickBird, (b) IKONOS.

Figure 3 shows the binary images of the road network extracted from the segmentation images shown in Figure 2. It can be observed from Figure 3 that the extracted roads are corrupted by other objects with similar colours to objects.

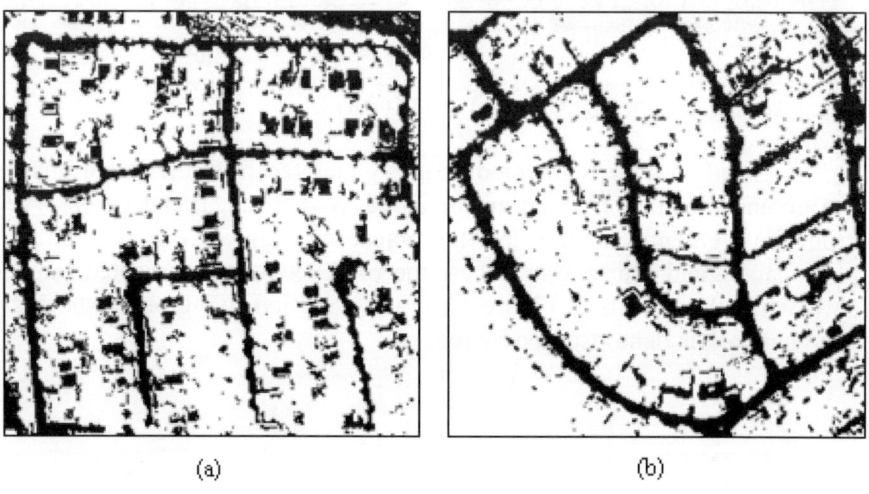

Fig. 3 Binary road net images: (a) QuickBird, (b) IKONOS.

Figure 4 shows the road regions obtained after filtering the road images depicted in Figure 3 using the binary morphological operators. A visual comparison of the images clearly favours the filtered images (see Figure 4) over the road images shown in Figure 3. Figure 4 shows the results obtained by filtering Figure 3 using binary direct dilating with a structuring element of 1 × 20 which eliminates the objects with length less than 20 pixels (20 *m* in ground distance).

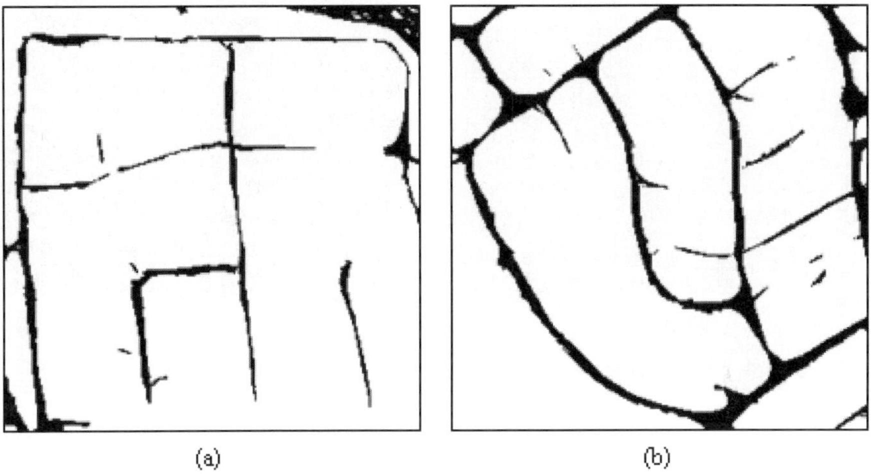

Fig. 4 Filtered binary road net images: (a) QuickBird, (b) IKONOS.

The road centrelines are delineated using the thinning algorithm (Zhang et al.; 1984) and the results are shown in Figure 5.

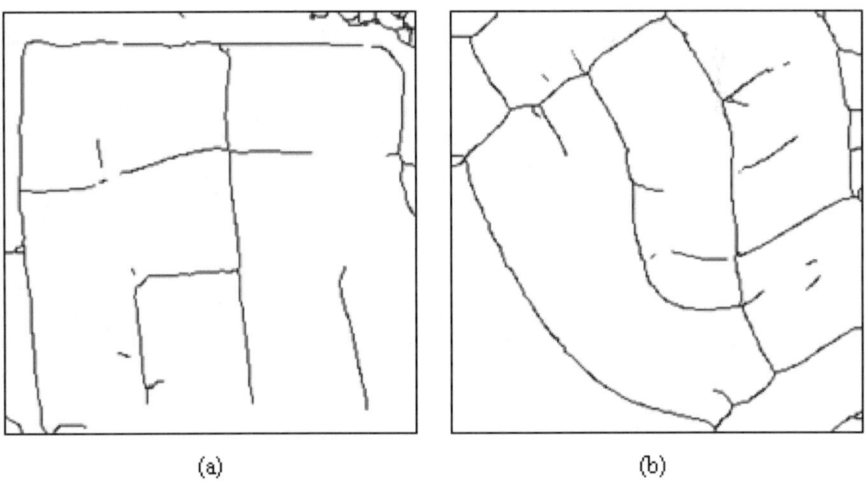

Fig. 5 Road centreline images: (a) QuickBird, (b) IKONOS.

In order to illustrate the accuracy, the extracted road centrelines are overlaid on the original image, see Figure 6. In the overlying images the thin red lines indicate the road centrelines. It can be obverted in Figure 6 that the road centrelines match the roads well.

Fig. 6 Overlay testing images and road centrelines: (a) QuickBird, (b) IKONOS.

5. Conclusions

This paper proposed a new method for extracting roads from high resolution remote sensing images such as QuichBird and IKONOS. This method employs the fuzzy relation based segmentation algorithm and works in three steps: (1) calculating colour similarity measurement in the RGB colour space, (2) segmenting colour satellite imagery using fuzzy relation based segmentation algorithm, (3) extracting central lines of roads by a post-processing procedure.

The contributions in this study are: (1) extent the fuzzy relation to solve colour image segmentation problem, (2) propose a method to induce a fuzzy relation from colour similarity point of view, (3) develop a road extract method which use colour and shape features. The advantage of this method is to automatically determinate the number of classes in segmentation. In most of situations, it is difficult to specify any desired number of clusters. For example, the situations often happen in the segmentations of remote sensing images, because the ground truth is always not available for these images.

The proposed method is examined by extracting road networks from QuickBird and IKONOS imagery. The results show that the proposed method for road extraction is very effective. In order to illustrate the accuracy, the extracted road centrelines are overlaid on the original images.

Acknowledgements

The research was partially supported by a Natural Sciences and Engineering Research Council of Canada (NSERC) discovery grant. The PhD research scholarship provided by the School of Graduate Studies at Ryerson University is also acknowledged.

References

Agouris, P., Stefanidis, A. (Eds.), 1999. Integrated Spatial Databases- Digital Images and GIS. Springer-Verlag, Berlin.

Baltsavias, E., Gruen, A., van Gool, L. (Eds.), 2001. Automatic Extraction of Man-Made Objects from Aerial and Space Images (III). Balkema, Lisse, The Netherlands.

Dubois, D. and H. Prade, 1980. Fuzzy Sets and Systems: Theory and Applications, Academic Press, New York, USA.

George, J. K. and Y. Bo, 1995. Fuzzy Sets and Fuzzy Logic: Theory and Applications, Prentice

Hall PTR, Upper Saddle River, New Jersey, USA.

Gottwald, S., 1979. Set theory for fuzzy sets of higher level. Fuzzy Sets and Systems, 2, 125-151.

Gruen, A., Kuebler, O., Agouris, P. (Eds.), 1995. Automatic Extraction of Man-Made Objects from Aerial and Space Images. Birkha¨user Verlag, Basel.

Gruen, A., Baltsavias, E., Henricsson, O. (Eds.), 1997. Automatic Extraction of Man-Made Objects from Aerial and Space Images (II). Birkha¨user Verlag, Basel.

Heller, A., Fischler, M., Bolles, R., Connolly, C., 1998. An integrated feasibility demonstration for automatic population of spatial databases. Proc. of Image UnderstandingWorkshop. Morgan Kaufmann, San Francisco, CA, pp. 403–426.

Li, J., Y. Li and H. Dong, 2002. Automated extraction of urban road networks from IKONOS imagery using a fuzzy mathematical morphology approach, International Archives of Photogrammetry, Remote Sensing and Spatial Information Sciences, 34(2II): 259-263.

Li, Y., 2004. Fuzzy similarity measure and its application to high resolution colour remote sensing image processing, Master Thesis, Ryerson University.

Wang, Y., Trinder, J., 2000. Road network extraction by hierarchical grouping. International Archives of Photogrammetry and Remote Sensing, vol. 33, Part B3/2, pp. 943–950.

Wiedemann, C., Ebner, H., 2000. Automatic completion and eval uation of road networks. International Archives of Photogrammetry and Remote Sensing, vol. 33, Part B3/2, pp. 979–986.

Zhang, T. Y. and C. Y. Suen, 1984. A fats parallel algorithm for thinning digital patterns. Communications of the ACM, 27(3), 236-239.

Development Of Processing Chains For Rapid Mapping With Satellite Data

Yves A. Buehler and Tobias W. Kellenberger

Remote Sensing Laboratories, Department of Geography, University of Zurich, Winterthurerstrasse 190, 8057 Zurich, Switzerland ybuehler@geo.unizh.ch

Abstract

In case of a disaster on a large scale the International Charter of Space and Major Disasters (Charter) provides satellite data from different sensors. Since 2000 more than 140 Charter Calls have proven the great value of this fast disaster response instrument but have also shown different handling problems causing time delay and downgrading the value of rapid mapping products. This paper proposes a framework for a rapid mapping processing chain based on the experiences gathered during the 2005 flood crisis in Switzerland. It focuses on the pre-processing of satellite data, the value-adding step and the visualization of the results. Critical elements are hereby the availability of essential datasets, the processing time, the information extraction and the usability of the products for the end-users. International programs containing rapid mapping elements as PREVIEW or GMES RESPOND and RISK-EOS need further research on the above-mentioned topics to improve the usability and significance of their products and to improve the benefit of satellite data for disaster management support.

1 Introduction

The great value of earth observation data to map the extent and damage of a disaster has been shown in different cases (e.g. Voigt et al. 2005, Allenbach et al. 2005). The International Charter on Space and Major Disasters (Charter) urges the member Space Agencies to acquire satellite images with highest priority and to deliver the datasets to Authorized Users (AU) as fast as possible. The datasets are delivered at no charge during a natural or man-made disaster on a large scale. Table 1 is a survey of the current members of the Charter with its earth observation satellites resources.

Table 1 The Charter members and their space resources (up to January 2007)

Member	Space Resources
European Space Agency ESA	ERS ENVISAT PROBA
Centre national d'études spatiales CNES (France)	SPOT
Canadian Space Agency CSA	RADARSAT
Indian Space Research Organisation ISRO	IRS
National Oceanic and Atmospheric Administration NOAA (USA)	POES GOES
Argentina's Comisión Nacional de Actividades Espaciales CONAE	SAC-C
Japan Aerospace Exploration Agency JAXA	ALOS
National Aeronautics and Space Administration NASA (USA)	ASTER TERRA AQUA
United States Geological Survey USGS	LANDSAT
Disaster Monitoring Constellation DMC (led by UK with the partners Algeria, Nigeria, and Turkey)	UK-DMC ALSAT-1 NigeriaSat BILSAT-1
China National Space Administration	Beiijng-1

Furthermore, Taiwan's National Space Organization (Formosat-3), commercial companies such as Digital Globe (QuickBird) and GeoEye (Ikonos, OrbView-3) have agreed to make their data available through France's Centre National d'Etudes spatiales (CNES) and U.S. Geological Survey (USGS), respectively. In the near future, German TerraSAR, Italian Cosmo and the Chinese/Brazilian CBERS Satellites will join the Charter too. These satellites resources have increased the temporal resolution of appropriate data during the last years substantially. The availability of valuable data with spatial resolution better than 10 m is adequate today and will further improve. Especially in remote regions with poor accessibility, satellites are the best and in most cases only source available for up to date and reliable geo-information on a large scale. Independent of the remoteness and developing state of a country, disasters (e.g. strong earthquakes or floods) can destroy essential traffic lines and communication networks. Geo-information derived from satellite data is of great value for the planning and accomplishment of the disaster response actions. To further improve the benefit from satellite data for disaster management, research into data pre-processing, the information extraction and the appropriate visualisation under pressure of time has to be intensified.

In August 2005, a stationary Vb cyclone transported very humid air from the Adriatic Sea towards the mountain range of the Alps and caused extremely heavy precipitation in the central part of Switzerland. It was the highest amount of rainfall ever measured since more than 100 years in the Alpine area. The subsequent floods, landslides and mudflows (Fig. 1) affected large parts of Switzerland, killed six people and caused damage of more than €1.5 billion, mainly on traffic-lines, buildings and farmland (FOWG 2005). The National Emergency Operations Centre (NEOC) of Switzerland as AU invoked the Charter to get up-to-date satellite data to support the disaster management process. This Call was the 100th invoking of the Charter since it's operational status in 2000 and is known as Charter Call 100. The Swiss National Point of Contact for Satellite Images (NPOC) was responsible as Project Manager (PM) and Value Adding Agency (VAA) to process and interpret the delivered data. The first dataset arriving at the NPOC was a SPOT5 scene acquired three days after the peak of the flooding (Seidel et al. 2006).

Fig. 1 Left image: flooded campground at the shoreline of lake Brienz BE, right image: large mudflow in the village of Brienz BE; images © Swiss Air Force

A variety of issues and bottlenecks concerning the application of satellite data for disaster management support occurred at the NPOC along with this crisis event. The experiences within this case of emergency showed us the need for more research into processing of earth observation satellite data for rapid mapping applications. This paper deals with the main problems tackled during the 2005 flooding and proposes a framework for an improved processing chain.

2. Experiences within Charter Call 100

Experiences from earlier Charter Calls point out the response time, availability and usefulness of the data, the narrow scope of the Charter and legal problems as main issues (Ito 2005). During the chain of Charter Call 100 we also faced other problems that have important impact on the value of the rapid mapping products.

2.1 Data acquisition and delivery

The point in time when the Charter is invoked is crucial. In case of Call 100, the procedure was started not until three days after the disaster peak and therefore caused a loss of valuable time. This time-delay can be avoided if the AU is better aware of the possibilities and limits of earth observation data. During the Call 100, first SPOT5 satellite data arrived at the VAA six days after the disaster peak. The availability of trained and well-prepared VAA that can advise the AU and can take over the project management, data pre-processing and value adding steps is very important. To change the operational tasking of an earth observation satellite is a difficult and time-consuming challenge that needs up to 24 hours time depending on the satellite system and the available ground stations. With the establishment of fast and easy to use data distribution networks and 24/7 operational distributors, project managers and VAA, time can be gained within each step of data delivery chain. The selection of appropriate sensors, which depends on the requirements of the affected region, is fundamental. Especially the spatial resolution but also other sensor specific properties as the spectral- and the radiometric resolution can affect the suitability of a sensor. Meteorological conditions and other external factors such as terrain-effects can additionally cause problems for the data processing. Not all satellite data is adequate for every disaster-type. During the Charter Call 100, RADARSAT-1 data acquired in standard beam mode with a spatial resolution of ca. 25 × 28 m was delivered. Radar data, generally not affected by clouds, is a good option to monitor precipitation-triggered disasters. But within Call 100 the spatial resolution was not sufficient to classify the flooded areas of small spatial extent typical for the rough terrain of the Swiss Alps. A careful selection of adequate sensors and sensor-modes is important for achieving optimal results.

2.2 Data processing and information extraction

Data processing has to be well planned and organized before a disaster strikes. To achieve a fast and precise orthorectification it is essential to have reference datasets ready to use. Orthorectification is a prerequisite for combining satellite data with other geo-information and performing change detection analysis. In regions with a poor availability of high quality geo-data this can be a difficult and time-consuming prearrangement. Even in regions with sufficient reliable geo-information the data must be easily accessible and pre-processed in the appropriate format. High-capacity hardware, suitable image processing software and skilled staff members have to be available to avoid an additional time delay.

Optical satellite data is highly dependant on cloud cover. In the case of Call 100, dense clouds covered for several days after the heavy rainfall more than 80% of the affected area. Luckily the main valleys containing the lakes and rivers causing the big part of the damage were partially cloud free at the time the data was acquired. Nevertheless, cloud- and cloud-shadows are major problems for the extraction of the flooded areas due to spectral similarity in most bands compared to water (Fig. 2). Additionally, illumination effects and the geometric and radiometric influences of the rough terrain aggravate the classification. Methods to handle these problems should be developed and tested in advance to gain valuable time in a disaster case.

Fig. 2 A section of the SPOT5 near-infrared band showing the spectral similarity of flooded areas and shadows

2.3 Visualization and product delivery

The link between the VAA and the end-users of rapid mapping products is not established well enough in Switzerland and other countries today. There are only very few research studies published dealing with the needs of end-users as disaster response teams and decision makers. Most persons in these positions are not used working with satellite data. Critical information should be visualized in a manner that can easily be interpreted by non-professionals. Further feedback from end-users has to be gathered and analyzed and investigations about the most needed information have to be accomplished. New technologies such as the World Wide Web, Google-Earth and focused geo-information systems may be used to improve the usability of rapid mapping products for end-users by better accessibility, faster updating and improved availability of datasets from remote regions.

3 Rapid mapping processing chains

Predefined processing chains are tools to optimize the process of the Charter Call. They help to mitigate some of the main problems discussed above. Though data from different sensors and distinct disaster types have to be processed differently, the framework persists in most cases. Fig. 3 shows a processing chain proposed in a case where optical- and/or radar satellite data is applied. This proposition is mainly based on the experiences gathered during the Charter Call 100 in Switzerland.

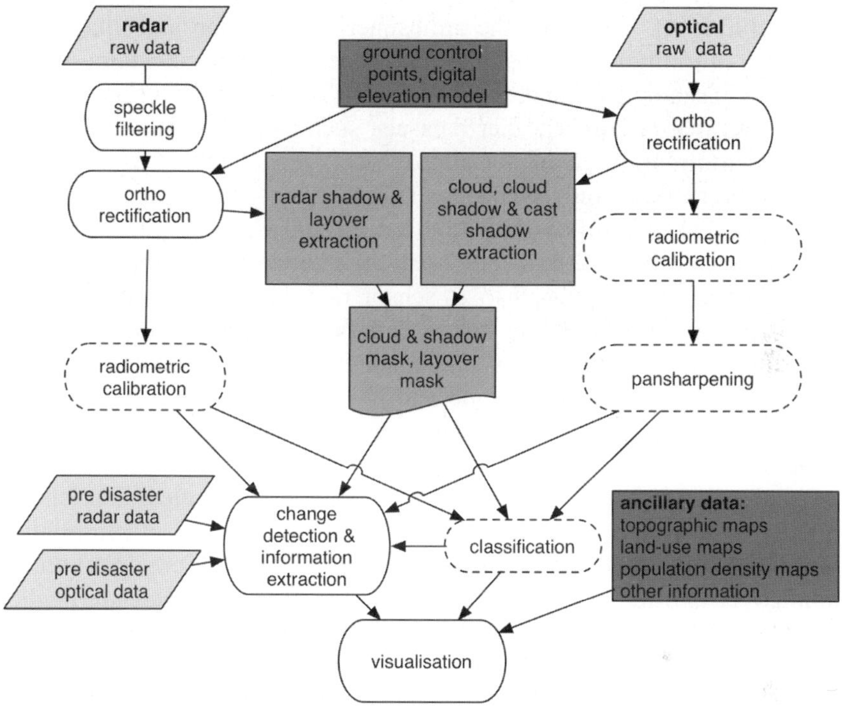

Fig. 3 Framework for a rapid mapping processing chain. Dashed boxes indicate processing steps that are optional.

3.1 Data pre-processing

The data delivered by the Charter is usually not or only inaccurately orthorectified. Precise orthorectification is crucial for applying change detection algorithms and combining the satellite images with ancillary data. It is very important to store a reliable set of ground control points for the potentially affected regions. These dataset should be available, ready to use and well maintained before a disaster event. Especially in regions with poor geo-data coverage, this is n important point. If no reliable sets of geo-data are available, the pre- and post disaster satellite data may also be orthorectified by an image-to-image matching algorithm. But the subsequent application of ancillary datasets will be more difficult and inaccurate.

In case of change detection algorithms it is important to process the images acquired at different points in time under various conditions to a comparable level. Therefore the atmospheric calibration and illumination correction are important steps within a processing chain including change detection. On the other hand, the radiometric calibration is a time consuming process that requires further datasets, such as sensor- and orbit metadata and information about the atmosphere. In a Charter Case it is very likely that not all information needed is immediately available. The extraction of flooded areas in case of Charter Call 100 was possible without radiometric calibration of the satellite data. For better comparability, the data sets may also be normalised to at sensor radiance using gain and offset calibration values. Depending on the data type, the image quality and the available time, an illumination- and/or an atmospheric correction can further improve the results (Lillesand and Kiefer 2000). The optimisation of radiometric calibration within rapid mapping chains should be a topic of further research.

Clouds, cloud-shadow and cast-shadow may affect optical datasets and partially aggravate or even prevent the extraction of critical information. Unfortunately in the case of a disaster caused by heavy precipitation, the cloud cover is often stationary for many days. Radar datasets are negligibly affected by clouds but are affected by layover, foreshortening and shadow, particularly in alpine terrain. Information extraction in these areas can cause loss of time and downgrade the results. The implementation of automatic cloud- and cloud-shadow extraction based on cloud ratios, geometric information and digital elevation models as described in (Li et al. 2003) help to improve the results considerably. Further research is needed to find the most suitable algorithm for the special requirements of rapid mapping applications.

3.2 Change detection and information extraction

The feasibility for a fast and reliable classification of the crucial land-cover items is highly dependant on the spatial pattern of the disaster and the affected area. Detection of flooded areas especially in flat terrain is simpler than the recognition of a damage caused by small landslides and mudflows. Mountainous terrain causes several additional problems such as higher probability of cloud-cover, cast-shadow, snow- and ice cover, geometric- and scale effects that have to be considered (Buchroithner 1995).

An object-based classification approach enables the implementation of additional information as intrinsic, topological and contextual information compared to the traditional pixel-based approach. Furthermore, the integration of thematic data from non- remote sensing sources is simpler (Benz et al. 2004). Our investigation into object-based classification for rapid mapping of flooded areas in mountainous terrain has yielded promising results (Buehler et al. 2005). The classification procedure in the object-based approach can be filed as a processing scheme and reactivated during a future Charter Call with comparable sensors fitted with only few adaptations. These processing schemes are developed, evaluated and enhanced prior to a disaster event. This saves a lot of handling time for a future emergency. A disadvantage of the object-based method is the high amount of processing power and memory needed, slowing down the classification process. Research on object-based classification with special consideration of rapid mapping demands should be intensified and expanded to further disaster types such as landslides, mudslides, avalanches, etc.

3.3 Visualisation

Decision-makers and civil protection staffs in the field are typically working over capacity during an emergency. In these situations the critical information has to be visualised in a coherent manner and on appropriate background to cover the needs of the disaster management. Only very few investigations into end-user requirements during a disaster event are published today (e.g. Koehler 2005). To be effective, the rapid mapping products have to consider the information network of the disaster management organisation. Fig. 4 shows an enhanced final rapid mapping product developed at the NPOC implementing the experiences made with the Charter Call 100. The information about the flooded areas has been extracted from a SPOT5 dataset and was combined with the familiar layout of the digital topographic map of Switzerland. The two water level classes are identified due to the different spectral reflection of deep and shallow water. The map allows identifying disconnected traffic routes and affected general infrastructure. In order to improve these products and to investigate the essential information for the different end users, further case studies have to be accomplished. Feedback from the different Charter Calls have to be collected, evaluated and published.

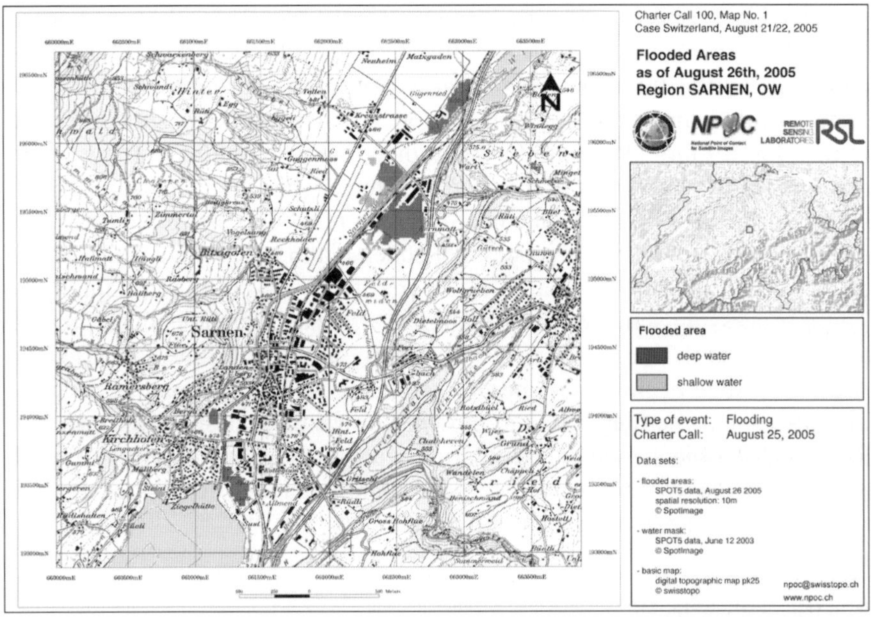

Fig. 4 Enhanced rapid mapping product showing flooded areas extracted from satellite data superposed on top of the topographic map of Switzerland

4 Discussion and Conclusions

Our experience with Charter Call 100 during the flooding 2005 in Switzerland revealed various critical elements of the processing of earth observation satellite data for rapid mapping. Considering the current technical possibilities it needs to be possible to deliver the first appropriate satellite datasets to the VAA within 24 hours after a disaster event. In another six to twelve hours, first rapid mapping products need to be available. Invoking the Charter, the tasking of appropriate satellites and the data delivery may result in major loss of time. The lack of important ancillary data, lack of experience with the data pre-processing and information extraction and unsuitable hard- and software are time constraining but preventable elements. With the predefinition of the data pre-processing approach, information extraction and the visualisation within a processing chain, the rapid mapping workflow can be optimised and accelerated. We propose a framework for a processing chain based on the experiences gathered with Charter Call 100. The timely availability of different base-datasets in the proper format is very important for rapid processing. Table 2 specifies the most important datasets required to be available in high quality prior to a disaster event. We are prepared to deliver first rapid mapping products for Switzerland within four to six hours after the arrival of the raw satellite data. The availability of the basic datasets, the preparation of optimized processing chains and the training of the staff members are identified to be the most important preconditions.

Table 2. Most important basic datasets for the rapid mapping chain

Dataset	Application in the processing chain
Digital Elevation Model DEM	Orthorectification, radiometric correction, cast shadow extraction, classification, information extraction, visualisation
Satellite orbit information	Orthorectification, radiometric correction
Topographic maps, aerial images	Orthorectification, visualisation
Ground control points GCP	Orthorectification
Archive satellite data (cloud-free, same season)	Change detection, visualisation
Ancillary data (land-use maps, population density maps, map of sensible infrastructure etc.)	Visualisation, information extraction

Within the rapid mapping processing chain, we observed a major lack of research on data pre-processing, information extraction and end-user requirements. The importance of rapid mapping for disaster management support has been recognized and different international initiatives for operational applications have been initiated. UNOSAT and GMES RESPOND are already well-established programs with the aim of making rapid mapping products accessible for the humanitarian community. PREVIEW was founded recently to develop and produce enhanced information service for risk management on a European level. These programs require further research on rapid mapping processing chains in order to improve the suitability and significance of their products for disaster management. The chance of getting appropriate satellite data for rapid mapping applications improves with every Earth observation satellite that is launched. To improve the benefit from the rising availability of earth observation data, the research into the data pre-processing and the information extraction has to be intensified.

References

Allenbach B, Andreoli R, Battiston S, Bestault C, Clandillon S, Fellah K, Henry J-B, Meyer C, Scius H, Tholey N, Yésou H, de Fraipont P (2005) Rapid EO Disaster Mapping Service: Added Value, Feedback and Perspectives after 4 Years of Charter Actions. In: IGARSS'05 Proceedings pp 4373 – 4378

Benz U, Hofmann P, Willhauck G, Lingenfelder I, Heynen M (2004) Multi-resolution, object-oriented fuzzy analysis of remote sensing data for GIS-ready information. ISPRS Journal of Photogrammetry & Remote Sensing, 58: pp 239 – 258

Buchroithner M (1995) Problems of mountain hazard mapping using spaceborn remote sensing techniques. Advances in Space Research, Vol. 15, pp 57 - 66

Buehler Y, Kellenberger T, Small D, Itten K (2005) Rapid mapping with remote sensing data during flooding 2005 in Switzerland by object based methods – a case study. In: Martin-Duque J, Brebbia C, Emmanouloudis D, Mander U (Eds) Geo Environment & Landscape Evolution II. WIT Transactions on Ecology and the Environment, pp 391 – 400

Charter (2007) International Charter on Space and Major Disasters. www.disasterscharter.org (access on 18 Jan 2007)

FOWG (2005) Federal Office for Water and Geology report on the flood event 2005. www.uvek.admin.ch/themen/umwelt/00640/00815 (access on 17 Jan 2007)

Ito A (2005) Issues in implementation of the International Charter on Space and Major Disasters. Space Policy Vol. 21, pp 141 – 149

Koehler P (2005) User-Oriented Provision of Geoinformation in Disaster Management: Potentials of Spatial Data Infrastructures considering Brandenburg/Germany as an Example. In: van Oosterom P, Zlatanova S, Fendel E (eds) Geo-information for Disaster Management, Springer, pp 171 – 180

Li M, Liew S, Kwoh L (2003) Producing cloud free and cloud-shadow free mosaic from cloudy IKONOS images. IEEE International Geoscience and Remote Sensing Symposium Vol. VI, pp 3946-3948

Lillesand T, Kiefer R (2000) Remote Sensing and Image Interpretation. John Wiley & Sons, Inc. New York

Seidel S, Kellenberger T, Buehler Y, Perret J-P (2006) Satellite response to the August 2005 floods in Switzerland: Charter Call 100. In: Ammann W, Haig J, Huovinen C, Stocker M (eds) Proceedings of the International Disaster Reduction Conference, Davos, Switzerland, pp 531 – 533

Voigt S, Riedlinger T, Reinartz P, Künzer C, Kiefl R, Kemper T, Mehl H (2005) Experience and Perspective of Providing Satellite Based Crisis Information, Emergency Mapping & Disaster Monitoring Information to Decision Makers and Relief Workers. In: van Oosterom P, Zlatanova S, Fendel E (eds) Geo-information for Disaster Management, Springer, pp 519 – 531

Automatic Generation of Remote Sensing Image Mosaics for Mapping Large Natural Hazards Areas

Yubin Xin [1,3], Jonathan Li [2], and Qiuming Cheng [3]

[1] PCI Geomatics Inc., 50 W. Wilmot Street, Richmond Hill, Ontario, L4B 1M5, Canada, xin@pcigeomatics.com

[2] Department of Geography, University of Waterloo, 200 University Avenue West, Waterloo, Ontario N2L 3G, Canada, junli@fes.uwaterloo.ca

[3] Department of Earth and Space Science and Engineering, York University, 4700 Keele Street, Toronto, Ontario M3J 1P3, Canada, {yxin, qiuming}@yorku.ca

Abstract

Remotely sensed data, ranging from satellite imagery, airborne laser scanner data, and aerial photograph, play more and more important roles in environmental monitoring, emergency response, and disaster assessment. Among the products of the broad range of applications, raster base maps, which are generated from various sources of remote sensing data, are becoming very critical for effective and efficient disaster management. The raster base maps can provide detailed topographic, land-use and land-cover information on the earth's surface in a short period or near real time. With the growing requirements of such raster base maps, the techniques which can be used for automatically correcting raw data and generating digital maps are urgently required.

This paper presents a system that consists of a set of processing steps to georeference and merge many satellite or aerial images together in order to quickly map a large geographic region. The periodic processing results can be compared and analyzed for monitoring a large emergency area. The technique makes full use of georeference and sensor model information, such as ephemeris data, geometric model, and/or GPS/INS navigation and positioning information, to automatically register and orthorectify the raw image data. Through mosaicking process, a seamless mosaicking image or image tiles is produced, which will be in a selected map projection system with consistent spatial resolution. Additionally, semi-automatic and manual editing can be performed to produce a standard map to satisfy the requirements of mapping agencies.

1 Introduction

Natural hazards can be described by two themes. The first theme can be categorized as chronic while the second is episodic or periodic (Bryant, 2005). Chronic hazards include desertification, soil degradation, and melting of permafrost. The cause could be due to humans or global warming. Peridic hazards are large magnitude events that appear over a short time period. They include phenomena such as earthquakes, tsunami, volcanic eruptions and flash floods. Human choose hazardous areas because they often offer benefits. For example, floodplains provide easily cultivated agricultural land close to a water supply. Humans in these environments are forced to predict and avoid natural calamities such as landslides, cyclones, earthquakes, and drought. A recent example is the December 2004 Tsunami, which caused a large area coastal flooding in various countries namely Sri Lanka, India, Indonesia, Bangladesh, Malaysia, and Thailand, etc. In Thailand alone, the affected area by the tsunami was covering in southern part, six provinces including Ranong, Phang-nga, Phuket, Krabi, Trang and Satun, 5,393 people were killed, 8,457 injured and 3,062 missing, approximately 58,000 people or 12,000 households affected, 4,800 houses destroyed wholly or partially, 5000 fishing villages affected, 6,000 fish vessels destroyed. The environment has also greatly affected, marine and coastal parks were damaged, some coral reefs were also destroyed. In addition, coastal flood plain which is mostly narrow, caused damage to buildings, road networks, bridges, bay or inlets, coastline, etc. Moreover, an electricity supply and telephone lines were disrupted for a couple of days (Sanguantrakool et al. 2005).

Earth observation (EO) through imaging sensors is the most important method for monitoring natural hazards. Although platforms carrying EO instruments could be space-, air-, land-, or ocean-borne, the satellite remote sensing provides a unique way for collecting the information about natural hazards around the globe. Currently satellite remote sensors, in combination with numerous other sensors, are monitoring all aspects of the Earth. Most nature hazards, such as volcanoes, landslides, severe storms, floods, wildfires, and tsunamis, are rapid-changing phenomena of the Earth system, happening at regional or local scales and scattering around the world. In order to support decision makings on hazard prevention and migration, targeted observations at high spatial/temporal resolutions are required when such a hazard is forecasted and/or detected. Given the limit swath width of the medium-to-high resolution satellite and airborne imaging sensors, such as 185 km for Landsat-7 Enhanced Thematic Mapper (ETM+), 60 km for SPOT-4 and -5 High Resolution Visible (HRV) and High Resolution Visible and Infrared (HRVIR) sensors as well as NASA Advance Spaceborne thermal Emission and Reflection Radiometer (ASTER), 70 km for IRS-1C and 1D (panchromatic), 11 km for Ikonos, 8 km for OrbView-3, and 16 km for QuickBird (Jensen 2007), mosaicking is required to georeference and merge many satellite or aerial images together in order to cover a large geographic area affected by the natural hazards

This paper presents a mosaicking system that consists of a set of processing steps to georeference and merge many satellite or aerial images together in order to quickly map a large geographic region. The periodic processing results can be compared and analyzed for monitoring a large emergency area. The technique makes full use of georeference and sensor model information, such as ephemeris data, geometric model, and/or Global Positioning System and Inertial Navigation System (GPS/INS) navigation and positioning information, to automatically register and orthorectify the raw image data. Through mosaicking process, a seamless mosaicking image or image tiles is produced, which will be in a selected map projection system with consistent spatial resolution. Additionally, semi-automatic and manual editing can be performed to produce a standard map to satisfy the requirements of mapping agencies.

The paper first briefly describes the mapping system concept and the production mapping flow. It then addresses the image correction and merging methodology including automatic image to image registration with initial direct georeferencing and generation of digital elevation models (DEMs), automatic ground control point (GCP) error detection and sensor model refinement, orthorectification, seamless cutline selection and mosaicking. The proposed image correction system is an automated system, which consists of multiple processing component functions in a scripting environment. The scripting work flow is highly customizable to deal with a large data set. It is also capable of data retrieval and delivery to enterprise database, such as Oracle 10G, or the Web via complementary subsystems. All the algorithms developed are implemented in the PCI Automatic Mapping environment.

2 Methodology

2.1 Pre-processing Data

The image import function automatically identifies the file structure of a stored set of raw image files and extracts/converts relevant information to a known format. A DEM can be obtained in various ways, for example it can be acquired directly from third party providers, or it can be generated by software with stereo image data sources.

The approximate sensor geometric models are derived either from the ephemeris/orbital data for satellite images or from the GPS/INS data for aerial photographs. GCPs, which inherently contain both image and ground coordinates, are required to detect blunders, to control the quality of geometric model, and to refine the approximate model to achieve a high-precision georeferenced result. For the purpose of improving efficiency and for automating the procedure, we select and store a set of GCPs in a database structure. All the relevant information about a GCP and the neighboring pixels that surround the GCP are included as the structure data. This structured data is called a GCP chip, while the database is called the GCP chip database. The purpose of using a GCP chip database is to permanently store GCP chip information that can be easily accessed and reused with new images in an automated procedure. We have various options to create a chip database: manually extract in an interactive environment, automatically extract from orthorectified images, or from georeferenced raw images with precise math models and corresponding elevation information using various PCI software tools.

2.2 Automatically Registering Raw Imagery

The sensor geometric model relies on two sets of data: camera/sensor information to set up the mapping equation, and GCPs to refine the parameters in the model. During the modeling stage, both forward and backward transformations between the raw pixel and line coordinates and the georeference coordinates can be established. The GCPs are used to refine the approximate model to achieve a precise georeferencing result and to detect blunders. Fig. 1 shows a testing workflow of the system.

In automating the procedure, the main problems we met with were geometric deformation and/or clouds on the raw image. Clouds on the image cause mismatching in the correlation step and also affect the radiometric balancing algorithms. The geometric deformation in some extreme cases is so serious that the pixel ground resolution at the edge of a raw image is almost two times as that of the center, and the orientation for each pixel is dependent on its position in the image. To solve these issues, a special procedure was developed with two options. The first option is to directly register each raw image onto a georeferenced image that has been orthorectified in a previous processing case with possibly various resolution and common overlapping area. Depending on the approximate camera/sensor model and DEM, a roughly orthorectified image patch surrounding the estimated position on the raw image is generated on the fly, and a correlation processing is then applied between that patch and the projected template area on the georeferenced image. Each successful correspondence will be reported and collected as a new GCP point of that raw image. Furthermore, a set of these newly collected GCP points with relevant sub-orthoimage may be stored in the format of GCP chip database for repeat use in future. Fig. 2 shows a result of image to image registration. We registered an Ikonos raw image to an orthorectified airphoto image, then orthorectified the Ikonos image. The result indicates a good overlap of the two orthorectified images to prove the method. The usage of chip database in fact is our second option. In this second option, a newly generated or existing GCP chip database will be used as precise georeference base. Similarly, we orthorectify a sub-area from the raw image by using the initial georeferencing data, the elevation, and the pixel sampling distance (PSD) from the GCP chip structure to make both the sub-orthoimage and the chip images in the same reference surface and resolution. This process can greatly reduce the affects of geometric deformation. After this preparation, we are ready to perform the critical matching procedure to register each individual GCP chip onto the sub-orthoimage, and project into the raw image. Assuming all the GCPs are located on edge corners, we set up the following hypothesis between the sub-orthoimage and the chip image:

- They have same edge features and similar corner points.
- The GCP should be located on or very close to one of the extracted corner positions.
- There is the same number of intersected edge lines crossing the corresponding GCP point. And, those conjugate linear edge features have similar geometric and radiometric characterizations.
- Violation to any one of the matching criteria leads to an incorrect match and the candidate position should be filtered out from the matching list.
- Successfully passing all the matching criteria results in a correct position.

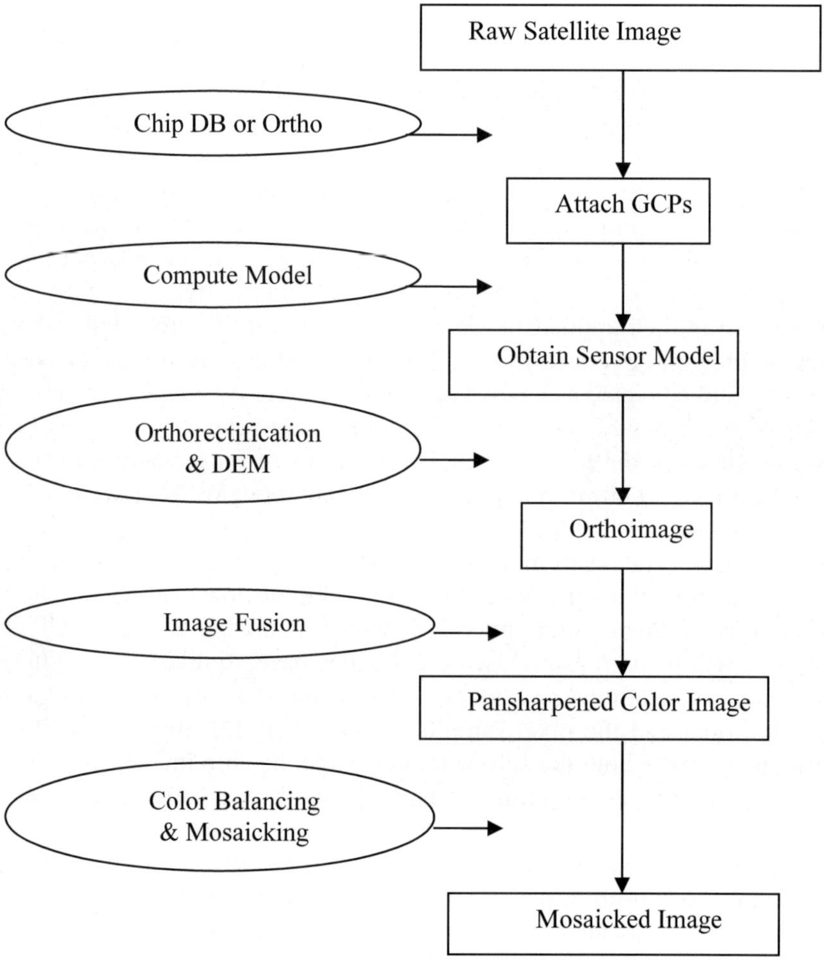

Fig. 1 Diagram of the system workflow

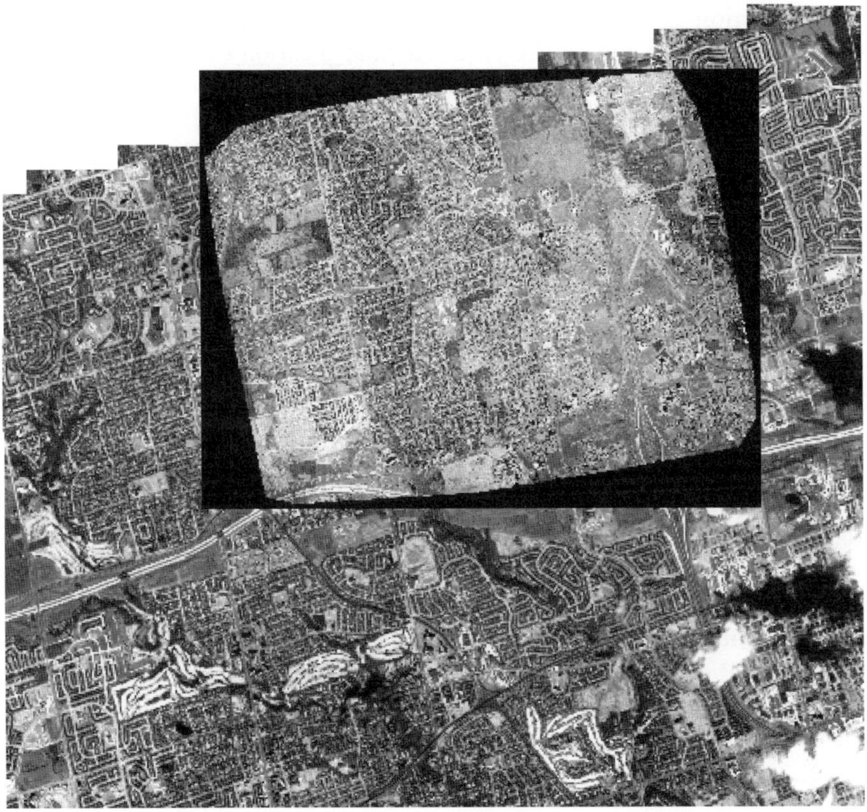

Fig. 2 Orthorectified result of image to image registration with different resolutions and sensor types – Ikonos raw image to aerial orthoimage.

3 Implementation Procedures

3.1 Edge detection

To locate each edge precisely, we firstly check and preprocess the raw image quality. If a raw image data contains noise and blurriness, a Gaussian smoothing is performed on the sub-area. Then the gradient in horizontal, vertical, and diagonal directions are calculated and stored for both edge detection and corner detection. In the edge detection step, weak edges (and also weak corners) always exist. To avoid skipping those features, we consider about all the four directional gradients and compare them with corresponding thresholds to decide each edge pixel. The output of the gradient edge detection is a binary edge image but it may contain wide ridges around the local maxima. The required final position of the edge lies approximately in the middle of this wider edge (Habib et al. 2003). To extract the central position of the edge, a thinning (or skeleton) filter is applied. We have implemented the Kreifelts (1977) algorithm to filter out non-skeleton pixels.

3.2 Corner detection

Since a GCP is always located near a corner feature, we can capture corner structures in patterns of intensities (Watson 2006). Using the gradient images, we immediately calculate the eigenvalues of a sum-gradient matrix C. Comparing the minimal of the two eigenvalues with a pre-defined threshold, we may decide the candidate corner position.

Azimuthal characterization matching is an algorithm that was first proposed by Motrena and Rebordao (1998). The approach is to solve the rotation-variant problem existing in area-based correlation method. It is based on the autocorrelation function of one azimuthal projection around a candidate GCP and has rotation and contrast invariances. If an orientation variance (rotation) exists, the azimuthal projections of the sub-orthoimage and the GCP chip image about the corresponding central pixels will differ by a phase factor: the sets are the same but the initial elements are different. We may apply the autocorrelation function to determine the rotation angle.

3.3 Linear feature extraction and comparison

To improve the robustness, we add line comparison in the neibourhoods of the candidate point and of the GCP position in the chip. The Hough Transform line detection is performed in the thinned edge image surrounding those two positions. Usually there are two to four line sections in each patch that can be found. We can directly compare their slopes and/or the lengths and the brightness between the conjugate features to filter out mismatching.

3.4 Pyramid cross-correlation

Based on the rotation angle and the initial camera/sensor model, we may need to re-calculate the resampling to generate the sub-orthoimage for each GCP chip, then perform coarse correlation and fine correlation in a hierarchical matching process. The matched position is back projected onto the raw image (Xin and Parent 2004). Fig. 3 shows raw/orthorectified image patches and a GCP chip.

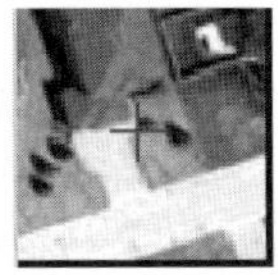

Fig.3a (AP) Fig.3b (AP) Fig. 3c (AP)
Sub-raw image Sub-ortho image GCP chip image

3.5 Recovering and Refining Sensor Model

The mathematical model relies on two sets of data: orbit/sensor information to set up the mapping equation, and ground control points to refine the parameters in the model.

The structure of the math model varies with the different sensors. The sensor geometric models define the mapping between raw image coordinates and a reference coordinate frame of the sensor, together with the imaging time. The orbital model encapsulates all knowledge about the characteristics of the satellite position and attitude in its orbit, over the image of interest. The model corrected or constructed with the ground control points forms the basis of precision mapping. In the implementation, a sensor model developed by Toutin (1995) was integrated into the PCI Geomatica software system.

3.*6* Generating Ortho and Mosaic

With an accurate georeferencing model for each image and a DEM for the working area, we can generate orthorectified images. Because of the existence of inconsistent radiometry, we may perform hot spot removal and radiometric balancing on the orthorectified images, and then merge them into an image map.

Hot spots can be described or represented as a 3D hemi-ellipsoidal surface located on an image. The third dimension represents the intensity variance. The parameters of this model can be calculated through a least squares algorithm. Using the obtained parameters we can reduce or eliminate the hot spot effect. When inconsistent radiometry exists between images, we have to apply radiometric balancing. A global bundle calculation is performed based on overlapped areas to figure out a set of correction parameters or lookup table for each image. Cutline selection is one of the most important steps. An optimization model for computing cutlines can be set up, which includes a cost function of a criteria and a set of constraints on a set of overlapped ortho images. We use optimal programming algorithm to solve the model and get polygon shaped cutlines for each ortho image. Pixels inside the cutline will contribute to the final mosaic image.

3.7 Accessing and Storing Data

The objective of the set of steps is to build a fully automatic image correction system. We are targeting on processing large data set for enterprise level usage. Beside the correctness of the algorithms, the critical consideration includes efficiency of data accessing and storage capability. In our system, Oracle 10G has been used as the central data storage and all the raw, intermediate and final results of the procedure are stored in the Oracle database and accessed through PCI vs Oracle interface engine.

4. Results

We have developed an automatic image correction and mapping system in a Python PPF (PCI Pluggable Functions) scripting environment. This is a fully automatic processing system with Python script batch processing environment supported by PCI Dynamically Linked Library (DLL). One study area is in Canada. The satellite images selected are Landsat-7 ETM+ with 15 m panchromatic (PAN) and 30 m multispectral (MS) imagery. To cover the entire country, we need more than 700 Landsat-7 image scenes. A Chip database contained 50k of various resolution image chips in this area has been used as the georeference base. The 90 m USGS DEMs were also used in this study.

After automatically collecting GCPs, a rigorous sensor geometric model was constructed for each image scene. The images were then orthorectified and a Pan Sharpening process was performed to fuse the multispectral and panchromatic imagery to obtain Pan-Sharpened orthorectified images. Finally, we mosaicked the orthorectified images into a large raster mosaic map, with hot spot removal and histogram color balancing processing. Fig. 4 shows the mosaic, which consists of four pan-sharpened images. Fig. 5 shows the mosaic generated from over seven hundreds of Pan-sharpened scenes. In this testing data set, the images were taken from different seasons of various years. So the color matching in the mosaic is not perfect. Also, some scenes are not available because of the lack of the relevant raw imagery. This example is to show the concept of our automatic procedure and some tests on the system. To obtain map-quality mosaicking result, a good quality raw images with more consistent radiometric characteristics, i.e. taken within a bounded period, is recommended. A research direction is to construct a parameter configuration matrix on each processing component based on various input/output data and their configuration parameters.

Fig. 4 Mosaic of four Landsat-7 scenes

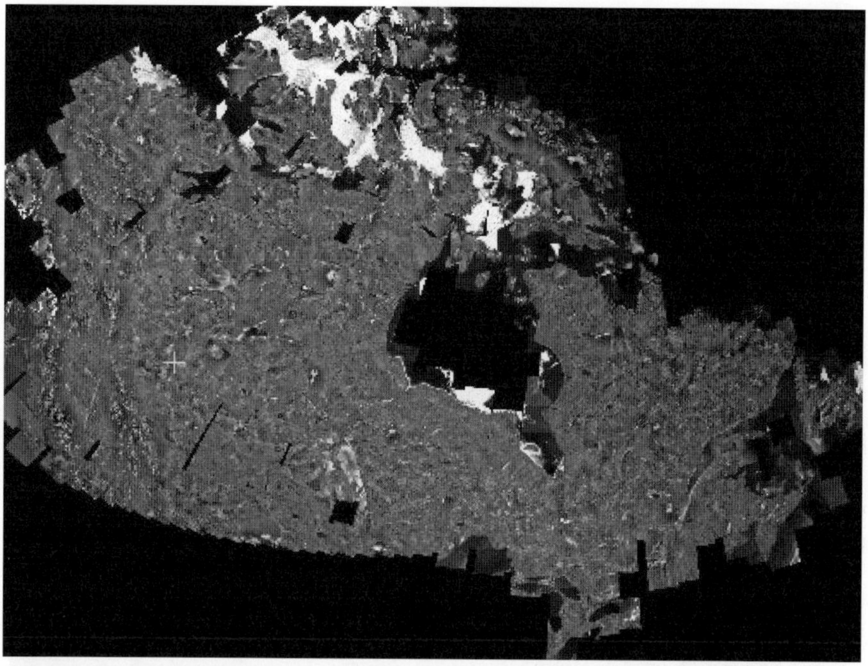

Fig. 5 Mosaic of the entire Canada with 700+ Landsat-7 images taken at different times.

5. Conclusions

The automatic mosaicking procedure described above has been developed and implemented in PCI's ProPack software product. The experiment shows very positive result in terms of efficiency, reliability and accuracy. Through the development, implementation, and analysis of the results, the following conclusions are made:

- The presented automatic procedure is implemented for fast, accurate, efficient and reliable image map generation, in a Python environment.
- The methodology and results introduced in this paper can be applied and used in various applications in environment monitoring, disaster analyzing and prediction, and crop disease detection.
- Oracle 10G is a powerful data storing and accessing system in handling raster, vector, and attribute types of data through PCI-Oracle interface software.
- The technique has been successfully applied on various types of remote sensing imagery.

- The system is able to process a large data set fully automatically.
- Although the whole procedure is highly automatic, some options and interaction are still required to guarantee the product quality.

Acknowledgments

This work was done in the development environment at PCI Geomatics, The authors would especially thank Dr. Philip Cheng and Mr. Doug Parent for their precious advices and valuable discussions.

References

Bryant E (2005) Natural Hazards, Cambridge University Press, New York, 551 pp

Habib AF, Lee Y, and Morgan M (2003) Automatic Matching and Three-Dimensional Reconstruction of Free-Form Linear Features from Stereo Images, *Photogrammetric Engineering & Remote Sensing*, 69(2): pp 189-197.

Jensen JR (2007) Remote Sensing of the Environment, Pearson Prentice Hall, Upper Saddle River, NJ, 592 pp

Kreifelts T (1977) Skelettierung und Linienverfolgung in raster-digitalisierten Linienstrukturen, in Nagel, H.-H. (Ed.), Digitale Bildverarbeitung, pp 223-231

Motrena, P. and Rebordao, J.M., 1998. Invariant models for ground control points in high resolution images, *International Journal of Remote Sensing*, 19(7): 1359-1375.

Sanguantrakool T, Pricharchon E, Phoompanich S (2005) Remote sensing technology for Tsunami Disasters Along the Andaman Sea, Thailand, 3 rd International Workshop on Remote Sensing for Post-Disaster Response, September 12-13, 2005, Chiba, Japan, 16 pp

Toutin Th (1995) Multisource data fusion with an integrated and unified geometric modelling. EARSeL Journal Advances in Remote Sensing, 4(2): 118-129.

Watson W (2006) Automated Georeferencing for Rapid Data Production, Photogrammetric Engineering and Remote Sensing, 72(4): pp 337-338.

Xin, Y. and Parent, D., 2004. Automated Procedure of a Prototype Mapping System, *Proceeding of the 25st Asian Conference on Remote Sensing,* pp *454-45*

Mapping Hazardous Slope Processes Using Digital Data

John Barlow and Steven E. Franklin

Department of Geography, University of Saskatchewan, Kirk Hall Rm 2, 117 Science Place, Saskatoon, SK, S7N 5C8
John.Barlow@usask.ca

Abstract

The use of satellite sensor data can be used to detect discrete slope processes and landforms with a high degree of accuracy. Whereas previous attempts to classify slope features using per pixel spectral response patterns have provided classification accuracies that are less than 60%, it is demonstrated that a combination of high resolution optical imagery, image segmentation and ancillary data derived from a digital elevation model can discriminate some types of mass wasting processes with accuracies of 80% or higher. The spatial resolution of the imagery is critical to the successful classification of such features both in terms of information derived from textural analysis and in the ability to successfully segment landslide features. Furthermore, the data generated in this manner can be used for geomorphic research in terms of characterizing the occurrence of mass wasting within the bounds of the image scene.

1 Introduction

This chapter deals with the creation of inventory maps for slope processes using remotely sensed data. For research being undertaken over very large areas in difficult mountainous terrain, the remote sensing approach to data collection becomes the only viable option (e.g. Malamud *et al.*, 2004; Hovius *et al.*, 1997; Rood, 1984). The accuracy of methods used in the mapping and digitizing of landslide inventories has a strong influence on the statistics generated from the inventory such as event magnitude, shape, and location (Malamud *et al.*, 2004). For this reason, such inventories have typically been generated from the manual interpretation of stereo aerial photographs with a program of ground validation. Such a process is necessarily empirical and subjective (Wills & McCrink, 2002) and is greatly influenced by the resolution of the aerial photographs (Stark & Hovius, 2001), and by the level of experience of the analyst.

Maps generated from satellite sensor data using traditional methods of image classification have produced less than satisfactory results (e.g. Epp & Beaven, 1988). Indeed, Brardinoni *et al.*, (2003) state that despite the huge advances that have been made in remote sensing, no reliable method had been developed to identify mass movements using digital image interpretation. Manual inspection of optical satellite imagery reveals a great many geomorphic forms and features. However, an automated approach to extracting these data over large areas has proven problematic. Early work indicates per pixel spectral response patterns, used in conjunction with maximum likelihood classification methods are unreliable in discriminating landslide scars from other barren areas on the landscape (e.g. Sauchyn & Trench, 1978; Connors & Gardner, 1987; Epp & Beaven, 1988). This inability to directly classify landslide features using per pixel spectral response patterns resulted in attempts to identify landslide prone sites using a combination of digital imagery and ancillary data such as DEM derivatives, soil maps, image textural analysis and digital slope profiles. While each data source has proven inadequate in classifying geomorphic features on its own, the combination of data sources provides a great deal of synoptic information regarding the surface characteristics of a given area and, consequently the geomorphic processes that are in operation (eg McDermid & Franklin, 1995; Shih & Schowengerdt, 1983)

A different approach to the automated identification of slope processes using digital data is to make use of associations between slope stability and land surface cover that occur in some slope systems (Warner *et al.*, 1994). McKean *et al.* (1991) noted relationships between vegetation, soil moisture, and bedrock morphology that were conducive to slope failure. They hypothesized colluvial deposits that provide the primary source areas for debris flows were located primarily in bedrock hollows and possessed different soil moisture characteristics and, consequently, differing vegetative properties.

More recently, Barlow *et al.* (2003) used a combination of image segmentation with Landsat ETM+ and DEM data to identify rapid mass movements; the overall accuracy was 75%. The detection and classification of individual process types (Cruden & Varnes, 1996) using an automated approach has been less successful using ETM+ data. Martin &

Franklin (2005) demonstrated that textural analysis of landslide scars may be capable of discriminating between rock slides and debris slides although the spatial resolution of image data was a limiting factor. Barlow *et al.* (2006) used a similar approach using high-resolution data and obtained classification accuracies of 80% or higher for debris slides and rock slides. A similar approach has also proven successful for the mapping of snow avalanche tracks in the Canadian Rockies (Barlow & Franklin, 2007). The innovation provided in the use of image segmentation is the ability to assign specific geomorphometric properties to differing image objects in order to place them within their geomorphic context. Both the landslide and avalanche studies will be discussed in more detail below though the use of two case studies: (1) the identification of mass movements using high resolution SPOT 5 data and (2) mapping snow avalanche tracks using Landsat TM data. The value of such data will then be demonstrated by applying the results to an assessment of slope stability and activity using the debris slide database

2 Case Studies

The following two case studies are used to demonstrate the ability of satellite sensor data to accurately map slope processes if used in combination with image segmentation and ancillary data. In both cases, only those failures with an area of over 1 ha were considered in the accuracy assessment as many smaller failures do not disrupt the forest canopy or are not resolvable by the imagery. Both examples rely heavily on image segmentation. This is a process whereby a digital image is subdivided into numerous subsets that correspond to areas of constant or smoothly varying spectral response in the scene (Nevatia, 1986). Jensen (2005) suggests that the segmentation method produced by Baatz & Schape (2000) produces representative and reproducible results. Within this method, image segmentation is accomplished through the use of a pixel heterogeneity criterion which is calibrated by the user in terms of the image channels to be used, the desired scale of the polygons, the pixel neighbourhood to be used, and the relative importance of pixel values over the smoothness of polygon boundaries (Baatz & Schape, 2000). The latter of these is of importance when using highly textured data as polygon branching can become problematic in the classification of the image objects. This is particularly true if shape criteria are incorporated into the class descriptions. This method is widely used in the eCognition software environment and was adopted for the case studies below.

2.1. Mapping Mass Movements using SPOT 5 Data

Barlow *et al.* (2006) demonstrated the utility of image segmentation in the detection of rapid mass movements over a portion of the Chilliwack Basin in southern British Columbia. The area is mountainous with elevations ranging from about 13 masl in the valley bottom to 2 283 m. The basin was heavily glaciated during the Pleistocene (Saunders *et al.*, 1987) and shows all of the features associated with alpine glaciation. Oversteepened and drift mantled slopes make debris slides, rock slides, and debris flows (Cruden & Varnes, 1996) fairly common occurrences within the basin. The mapping of these features using satellite imagery has been investigated by Barlow *et al.* (2003), Martin & Franklin (2005), and Barlow *et al.* (2006).

The data layers used by Barlow *et al.* (2006) and the classification hierarchy are illustrated in figure 1. The multispectral channels were fused with the panchromatic channel using IHS (Intensity-Hue-Saturation) in order to create a 2.5 m multispectral database (Pohl & Van Genderen, 1998). The imagery was segmented using a ratio of 8:2 for the relative importance of reflectance vs. shape and 7:3 for compactness vs. smoothness. This combination of parameters was established after several iterations in which the produced segments or objects appeared to capture the shape and size of the landslides within the study area. The process made use of an equal weighting of the spectral bands as well as the plan curvature layer.

The classification worked by progressively eliminating image objects through four boolean decision criteria. The first step divided objects into vegetated/unvegetated classes based on an NDVI (Normalized Difference Vegetation Index) threshold of 0.15. NDVI values have been demonstrated to have a high correlation with green leaf area and biomass (Kidwell, 1990). This accounts for over half of the image objects in the study area. The next level assigns each of the unvegetated image objects to either flatland or steepland based on the slope layer. Here the threshold between the two was set at 0.27 (15 degrees), as no rapid mass movements were observed below this gradient. All of the objects that were classified as steepland were then evaluated based on a length to width shape criterion. Rapid mass movements are generally identified as long thin features. Empirical inspection of the aerial photographic inventory demonstrated that mass movements had a length to width ratio of 2.5 or higher. Therefore, this threshold was required to be classified as a thin feature, whilst the remaining objects were classified and labeled as 'square features'.

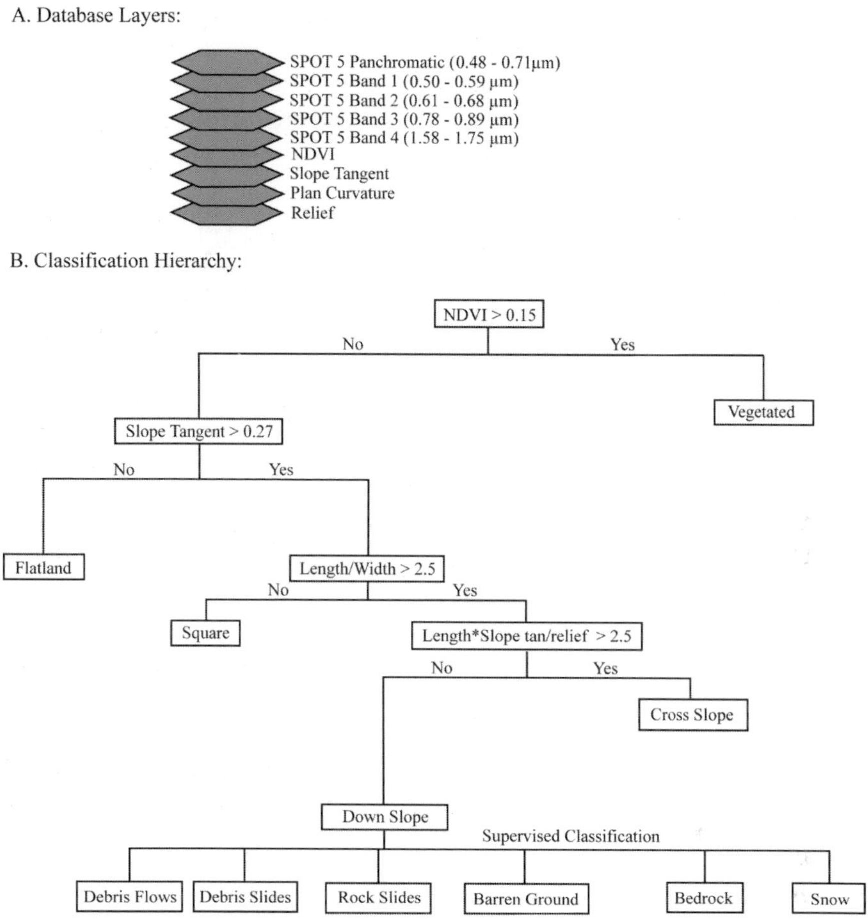

Fig. 1 Database and classification for the detection and mapping of rapid mass movements in the Chilliwack basin.

One of the most obvious characteristics of rapid mass movements is their dependence on gravity. Failure tracks tend to follow the path of steepest descent (fall line) down a given slope. The geomorphic context of an image object is therefore a useful tool in the classification. The orientation of the long axis of an object on the slope was used to separate those that ran roughly parallel to the fall line to those that extend across the slope. Theoretically, for a landslide on an infinite slope (Barlow *et al.*, 2006):

$$\frac{L \tan \beta}{\Delta z} = 1 \tag{1}$$

Table 1. Classification accuracy assessment for SPOT 5 errors of omission (producer's accuracy). The accuracy was determined by randomly selecting 20 failures of each type from the aerial photographs and comparing them to the classified image.

	Debris Slide	Rock Slide	Debris Flow
Debris Slide	18	2	2
Rock Slide	-	16	2
Debris Flow	1	-	12
Barren Ground	1	1	-
Bedrock	-	1	2
Snow	-	-	-
% Accuracy	90	80	60

Table 2 Classification accuracy assessment for SPOT 5 errors of commission (user's accuracy). The accuracy was determined by taking 20 failures of each type from the classified image and comparing them to the aerial photographs.

	Debris Slide	Rock Slide	Debris Flow
Debris Slide	16	-	1
Rock Slide	1	16	-
Debris Flow	1	2	16
Barren Ground	2	1	-
Bedrock	-	-	2
Road	-	-	1
Vegetated	-	1	-
% Accuracy	80	80	80

Classification of debris flows was less successful that that of the other types of failure. They generally occurred in deep mountain gully systems that were often obscured by shadow. This has obvious implications for the image segmentation and the debris flows were often broken up into several shorter segments. While the inclusion of plan curvature in the image segmentation alleviated this to some degree, many were designated as 'square features' (as opposed to the long thin features in the classification hierarchy). Additionally, the shadow was often misclassified as vegetated due to a higher NDVI measure compared to observations on exposed bare slopes. However, the classification accuracy of 80% for errors of commission indicates the criteria set forth in the supervised classification were effective in discriminating between the process types. In particular, curvature was an excellent class descriptor for debris flows due to their occurrence in areas of high concavity.

2.2. Mapping Snow Avalanche Tracks using Landsat TM

Snow avalanches are a common occurrence within the Canadian Rockies (Luckman, 1978). During winter months, this process represents an increasingly serious hazard as larger human populations use the region for recreational activities. Aavalanche activity results in a distinctive biogeographic response that can be associated to characteristic land cover patterns (Walsh *et al.*, 1990). Avalanche tracks typically manifest themselves within a forest matrix as a non-forested strip of meadow, rocky ground, willow shrubs or similar vegetation running vertically through the forest of a mountain valley side (Suffling, 1993). The recurrence of avalanches in the same location perpetuates the disruption of the forest canopy leaving more avalanche resistant shrubs and herbs to colonize these areas. Such areas are known as constrained avalanche tracks (McClung, 1993)

As with other slope processes, the use of spectral data alone has proven ineffective at mapping snow avalanche tracks as the vegetative communities that are common in these areas are also found in other portions of the landscape (Connery, 1992). However, the use of landcover with specific geomorphology has proven more favorable. Barlow & Franklin (2007) used a combination of Landsat TM data, image segmentation and DEM derivatives to map such features in a portion of the Canadian Rockies lying mostly within the boundary of Banff National Park. The mountains within this area were heavily glaciated during the Pleistocene and exhibit all of the features associated with alpine glaciers such as U shaped valleys, cirques, arêtes, and horns. Many of the peaks exceed 3 000 m and snow avalanches are common. The data layers and classification hierarchy used in the analysis are shown in figure 2.

The imagery was segmented using a ratio of 8:2 for the relative importance of reflectance vs. shape and 1:9 for compactness vs. smoothness with a scaling parameter of 10. As with the debris slide study, this combination of parameters was established after several iterations. The classification strategy differs somewhat from the identification of mass movements discussed above in that the spectral data layers and the DEM are used first to create a land cover map as shown in figure 2. As the tracks are typically associated with shrub or herb type vegetation, these classes are subjected to a series of decisions based on their shape and specific geomorphometry in a similar manner to that described above in the Chilliwack study. The accuracy of the automated system was assessed using a database derived from manual inspection of 1:50 000 stereo aerial photographs. A sample of 45 tracks was taken from the aerial photographic inventory in order to test for errors of omission and 45 events identified by the automated classification were tested against the air photos for errors of commission. The results of this analysis are shown in Table 3.

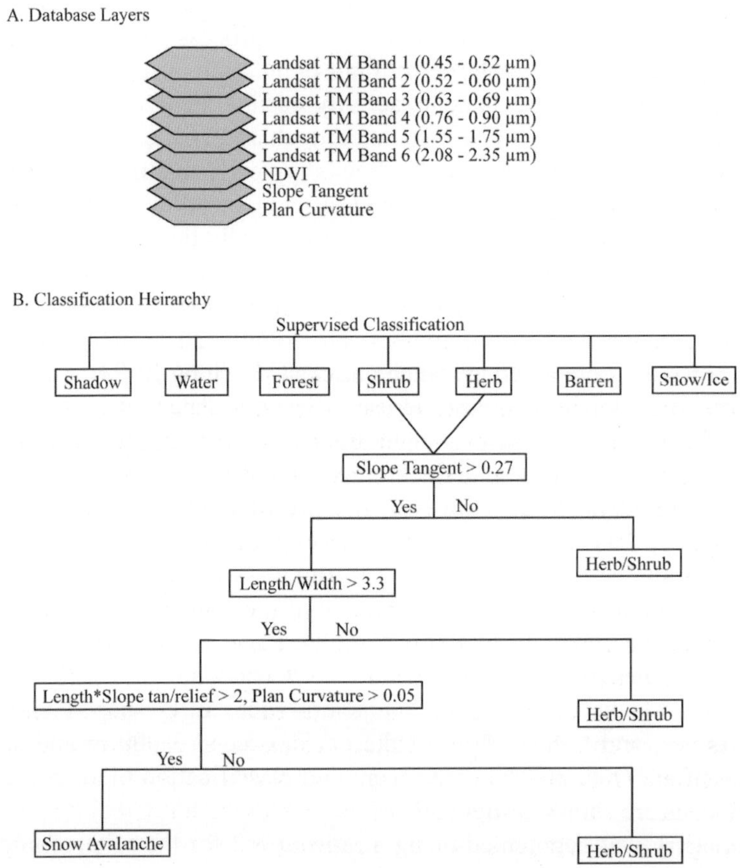

Fig. 2 Database layers and classification hierarchy for the Landsat TM classification and mapping of snow avalanche tracks in the Banff area.

The classification achieved an accuracy of 82% for errors of omission. The primary source of error involved avalanche tracks being misclassified as forested areas. This is mainly attributable to a failure of the segmentation to capture the tracks and due to mixed pixels along the edges of the thinner shoots affecting the spectral response of the area. Misclassification as shrub also occurred in one instance and can be attributed to the shape properties of the image object failing to meet the criteria set forth on the contextual portion of the classification.

Table 3. Classification accuracy assessment

Land Cover Type	Omission (n=45)	% Total	Commission (n=45)	% Total
Avalanche	37	82.2	34	75.5
Forest	7	15.5	-	-
Shrub	1	2.2	3	6.6
Herb	-	-	3	6.6
Barren	-	-	5	11.1

Errors of commission are also reported in table 3. As is evident, the greatest source of error is the assignment of barren areas to the avalanche class. Such areas are therefore being incorrectly classed as either shrub or herb and then being sent through the contextual phase of the classification. Other errors include shrub and herb areas being misclassified as avalanche tracks. In these cases, the misclassified areas possess both the spectral and contextual traits of an avalanche track but were not formed by this process. Indeed, the problem of polygenesis is an old one in geomorphology (Scheidegger, 1991). A portion of the imagery draped over the DEM is shown above the same area of classified imagery in Figure 3. The figure shows a number of avalanche tracks on the eastern flanks of Mt Cascade. The avalanche tracks in upper part of the figure are clearly visible and the value of shape characteristics and specific geomorphometry in their identification should be apparent. Below, the results of the classification are displayed. As is evident, the majority of the avalanche tracks have been correctly classified. Only one track has been incorrectly assigned to the shrub class. This is due to the segmentation of the track which divided it into two polygons. These did not possess the shape properties of a thin object and were therefore eliminated as potential avalanche areas.

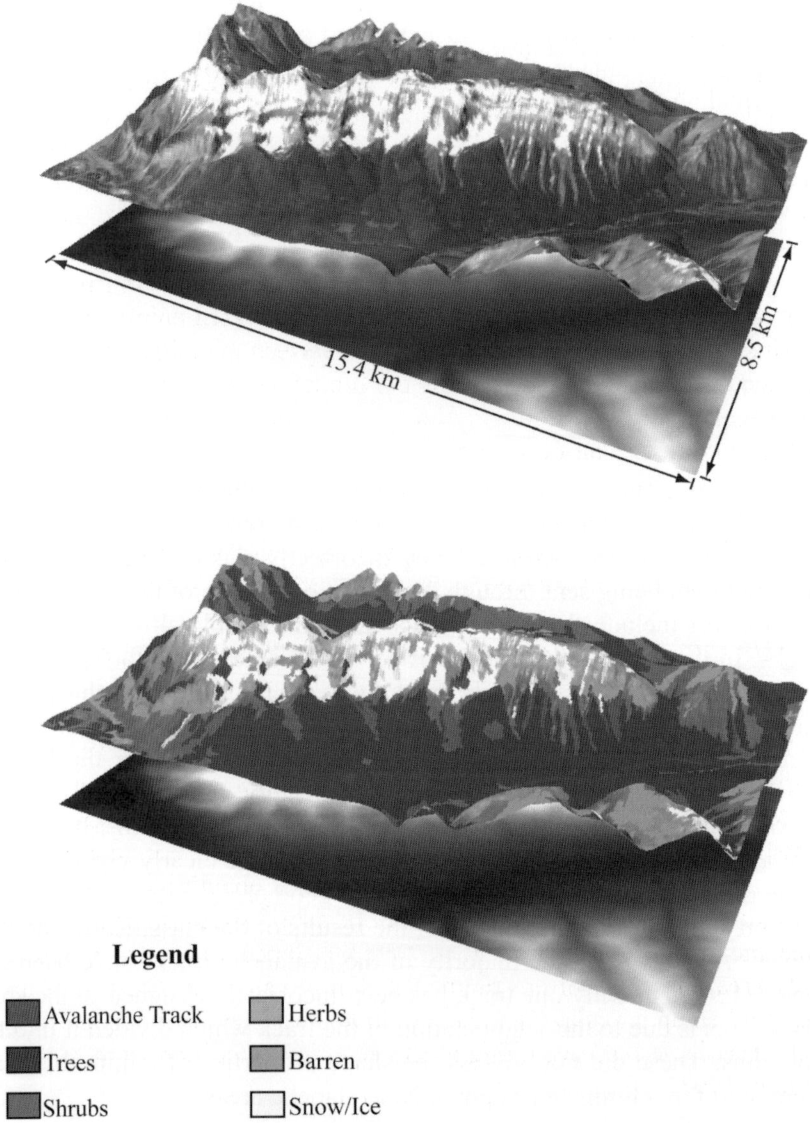

Fig. 3 Example of the Landsat TM imagery draped over the DEM, looking westward across the area around Mt Cascade. Top: several avalanche tracks are clearly visible descending the eastern slopes of the mountain. Bottom: The classified image draped over the same area.

3. Using Inventory Data for Slope Stability Assessment

The use of digital data sources to map slope hazards is beneficial in terms of data throughput into analyses of slope stability. One of the most powerful methods for the assessment of slope stability over large areas is the use of discriminant function analysis to define stability based on a series of independent variables (e.g. Baeza & Corominas, 2001). This method takes two populations and maximizes the space between them in n dimensional space (Manly, 1994). In example, the highest points within the debris slides mapped in the first case study were taken as initiation points and compared to a population of 2000 randomly generated points. A stepwise discriminant function was then performed between the initiation points and the random points generated across the model surface. For demonstration purposes, we only used geomorphometric parameters such as slope, aspect, profile curvature, plan curvature, and drained area. Aspect was converted into deviation from north and normalized to a value of between 0 and 1 due to the higher occurrence of debris sliding on south facing slopes within this area (Barlow *et al.*, 2007). Maximum separation was accomplished using only slope tangent, plan curvature and deviation from north such that:

$$DF = 0.469(slope\ tan) - 0.432(plan\ curve) + 0.727(deviation\ from\ north). \quad (2)$$

One problem with assigning slope stability in this way is the fact that debris slides tend to occur on metastable slopes (Scheidegger, 1991). These are slopes that are stable until a triggering event such as heavy precipitation or seismic shaking causes failure. Unstable slopes are unable to retain unconsolidated material in any significant quantities and will therefore experience a much lower incidence of debris slides. In order to determine the spatial distribution of metastable slopes within the Chilliwack study area, the range of discriminant function values obtained from Eq. 2 were broken up into classes and the sediment transport rate associated with each class is derived according to the equation (modified from Martin & Church, 1997):

$$q_{dc} = \sum_{i=1}^{n} \left[\frac{v_i}{a_{dc}} d_i \right] \quad (3)$$

where q_{dc} is the volumetric transport rate of sediment to be applied to the discriminant function class, v is the volume (calculated using a series of depth assumptions), a_{dc} is the area of the discriminant function class, d_i is the runout distance taken from the midpoint of the initiation zone to the midpoint of the deposition zone (Martin, 1998), and n is the number of landslides. The result is divided by 10 years, corresponding to the time-scale of the landslide inventory (Barlow *et al.*, 2006). The values of q_{dc} for the range of discriminant function values are plotted in figure 4. Based on these data, a best fit relation was derived using least squares regression such that:

$$q_{df} = 0.55\ DF^{5.50} \tag{4}$$

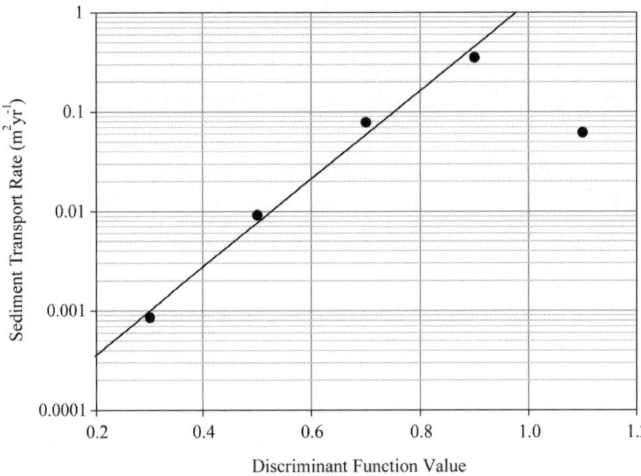

Fig. 4 Volumetric transport rates associated with discriminant function values. The rollover associated with high discriminant function values is interpreted as the transition from weathering limited to transport limited slopes.

Equation 4 describes the best-fit line for the first four points on the plot and has a r^2 of 0.99. It must be noted that a rollover occurs between the last two points. This is interpreted as the transition between transport limited and weathering limited slopes (Martin, 2000). These data indicate that due to the occurrence of debris slides on metastable slopes, the probability of failure and therefore the potential hazard is reduced on the most unstable portions of the basin as these are transport limited. Fig. 5 shows a map of debris slide occurrence based on the disriminant function values calculated in equation 4. As is evident, the potential for debris sliding increases up to a threshold DF value of about 0.9 but is reduced in areas that posses a higher discriminate function value. However, whilst these areas may have a lower potential for debris slide occurrence, they are inherently unstable. The zone of metastable slopes susceptible to debris sliding is therefore those areas on Fig. 5 that possess a discriminant function value between roughly 0 and 0.9.

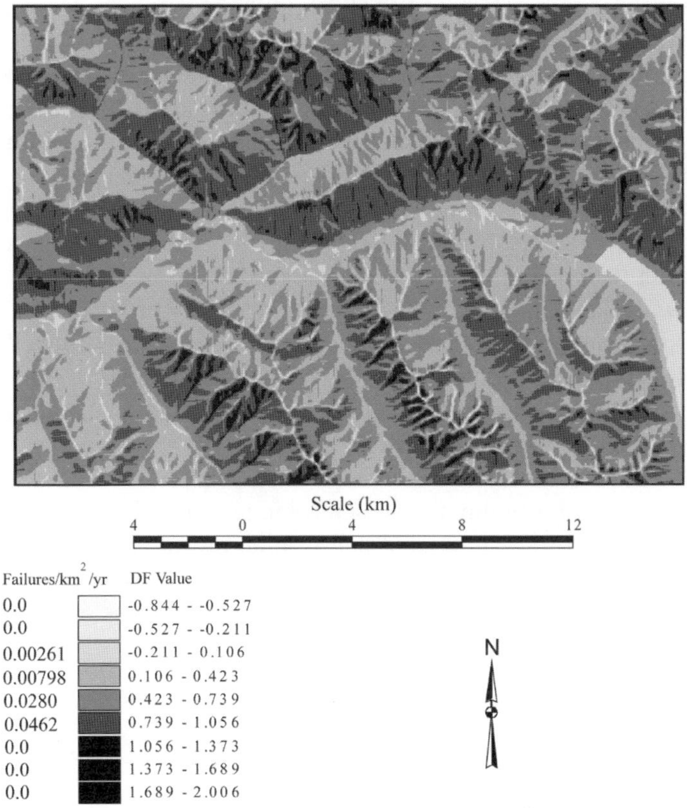

Fig. 5 Map showing the number of a debris slides/km^2/yr within the Chilliwack study area. The classes used are identical to those identified in Fig. 7. Note: the more unstable areas have a lower occurrence as these areas are transport limited.

4. Conclusions

Image segmentation, when used in conjunction with optical satellite imagery and ancillary data provides the necessary digital information to extract slope features from the image scene. Both of the mapping techniques discussed here are dependent on land cover changes that result from failure. The type of feature objects that can be detected and the accuracy of the classification are dependent upon the spatial resolution of the imagery used in the analysis. This research has investigated the applicability of two differing satellite sensor platforms in the automated detection of mass movements. A combination of SPOT 5 and DEM data was acquired to develop an automated system to detect and classify rapid mass movements that were fresh and over 1 ha in area in a high mountain region in British Columbia. The method yielded an overall accuracy of 77% for all rapid mass movements. These features were further divided according to the classification system of Cruden & Varnes (1996) into debris slides, debris flows, and rockslides; the process specific classification accuracies were 90%, 60%, and 80% respectively (Barlow *et al.*, 2006). The use of Landsat TM data with a similar set of DEM derivatives proved capable of mapping snow avalanche tracks with an overall accuracy of 79%. This was due to the distinctive land cover associated with these features as well as their shape characteristics and orientation on the landscape. These results strongly support the viability of using satellite remote sensing data and digital elevation models to map slope processes.

The ability to generate accurate maps of slope processes allows for great efficiency in data throughput to geomorphic research. Here we have demonstrated the use of an automated debris slide inventory in the quantification of slope instability in the Chilliwack Valley, British Columbia. Such research is critical in the identification of unstable slopes and the estimation of landslide hazard to human populations.

References

Baatz M, Schape A, 2000, Multiresolution Segmentation: An Optimization Approach for High Quality Multi-scale Image Segmentation, In: Strobl J., Blaschke T., Griesebner G. (eds) Angewandte Geographische Informationsverarbeitung XII. Wichmann-Verlag, Heidelberg, pp 12-23.

Baatz M, Benz U, Dehghani S, Heymen M, Holtje A, Hofmann P, Ligenfelder I, Mimler M, Sohlbach M, Weber M, Willhauck G, 2001, eCognition Users Guide, Definiens Imaging GmbH, Munich.

Baeza C, Corominas J, 2001, Assessment of shallow landslide susceptibility by means of multivariate statistical techniques, Earth Surface Processes and Landforms 26: 1251-1263.

Barlow J, Franklin S, 2007, Mapping snow avalanche shoots in the Canadian Rockies using digital data, Arctic, Antarctic and Alpine Research. Submitted

Barlow J., Martin Y., Franklin S., 2007, Characterizing Debris Slide Occurrence using Digital Data, Chilliwack Valley, British Columbia, Canadian Journal of Earth Sciences, Submitted.

Barlow J, Franklin S, and Martin Y, 2006, High spatial resolution satellite imagery, DEM derivatives, and image segmentation for the detection of mass wasting processes, Photogrammetric Engineering and Remote Sensing 72 (6):687-692.

Barlow, J, Martin Y, and Franklin S, 2003. Detecting translational landslide scars using segmentation of Landsat ETM+ and DEM data in the northern Cascade Mountains, British Columbia, Canadian Journal of Remote Sensing 29(4): 510-517.

Brardinoni F, Slaymaker O, Hassan M, 2003, Landslide inventory in a rugged forested watershed: a comparison between air photo and field survey data, Geomorphology 54: 179-196.

Connery D, 1992, Avalanche vegetation classification using Landsat imagery and a digital model in southwestern Alberta. MSc Thesis, University of Calgary.

Connors K, Gardner T., 1987, Classification of geomorphic features and landscape stability in Northwestern New Mexico using simulated SPOT imagery, Remote Sensing of Environment 22: 187-207.

Cruden D, Varnes D, 1996, Landslide types and processes, In: Turner K, Schuster R, (eds) Landslides Investigation and Mitigation, Transportation Research Board Special Report 247, National Academy Press, Washington D.C., pp. 36-75.

Epp, H, Beaven L, 1988. Mapping slope failure tracks with digital Thematic Mapper data. IGARSS'88, Proceedings of the International Geoscience and Remote Sensing Symposium, 13-15 September 1988, Edinburgh, Scotland, IEEE New York, vol. 3, pp 1649-1652.

Gao J, 1993, Identification of topographic settings conducive to landsliding from DEM in Nelson County, Virginia, USA, Earth Surface Processes and Landforms 18: 579-591.

Gardner J, 1970, Geomorphic significance of avalanches in the Lake Louise Area, Alberta, Canada, Arctic and Alpine Research 2 (2): 135-144.

Hovius N, Stark C, Allen P, 1997, Sediment flux from a mountain belt derived by landslide mapping, Geology 25: 231-234.

Haralick R, Shanmugam K, Itshak D, 1973, Textural features for image classification, IEEE Transactions on Systems, Man, and Cybernetics 6: 610-621.

Jensen J, 2005, Introductory digital image processing: a remote sensing perspective (3rd ed). Pearson Prentice Hall, Upper Saddle River.

Johnson E, 1987, The Relative Importance of Snow Avalanche Disturbance and Thinning on Canopy Plant Populations, Ecology 68 (1): 43-53.

Kidwell K, 1990, Global Vegetation Index User's Guide, U.S. Department of Commerce/National Oceanic and Atmospheric Administration/National Environmental Data and Information Service/National Climactic Data Centre/Satellite Data Services Division.

Luckman B, 1978, Geomorphic work of snow avalanches in the Canadian Rocky Mountains, Arctic and Alpine Research 10 (2): 261-276.

Malamud B, Turcotte D, Guzzetti F, Reichenbach P, 2004, Landslide inventories and their statistical properties, Earth Surface Processes and Landforms 29: 687-711.

Manly B, 1994, Multivariate Statistical Methods, A Primer. Chapman & Hall, New York.

Martin Y, 2000, Modelling hillslope evolution: linear and nonlinear transport relations, Geomorphology 34: 1-21.

Martin Y, Church M, 1997, Diffusion in landscape development models: on the nature of basic transport relations, Earth Surface Processes and Landforms 22: 273-279.

Martin Y, Franklin S, 2005, Classification of soil- and bedrock-dominated landslides in British Columbia using segmentation of satellite imagery and DEM data, International Journal of Remote Sensing 26 (7): 1505-1509.

McClung D, Schaerer P, 1993, The Avalanche Handbook. The Mountaineers, Seattle, Wa.

McDermid G, Franklin S, 1995, Remote sensing and geomorphic discrimination of slope processes, Zeitschrift fur Geomorphologie 101: 165-185.

McKean J, Buechel S, Gaydos L, 1991, Remote sensing and landslide hazard assessment, Photogrammetric Engineering and Remote Sensing 57 (9): 1185-1193.

Nevatia R, 1986, Image segmentation, In: T. Young & K. Fu (Eds) Handbook of Pattern Recognition and Image Processing. Academic Press, New York, pp 215 - 231.

Pohl C, Van Genderen J, 1998, Multisensor image fusion in remote sensing: concepts, methods, and applications, International Journal of Remote Sensing 19 (5): 823-854.

Rood K, 1984, An aerial photograph inventory of the frequency and yield of mass wasting on the Queen Charlotte Islands, British Columbia, BC Ministry of Forests, Land Management Report 34.

Sauchyn, D,Trench N, 1978. Landsat applied to landslide mapping. Photogrammetric Engineering and Remote Sensing 44: 735-741.

Saunders I, Clague J, Robets M, 1987, Deglaciation of Chilliwack River valley, British Columbia, Canadian Journal of Earth Sciences 24: 915-923.

Scheidegger A, 1991, Theoretical Geomorphology (3rd ed). Springer, Berlin.

Shih E, Schowengerdt R, 1983, Classification of arid geomorphic surfaces using Landsat spectral and textural features, Photogrammetric Engineering and Remote Sensing 49 (3): 337-347.

Stark C, Hovius N, 2001, The characterization of landslide size distributions, Geophysical Research Letters 28 (6): 1091-1094.

Suffling R, 1993, Induction of vertical zones in sup-alpine valley forests by avalanche-formed fuel breaks, Landscape Ecology 8 (2): 127-138.

Walsh S, Butler D, Brown G, Bian L, 1990, Cartographic Modeling of Snow Avalanche Path Location within Glacier National Park, Montana, Photogrammetric Engineering and Remote Sensing 56 (5): 615-621

Wills C, McCrink T, 2002, Comparing Landslide Inventories: The Map Depends on the Method, Environmental and Engineering Geoscience VIII (4): 279-29

Monitoring Xi'an Land Subsidence Evolution by Differential SAR Interferometry

Qin Zhang [1], Chaoying Zhao[1], Xiaoli Ding[2] and Jianbing Peng[1]

[1]College of Geological Engineering and Geomatics, Chang'an University, 126 Yanta Road, 710054 Xi'an, Shaanxi, China
[2]Department of Land Surveying and Geo-Informatics, HongKong Polytechnic University, Hung Hom, Kowloon, Hong Kong, China

Abstract

Differential SAR Interferometry (D-InSAR) technique has been used to monitor the land subsidence evolution in Xi'an, China during the period of 1992 to 2005. Three schemes have been made to detect the three subsidence stages, namely, stage I 1992-1993, stage II 1996-1997 and stage III 2004-2005. And annual subsidence rates have been calculated in three schemes which uncovered the land subsidence evolution in Xi'an from 1990s to now days. The D-InSAR results show that the maximum subsidence rate was up to 20m/a in the 1996, and decreased greatly from 20cm/a in 1996 to 5cm/a in 2005, which mainly owing to the controlling of underground water withdrawal policy in 1996s. For the lack of traditional monitoring results, only the first stage of D-InSAR result has been compared with leveling result, which demonstrated high consistence each other.

1 Introduction

Xi'an city, the capital of Shaanxi province, lies in the northwest China. Xi'an is famous worldly for its historic sites. However, with the rapid development of economy in recent decades, Xi'an has suffered serious land subsidence and land fissure since the end of 1950s mainly due to the uncontrolled groundwater withdrawal and widely high-tech zone construction. The leveling monitoring results showed that up to 1992, areas with cumulative subsidence of at least 100mm were over 105 sq. km, the maximum subsidence was 1940mm, and the average subsidence rate was 80-126mm/a, the maximum subsidence rate was up to 300mm/a, and eight subsidence funnels could be identified in the southern, eastern and southwest suburbs of Xi'an respectively in 1996 (Yan 1998).

Traditional ground monitoring methods e.g., high precise leveling had been used to collect precious land subsidence information. However, it is limited for its high cost, time consuming, and discrete benchmark surveying etc. The most adverse problem in Xi'an is that the wildly leveling surveying was carried out before 1990s, so almost no information could be found to research the recent subsidence. Fortunately, in the past two decades or so, InSAR technique has been widely used in the earthquake (e.g., Massonnet et al. 1993; Xia et al. 2003), volcano (e.g., Lu et al. 2003) and land subsidence (e.g., Li et al. 2004; Liu et al. 2001) monitoring. Thanks for the project on Xi'an land subsidence monitoring funded by the Ministry of Land & Resource, China, scores of ERS data and decades of Envisat data spanning from 1992 to 2005 have been acquired and decades of GPS benchmarks have been settled meanwhile. This contribution concentrates on Xi'an land subsidence evolution, where three schemes according to the ERS data with different tracks and Envisat data covering Xi'an and the surrounding areas have been made to calculate the vertical deformation. In order to calibrate the D-InSAR result, in situ leveling data with annual vertical contour map and some discrete leveling points have been carefully compared with D-InSAR results.

2 Review of the Principle of D-InSAR

D-InSAR is based on the phase difference which acquired from two different SAR acquisitions. The interferogram phase is proportional to the range difference in LOS (line of sight) as:

$$\varphi = -\frac{4\pi\Delta R}{\lambda} \qquad (1)$$

where φ denotes the phase of a given pixel in radar coordinate, λ is the wavelength, as for C band ERS and Envisat data λ is 5.66 cm. ΔR is the range difference in master LOS. The image usually expressed as a complex value $Ae^{i\varphi}$. When the master and slave images have been fine co-registrated, their coherence can be calculated as

$$|\gamma| = \left| \frac{\langle A_1 e^{i\varphi_1} \times A_2 e^{-i\varphi_2} \rangle}{\sqrt{\langle A_1^2 \rangle \times \langle A_2^2 \rangle}} \right| \tag{2}$$

where the brackets correspond to the mathematical expectation estimated in a window centered in a given pixel. Eq. 2 can be used to evaluate the interferogram quality. The coherence is high in urban areas and very low in vegetation areas. It has been clearly demonstrated in the test region of Xi'an city that nearly 80% are high coherence which mostly covers the interested areas.

The interferometric phase φ obtained with 2π ambiguity is the addition of different contributions:

$$\varphi = \varphi_d + \varphi_t + \varphi_a + \varphi_o + \varphi_n \tag{3}$$

where $\varphi_d = \frac{4\pi}{\lambda}\delta r$ is the deformation component, δr is the displacement in LOS which is occurred between the two SAR acquisitions. One fringe corresponds to half wavelength (2.8cm of ERS data) displacement. In the case of vertical motion, one fringe is associated with 3.0 cm of motion as the incidence angle of ERS is about 23° (Raucoules et al. 2003).

$\varphi_t = 2\pi \frac{h}{h_{2\pi}}$, is the topographic component, where h denotes the altitude of the target over the spherical surface, $h_{2\pi}$ is the height ambiguity which is given as (Massonnet et al. 1993)

$$h_{2\pi} = \left| \frac{\lambda R \sin\theta}{2B_\perp} \right| \tag{4}$$

where $B_\perp$ is the baseline perpendicular component and θ is radar's depression angle. According to the acquisition method of the φ_t, three D-InSAR methods have been deduced as: two-pass method (Massonnet et al. 1993), three-pass method and four-pass method (Zebker et al. 1994). The two-pass method has been applied in Xi'an test in following section.

φ_a is the atmospheric component. It is due to the incongruous atmospheric effect between the two SAR acquisitions (Zebker and Villaseno 1992). This effect is omitted in Xi'an study for the dry climate and mainly for the lack of auxiliary data to research it presently. φ_o is the orbital component, which is related to the relative position of satellite tracks. It can be modeled and removed using orbital data and/or by eliminating the residual orbital fringes by simulating them to a regular fringe pattern

(Scharroo et al. 1998). φ_n is the phase noise mainly due to the temporal and spatial de-correlation of the interferometric signal. In order to detect the deformation, the other components in Eq. 3 except φ_d must be calculated out.

Next, phase unwrapping must be taken into account to obtain the real phase which is related to true deformation. In Xi'an study phase unwrapping is a serious problem, because there exist decades of land fissures and great amplitude of land subsidence rate. SNAPHU phase unwrapping algorithm (Chen and Zebker 2001) has been used in this test. Fig. 1 shows the two-pass D-InSAR processing flow.

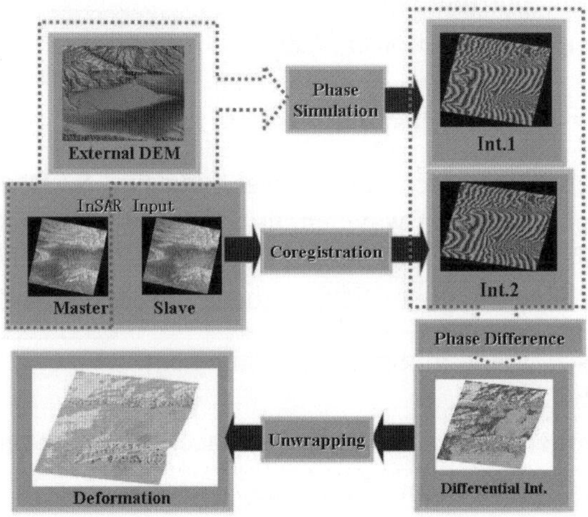

Fig. 1 Two-pass D-InSAR processing flow

3 Test Area and Data Sources

3.1 Location of the study area

The research area is 18×22km^2 (about 400 km^2) in Xi'an, shown in Fig. 2, which covers the main subsidence regions, where the city wall can be clearly identified which can be taken as a main geographic reference for the subsidence analysis. Also nine leveling bench marks are displayed by black triangle there.

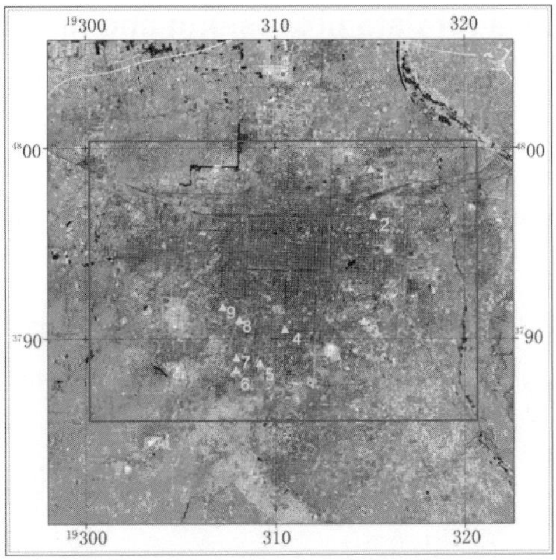

Fig. 2 Location of the study area in Xi'an (highlighted by red rectangle frame)

3.2 SAR data used in this research

Table 1 illustrates the SAR data including the ERS1/2 with different tracks and Envisat data, by using them, three schemes have been made to detect three main subsidence stages in Xi'an, namely, 1992-1993, 1996-1997 and 2004-2005 respectively. All InSAR results have been recalculated to annual subsidence rates, which same as the traditional leveling results for easy comparison.

Table 1. SAR data for Xi'an subsidence research

Scheme	date	Sensor	Track	$B_{//}$ (m)[a]	$B_{\perp}$ (m)[b]	Time span (days)
I	19920930 19930207	ERS1 ERS1	161	-29	124	140
II	19960107 19960701	ERS1 ERS2	390	47	-37	175
III	20040110 20050827	Envisat Envisat	2161	-32	118	595

[a] Baseline parallel component
[b] Baseline perpendicular component

3.3 Auxiliary data for data processing and result analysis

As mentioned before, two-pass D-InSAR method has been used for Xi'an land subsidence research, while the 3 arc second SRTM DEM shown in Fig.3 has been used as external DEM, which also re-sampled to 20m resolution to be subtracted from the interferogram during the differential data processing.

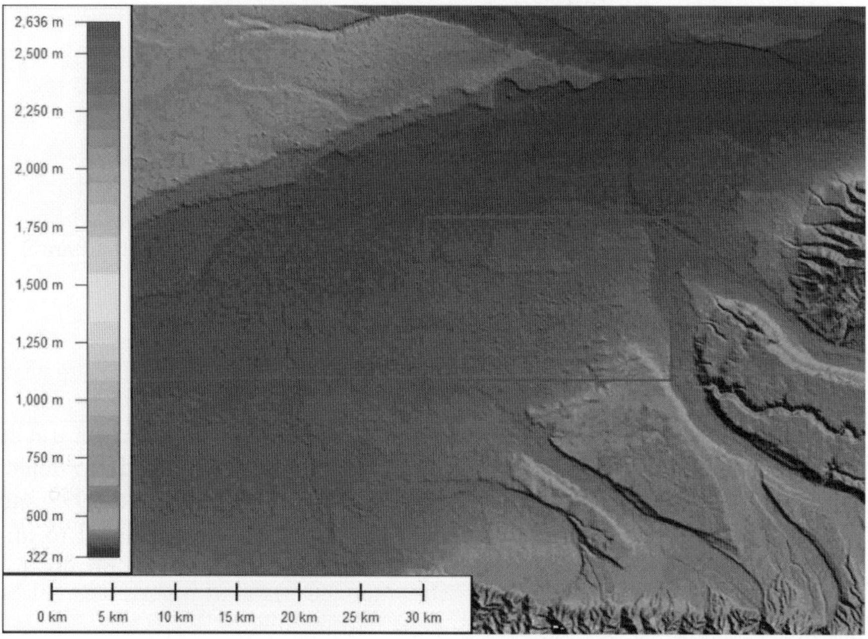

Fig. 3 SRTM DEM covering Xian area (highlighted by red rectangle frame)

Aiming at mitigating the orbital errors, precise ERS orbit data from the Delft University of Technology have been used in the processing of co-registration and flat-earth phase removing (Scharroo et al. 1998).

In order to calibrate the D-InSAR results, the high precise leveling contour map (Zhu et al. 2005) has been selected and shown as Fig. 4.

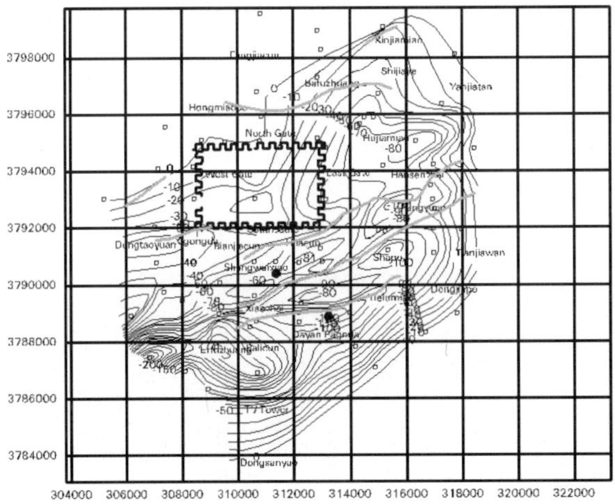

Fig. 4 Vertical deformation contour map by leveling in Xi'an (1988-1991)

4 D-InSAR Processing and Results

Earth-View InSAR software has been applied to process the standard D-InSAR data and all D-InSAR results of three schemes have been corrected to the Northeast point of Xi'an city wall, and recalculated to annual deformations and displayed as the overlay map with Landsat TM image shown as Figs. 5, 7, and 9, respectively. Also their corresponding contour maps have been displayed as Figs. 6, 8, and 10, respectively.

Fig. 5 Overlay map of Xi'an annual subsidence rate of Scheme I (1992-1993)

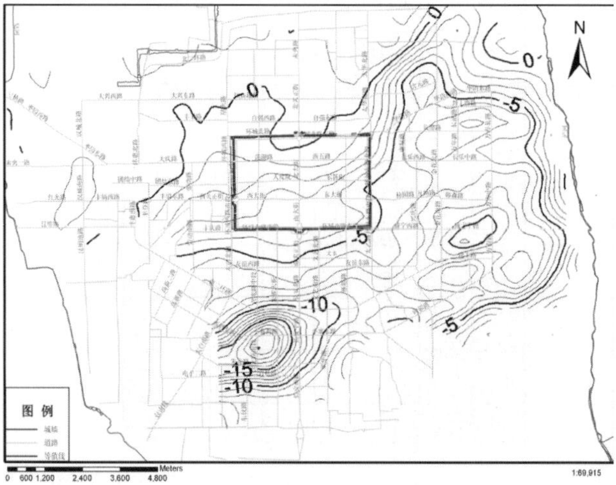

Fig. 6 Contour map of Xi'an annual subsidence rate of Scheme I (1992-1993)

By comparing two contour maps from leveling and D-InSAR techniques shown as Figs. 4 and 6, respectively, their subsidence funnels, effect areas and amplitudes are consistent very much by considering their time span difference. And the maximum subsidence rate was up to 22cm/a in the south of Xi'an city named Balicun funnel. Also nine leveling Bench marks (displayed in Fig. 2) have been collected to compare with D-InSAR results listed in table 2. The mean value was 1cm/a, and the root mean square (RMS) was 0.94cm/s.

Table 2. Comparison of leveling and D-InSAR results in stage I

Point No.	Leveling(cm/a)	D-InSAR(cm/a)	residue(cm/a)
1	-2.5	-1.5	1
2	-7.6	-7.2	0.4
3	-13	-7.8	5.2
4	-12	-13	-1
5	-14	-11	3
6	-11	-10.5	0.5
7	-15	-14	1
8	-12	-13	-1
9	-6	-6	0

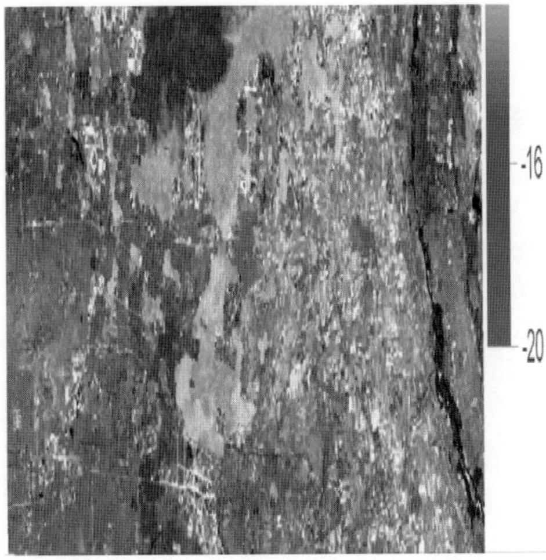

Fig. 7 Overlay map of Xi'an annual subsidence rate of Scheme II (1996-1997)

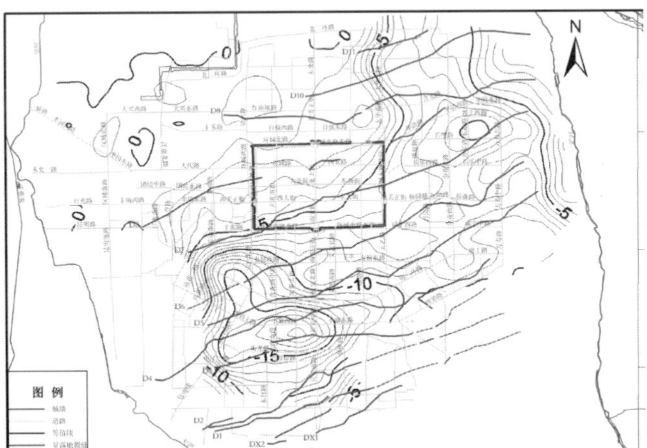

Fig. 8 Contour map of Xi'an annual subsidence rate of Scheme II (1996-1997)

It was clear that the subsidence effect area in the south of Xi'an city had been larger than that of stage I, and the maximum subsidence rate was up to 20cm/a, which was mostly related to the large scale high-tech zone construction in Xi'an since 1995.

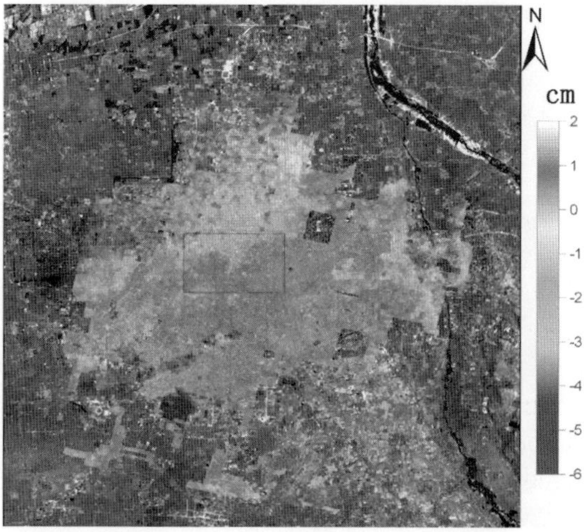

Fig. 9. Overlay map of Xi'an annual subsidence rate of Scheme III (2004-2005)

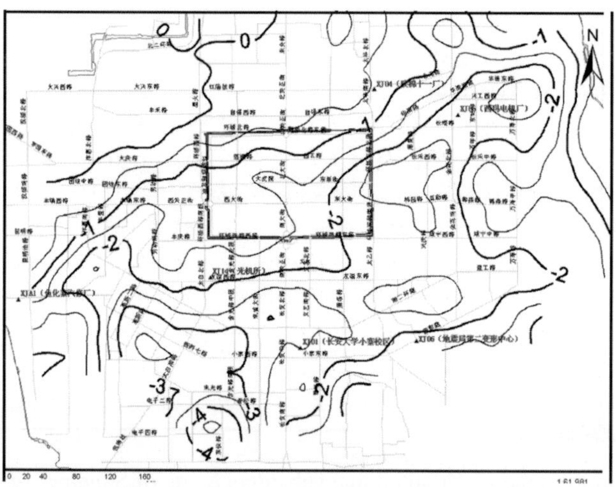

Fig. 10 Contour map of Xi'an annual subsidence rate of Scheme III (2004-2005)

From the Figs. 9 and 10 and the others series subsidence rate maps, it was evident to show that subsidence rate had been decreased greatly from stage II to stage III, exactly to say, the maximum subsidence rate had been decreased from 20cm/a in the 1996 to 5cm/a in the 2005, which were consistent with the policy of controlling underground water withdrawal by local government since 1996.

5 Conclusions

D-InSAR method has been successfully applied in the Xi'an land subsidence monitoring during the period of 1992 to 2005, and three schemes have been made to detect the subsidence evolution. Several conclusions could be clearly drawn from the D-InSAR results as: the subsidence rate amounted to the maximum in the subsidence funnel of high-tech zone with 20cm/a in the 1996, which was greatly related to the high-tech zone construction; Thanks for the policy of controlling of underground water withdrawal since 1996, the subsidence rate had been decreased greatly from 20cm/a to 5cm/a in 2005; as for the first monitoring stage, both the contour map and discrete bench marks comparison had shown the high consistence between leveling and D-InSAR results.

Meanwhile, the atmospheric effect in Xi'an should be considered quantitatively late, it is also necessary that the GPS monitoring data should be used to calibrate the recent D-InSAR results. How to consider the land subsidence and land fissure comprehensively are also need to be researched future.

Acknowledgements

This paper is funded by the key project of Nature Science Foundation of China (NSFC) (project No: 40534021), the key project of the Ministry of Land & Resources, China (project No: 1212010440410) and general project of Nature Science Foundation of China (NSFC) (project No: 40672173). The SAR images are provided by ESA and track 161 data are provided by ERS category 2 of Eurimage. We are also grateful for the provision of SRTM DEM, DELFT precise orbit data and SNAPHU phase unwrapping software, all of them are provided freely.

References

Atlantis Scientific (2003) EV-InSAR version 3.0. User's Guide. Atlantis Scientific Inc. Ontario, 257 pp

Chen CW, Zebker HA (2001) Two-dimensional phase unwrapping with use of statistical models for cost functions in nonlinear optimization. Journal of the Optical Society of America, 18 (2): 338–351

Li T, Liu J, Liao M. (2004) Monitoring City Subsidence by D-InSAR in Tianjin Area. In: IEEE IGARSS2004 Proceedings, Anchorage Alaska, pp. 3333–3336

Liu GX, Ding XL (2001) Settlement field of Chek Lap Kok airport, Hong Kong, detected by satellite synthetic aperture radar interferometry. Chinese Science Bulletin.46:1778–1782

Lu Z, Patrick M, Fielding EJ, Trautwein C (2003) Lava volume from the 1997 eruption of Okmok volcano, Alaska, estimated from spaceborne and airborne interferometric synthetic aperture radar. IEEE Transactions on Geoscience and Remote Sensing, 41: 1428–1436

Massonnet D, Feigl KL (1998) Radar interferometry and its application to changes in the Earth's surface. Reviews of Geophysics, 36: 441–500

Massonnet D, Rossi M, Carmona C, et al. (1993) The displacement field of the Landers earthquake mapped by radar interferometry. Nature 364:138–142

Raucoules D, Maisons C, Carnec C, Le Mouelic S, King, C, Hosford S (2003) Monitoring of slow ground deformation by ERS radar interferometry on the Vauvert salt mine (France): comparison with ground-based measurement. Remote Sensing of Environment, 4, 468–478

Scharroo R, Visser PNAM, Mets GJ (1998) Precise orbit determination and gravity field improvement for ERS satellites. Journal of Geophysical Research, 103: 8113–8127

Xia Y, Michel GW (2003) Seismic unloading and loading in northern central Chile as observed by differential Synthetic Aperture Radar Interferometry (D-InSAR) and GPS. International Journal of Remote Sensing, 24: 4375–4391

Yan W (1998) Analysis on the Origin of Land Subsidence and Its Countermeasures of Control in Xi'an (in Chinese). The Chinese Journal of Geological Hazard and Control. 119: 27–32

Zebker H, Villaseno J (1992) Decorrelation in interferometric radar echoes. IEEE Transactions on Geoscience and Remote Sensing, 30: 950–959

Zebker H, Rosen P, Goldstein R, Gabriel A, and Werner C (1994) On the derivation of coseismic displacement fields using differential radar interferometry: the Landers earthquake. Journal of Geophysical Research, 99: 19617–19634

Zhu Y, Wang Q (2005) A Study of the Space-Time Change Characteristics of Ground Subsidence in Xi'an and Their Mechanism (in Chinese). Acta Geoscientica Sinica, 26: 67-70

Evaluation of NARAD Precipitation Data for Rainfall Monitoring in Eastern Ontario, Canada

Dongmei Chen and Andrew Farrar

Department of Geography, Queen's University
Kingston, Ontario K7L 3N6, Canada
chendm@post.queensu.ca

Abstract

The development of NEXRAD weather radar products has greatly advanced the capacity to forecast and provide warnings of severe weather conditions over large areas in a time-efficient manner. However, most studies in the literature are conducted within the U.S. This study evaluates the reliability of NEXRAD precipitation data and rain gauge measurements in Eastern Ontario, Canada, for potential flood monitoring and water budget analysis. Five-month daily rainfall data from NEXRAD and rain gauge measurements were collected and generated for two Eastern Ontario conservation authority regions. The NEXRAD data was evaluated using rain gauge measurements as the reference. A good correlation (0.78) exists between the daily NEXRAD precipitation data and rain gauge measurements, especially for heavier rainfalls. The result also shows that 62% of radar precipitation data underestimates the daily precipitation. This underestimate is more common when the rainfall is small. The evaluation of spatial patterns of rainfall suggests that radar precipitation shows a more continuous pattern than the interpolated surfaces from rain gauges. Considering that small rainfall events contribute a relatively small portion of the total precipitation, NEXRAD products can play an important role in real-time flood monitoring and water budget analysis during heavy rainfall events in Canadian regions within the working range distance of the NEXRAD system.

1. Introduction

Accurate precipitation data are important for flood forecasting and regional water management. Traditionally, rain precipitation has been monitored at ground rain gauges placed at different locations. However, this type of monitoring network cannot capture spatial variation and patterns of rainfall, especially for regions with complex terrain (Young et al. 1999; Morin et al. 2003; Vieux and Bedient 2004). The development of remotely sensed weather radar data has greatly advanced the forecast of spatial pattern and rainfall quantity over large areas in a time-efficient manner (Maddox et al. 2002; Bedient et al. 2003).

In order to better detect severe weather and improve the accuracy of precipitation forecast, the U.S. National Weather Service (NWS) began installing the Next Generation Weather Radar WSR-88D (NEXRAD) system in 1988. NEXRAD comprises 158 weather surveillance radar–1988 Doppler (WSR-88D) radars across the United States. The WSR-88D radar sends out radar beams at several different altitudinal scanning angels using a 10 cm wavelength (S-band) and penetrates the atmosphere and rainfall with little attenuation over long distances. The ground-based radars then receive radar signals bouncing off of precipitation. Precipitation intensity is estimated by using information about the strength, velocity, and spectrum of the reflected beam (more details on how NEXRAD works can be found at www.wunderground.com/radar/). Under most conditions, its usefulness range is considered to be 180 km, even though it can produce precipitation estimates up to 230 km away (Vieux and Bedient 2004). The launch of the NEXRAD system has revolutionized the capability of the NWS to forecast and warn of severe weather conditions (Xie et al. 2006). Since the installation of NEXRAD system, NWS's precipitation products have been evaluated for their application to rainfall estimation (Johnson et al. 1999; Krajewski and Smith 2002; Morin et al. 2003; Xie et al. 2006), hydrologic modeling (Pereira Fo et al. 1999; Young et al. 2000; Vieux and Bedient 2004), and flood forecasting and validation (Vieux and Bedient 1998; Bedient et al. 2000; Bedient et al. 2003; Zhang and Smith 2003; Vieux and Bedient 2004).

NEXRAD precipitation products are categorized into four product levels according to the amount of preprocessing, calibration, and quality control performed (Reed and Maidment 1999). The lowest level NEXRAD product, Stage I, is the hourly digital precipitation (HDP) estimate directly derived from radar reflectivity. In Stage II, the HDP is calibrated by merging with surface rain gauge measurements with a mean field bias correction. The most commonly used is the NEXRAD Stage III data, in which, the

Stage II data from multiple weather radars covering the entire NWS River Forecast Center is combined and corrected using the average of all available rain gauge measurement (Young et al. 2000, Xie et al. 2006). Another NEXRAD product is the mosaicked Stage III precipitation product covering the entire Continental United States.

Despite the wide use of NEXRAD data, a key concern about these products is their accuracy and uncertainty (Anagnostou et al. 1999; Ciach 1999; Seo et al. 1999; Habib 2002; McCollum et al 2002). Because NEXRAD precipitation data are increasingly used as inputs for hydrological models of flood forecasting and warning, the accuracy of and uncertainty about this data needs to be evaluated. Many studies have compared the NEXRAD precipitation data with ground measurements at rain gauges in different regions of the U.S. (Johnson et al. 1999; Young et al. 1999; McCollum et al. 2002; Stellman et al. 2000; Krajewski et al. 2003; Xie, et al. 2006). The most common approach to evaluating accuracy is to compare the differences between radar estimates and rain gauge measurements through standard statistics, although the gauge measurements do not reflect the spatial scale of a NEXRAD pixel (Xie et al. 2006).

Since all NEXRAD radar stations are located within U.S., all previous studies related to NEXRAD precipitation data have focused on their application within the U.S. Although many populated regions of Canada are within 250 km of the border, within the usefulness range of radar located in the border areas of U.S., no studies have tested the reliability of NEXRAD data in Canada.

Since 2004, the Ontario government has adopted the Source Water Protection Act. This act requires all conservation authorities in Ontario to maintain detailed catchment-scale water budgets. Due to the low density of weather stations in Ontario, it is almost impossible for a conservation authority to declare a regional drought or a flood alert, because the precipitation varies so widely over the large, lake-influenced areas of Eastern Ontario. Several conservation authorities have been interested in using free NEXRAD precipitation data to track how much rainfall has fallen in their areas. They contacted us to conduct an evaluation study to determine the usability of free NEXRAD data for estimating rain fall in their areas.

The purpose of this study is to evaluate the usefulness of NEXRAD Stage III precipitation data to map the spatial distribution of rainfall for flood monitoring and water management in the area of Eastern Ontario, Canada, near the U.S. border.

2. Methods

2.1 Study area

Fig. 1 shows the two conservation authorities used in this study: the Quinte Conservation Authority and the Cataraqui Conservation Authority, both located in Eastern Ontario, Canada. The study area belongs to the ecozone of the mixedwood plains. The total area is about 8400 km^2. Elevation of the study area ranges from 58 m to 457 m. Most of the areas are relatively flat, with slopes of less than 30°. The north part of the region is mainly covered by forest, while agriculture dominates the south area. Largely influenced by the Great Lakes, the climate in the study area is highly changeable.

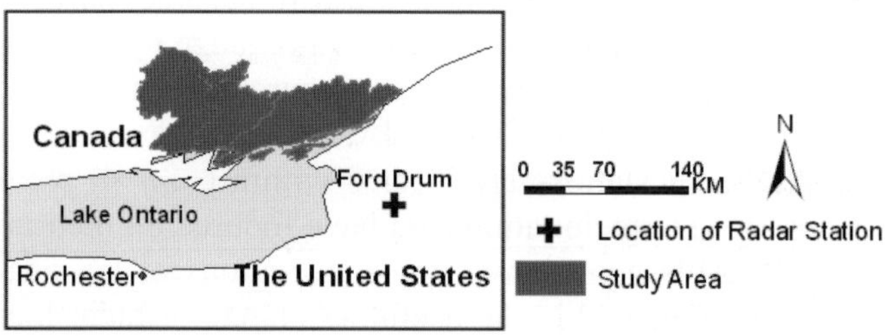

Fig. 1 The study area

Within the 8400 km^2 study area, there are only ten meteorological stations that have continuous daily precipitation records from Environment Canada (Fig. 2 and Table 1). Instruments at each station continuously record precipitation and other climate variables, such as air temperature, relative humidity, and wind speed and direction, on an hourly basis. Most rain gauges use the Geonor T-200B accumulating precipitation gauge. However, hourly precipitation data are not always available for all stations at the National Climate Data and Information Archive (www.climate.weatheroffice.ec.gc.ca) of Environment Canada.

Rates of rainfall can change dramatically over space and time, particular during convective events due to lake effect. The mean annual precipitation within the study area ranges from 892 mm at Belleville to 1027 mm at Glenburnie (Table 1). For the conservation authorities to declare a regional drought or flood alert using the rainfall data from the limited rain gauges is almost meaningless. For that reason, spatial distribution of the rainfall is needed. Conservation authorities have expressed a need for more precipitation gauges all over the region. However, that is financially impossible.

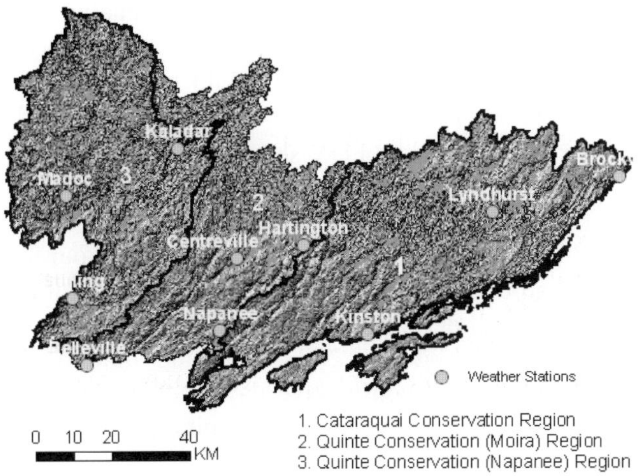

Fig. 2 Location of weather stations. The background image is a shaded relief DEM of the study area.

Table 1. Weather station information

Climate ID	Station Name	Elevation (m)	Average annual rainfall (mm)	Average annual precipitation (mm)
6104146	Kingston	93	795	968
6150689	Belleville	76.2	736	892
6100971	Brockville	96	784	983
6103367	Hartington	160	796	967
6104725	Lyndhurst	86.9	801	977
6152555	Centreville	114.3	778	901
6156533	Picton	76.2	759	964
6102808	Glenburnie	114.3	868	1027
6101265	Cararaqui	144.8	813	994

2.2 Radar data

The American radar station used in this study is located at Fort Drum, New York, and was chosen because of its proximity to the study area. The Fort Drum (FDX) station is situated at an altitude of 562 m above sea level and has been active since 1997. Fort Drum is within 70 km of the nearest border of the study area and within 200 km of the farthest border, which means both conservation areas are within the working range of the radar.

The altitude of the Fort Drum station is higher than the highest elevation in the study area. An elevation check shows that no mountains block the radar beams to the study area. NEXRAD precipitation products typically use data from the lowest of the four radar tilts with no significant beam blockage (Morin et al. 2003). Therefore, the first four tilts (0.5°, 1.5°, 2.4°, and 3.4°) of the radar beam from the FDX should be useful for the study area.

The initial intent of this study was to obtain the most recent rainfall data; however, due to the lack of rainfall prior to the start of our study, a different time period had to be selected. To find time periods with a relatively high number of rainfall dates, we searched the Environment Canada database. Five months (May and July 2002, May and June 2003, May 2004) with a high percentage of consecutive days with large amount of rainfall were selected to compare the radar precipitation data with rain gauge measurements.

The most time-intensive aspect of the analysis involved the preprocessing the radar data extracted from the NOAA website. The radar data used in this study had to be accessed through the NOAA archives, given that the NOAA website posts only records for the two most recent days. The variable examined was THP, the three-hour total precipitation. Once downloaded via ftp, the data were viewed and exported into ArcGIS shapefile using the "NEXRAD viewer" program. Fig. 3 shows an example of three-hour precipitation during a storm event on May 2005 as shown in the NEXRAD viewer.

The radar precipitation data exported using the NEXRAD viewer was displayed in a polygon shapefile in ArcGIS. We added up all THPs for each rainfall day to generate the radar rainfall layer for the study area. Then we extracted the daily radar precipitation for each station from the summarized radar rainfall layer for all days in ArcGIS and compared that data with the daily rainfall records from rain gauges, which were imported to point shapefiles in ArcGIS. All these processes were conducted by a Model built by Model Builder in ArcGIS.

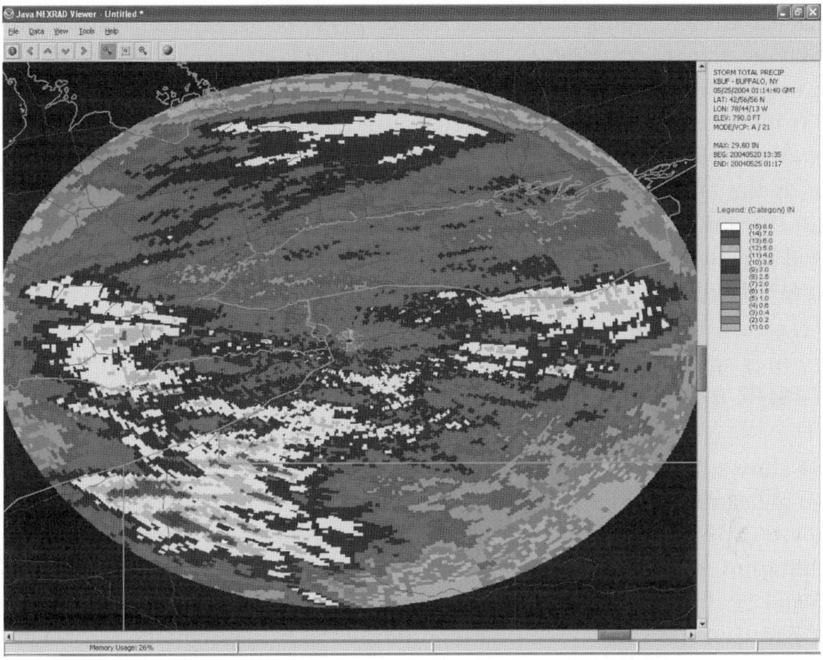

Fig. 3 An example of three-hour total precipitation during a storm event in May 2004. The center is Fort Drum, the location of the radar station.

2.3 Comparison of radar precipitation and rain gauge data

The comparison of radar precipitation and rain gauge data was conducted using two approaches. One was standard statistical analysis: comparing the mean, standard deviation, and the correlation coefficient of individual stations. A bias factor is also calculated by the ratio of radar precipitation and rain gauge measurement. Dates without rainfall were excluded from the analysis. If both radar precipitation and rain gauge measurements were zero for a particular station, that pair was excluded. In total, 177 pairs of NEXRAD precipitation and rain gauge measurement were used in the analysis.

The second comparison method used involved checking the spatial pattern of radar precipitation surface and interpolated surface from rain gauge measurements. The latter approach is commonly used to generate the rainfall surface. Three types of interpolation methods were tested: inversed distance weighting (IDW), kriging, and spline interpolation. Kriging is a geostatistical method that considers the spatial autocorrelation in the data, while IDW and spline interpolation are deterministic methods that use mathematical formula. IDW uses a linear combination of distance-based weights at known points to estimate values at unknown points. In spline interpolation, the interpolated surface is smoothed because the points do not have to pass through the original points of the data set (see Longley et al. (2003) for more details on each interpolation method). In order to generate a better interpolated surface, four additional weather stations outside but close to the study area were used together with the ten weather stations within the study area. The correlation coefficient (CV) between the interpolated surface and the radar rainfall surface was used to evaluate similarities and differences between the two data sets.

3. Result analysis

Fig. 4 shows the scatter plot of the NEXRAD daily precipitation and rain gauge measurements for all dates considered. Figure 4 shows that although there are differences between most of rain gauge measurements and radar precipitation values, the changes in the rain gauge measurements are proportional to the changes in the radar precipitation. The correlation coefficient between the radar precipitation and gauge measurements is 0.78, at the significance of 0.05. This correlation coefficient is much higher than those obtained in a similar study of a semi-arid region by Xie et al. (2006) and in other American studies. One reason for this high correlation is that only five months with consecutive daily rainfall were used in our study, while the other studies used several years of data. Also, Xie et al. used hourly precipitation, which fluctuates more than daily accumulation.

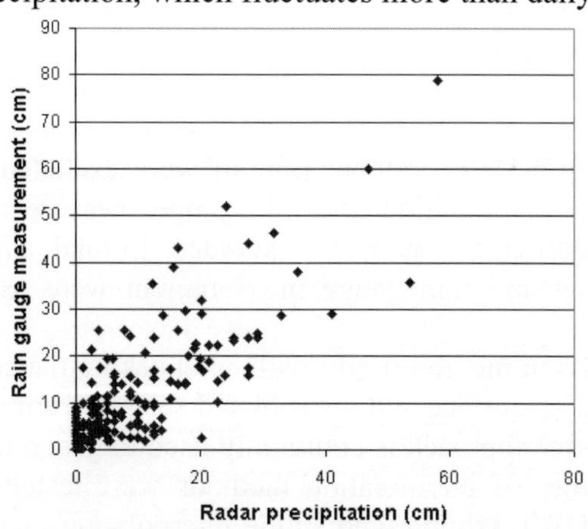

Fig. 4 The scatter plot of daily accumulated NEXRAD precipitation and rain gauge measurements for all stations and days.

Table 2 lists the basic statistics of radar precipitation and rain gauge measurements of weather stations for the study period. The average bias (the ratio of radar data and rain gauge data) is 1.16, showing, in average, the radar precipitation underestimates the actual rain fall. The worst underestimate case in this study is that radar estimate is 27.59 cm less than the rain gauge measurement in one day. However, the daily accumulated radar precipitation can also overestimate the actual rainfall as high as 18 cm. In this study, there are 62% of cases in which the radar precipitation value is smaller than the rain gauge measurement, indicating that NEXRAD precipitation underestimates the ground rainfall measurement, especially when the rainfall values are small.

Table 2. The basic statistics of radar precipitation and rain gauge measurements for weather stations

	Minimum	Maximum	Mean	Standard Deviation
Radar data	0.07	58.2	10.79	10.50
Rain gauge data	0.1	78.8	12.87	12.41
Difference between radar and rain gauge data (cm)	-27.59	18.06	-2.08	7.89
Ratio of radar and rain gauge data	0.07	10.16	1.16	1.29

Several previous studies (McCollum et al. 2002 and Xie et al. 2006) found that the 24-hour accumulation of NEXRAD precipitation is less than gauge measurements in the cold season, although, theoretically, the radar precipitation should be higher than the rain gauge measurements, due to radar's large detection area. There are two possible causes for this difference: an overshooting of the radar beam in stratified rainfall during cold seasons, or a truncation error in the NEXRAD processing. Our results seem consistent with the findings in these studies. Although June and June are considered the hot season in most of the U.S. region, the June-July temperature in Eastern Ontario is often only mild.

To further investigate the correlation between NEXRAD precipitation and rain gauge measurements, we also calculated the correlation coefficient for different rain gauge measurements of 0 to 5 cm, 5 to 15 cm, and over 15 cm. The correlation coefficients at these different rainfall ranges are 0.33, 0.44, and 0.62, respectively, indicating that NEXRAD precipitation may be more appropriate for forecasting a heavy rainfall than a small rainfall.

Considering the high correlation between the radar precipitation and rain gauge measurement, we can use the spatial pattern shown in the NEXRAD data to evaluate which interpolation method can generate the best pattern. Table 3 lists the correlation coefficients of the radar precipitation surface and the interpolated surfaces using different methods, from twenty randomly selected daily rainfall data in the data set. To our surprise, the spline interpolation method, not the Kriging, yielded the best interpolated rainfall surfaces, although the differences of the correlation coefficients were not significant.

Table 3. The correlation coefficient between the NEXRAD precipitation surface and interpolated rainfall surfaces from different methods

Surface	Minimum	Maximum	Average
NEXRAD vs. IDW	0.532	0.825	0.772
NEXRAD vs. Spline	0.612	0.892	0.787
NEXRAD vs. Kriging	0.587	0.886	0.782

Figs. 5 and 6 show examples of spatial distribution of rainfall from NEXRAD and interpolated gauge measurements using the spline interpolation method. Compared with Fig. 6, Fig. 5 has a more continuous (or realistic) spatial rainfall patterns and would better describe the spatial pattern of rainfall. In most places, the value in Fig. 5 is smaller than its corresponding value in Fig. 6, indicating that the radar data underestimated the rainfall amount.

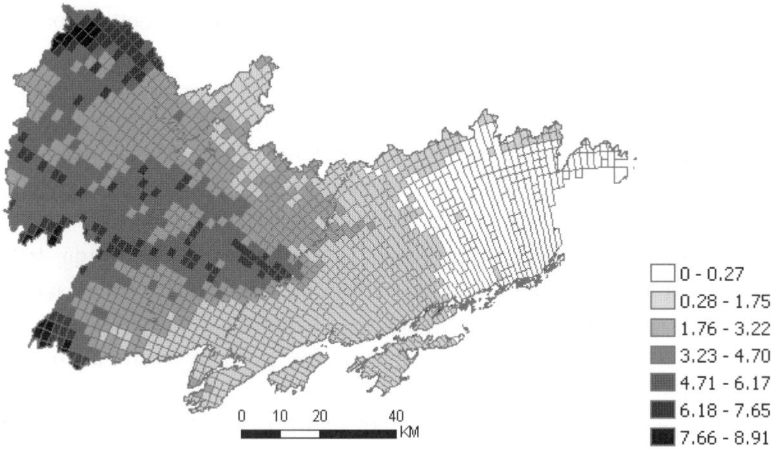

Fig. 5 Daily NEXRAD rainfall surface on July 22, 2002

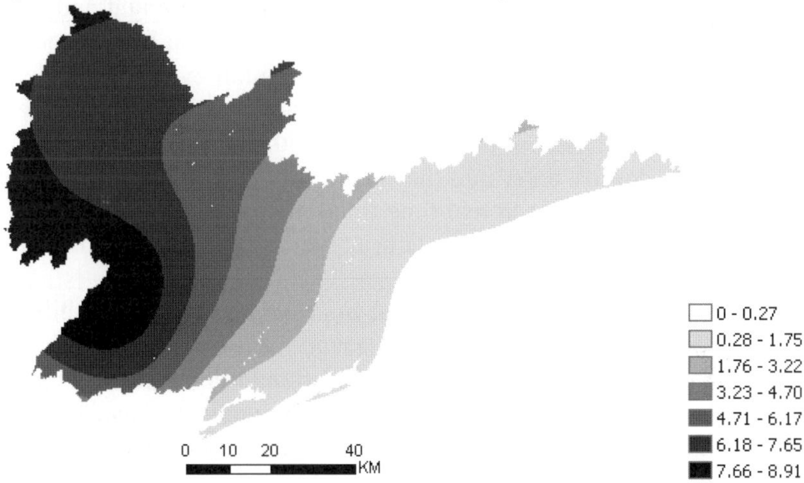

Fig. 6 Rainfall surface interpolated from rain gauge data on July 22, 2002 using the spline interpolation method

4. Summary and conclusions

This study compared the NEXRAD precipitation data and rain gauge measurements for five individual months in 2002 to 2004 for two conservation authority regions in Eastern Ontario. The NEXRAD data was evaluated using gauge data as the reference. The results show that a good correlation (0.78) exists between the NEXRAD precipitation data and rain gauge measurements. The NEXRAD precipitation has a better correlation with the rain gauge measurements for heavier rainfalls.

Statistical analyses suggest that radar precipitation underestimates the daily precipitation for 62% of rainfall days. This underestimation is more common when the rainfall is small, suggesting that there is more uncertainty in NEXRAD precipitation when it is used to estimate small rain events.

The evaluation of spatial patterns of rainfall suggests that radar precipitation shows more continuous patterns than the interpolated surfaces from rain gauges. Among the three interpolated methods, the spline interpolation method generated the surface with the highest average correlation with the radar precipitation.

Several recommendations have been made to the two conservation authorities regarding the implementation of NEXRAD in their flood monitoring, water budget analysis, and definitions of regional drought and flood. Firstly, based on our analysis, the NEXRAD precipitation has a good correlation with rain gauges, especially for large rainfall events. Considering that the small rainfall events contribute a relatively small portion of the total precipitation, and most of this rain usually either absorbs into the soil or evaporates and doesn't enter into the stream flow, the NEXRAD products can play an important role in real-time flood monitoring and water budget analysis during heavy rainfall events in regions within the working range distance of the NEXRAD system.

However, further research should be done regarding the reliability of NEXRAD by using more detailed rain gauge measurements. In addition, snowfall is not considered in this study and should be examined. Finally, complete hourly precipitation data for all Environment Canada stations must be collected for additional studies, as well as other meteorological data, such as wind speed and temperature, to test the sensitivity of different NEXRAD products under different seasons and climate conditions.

Acknowledgements

The financial support for this research came from NSERC (National Science and Engineering Research Council of Canada) Undergraduate Student Research Award (USRA) and a NSERC discovery grant. The authors would like to thank Luke Eades and Mara Shaw at the Cataraqui Regional Conservation Authority for their suggestions and assistance.

References

Anagnostou, E. N., W. F. Krajewski, J. Smith (1999) Uncertainty quantification of mean_areal radar-rainfall estimates. Journal of Atmospheric and Oceanic Technology, 16: 206-215.

Bedient, P. B., B. C. Hoblit, B.C Gladwell, and B.E. Vieux (2000) NEXRAD radar for flood prediction in Houston. Journal of Hydrologic Engineering, 5: 269-277.

Bedient, P. B., A. Holder, J. Benavides, B.E. Vieux. (2003) Radar based flood warning system - T.S. Allison. Journal of Hydrologic Engineering, 8(6): 308-318.

Ciach, G. J. and W. F. Krajewski (1999) Radar-rain gauge comparisons under observational uncertainties. Journal of Applied Meteorology, 38: 1519-1525.

Habib, E. and W. F. Krajewski (2002) Uncertainty analysis of the TRMM ground-validation radar-rainfall products: Application to the TEFLUN-B field campaign. Journal of Applied Meteorology, 41: 558-572.

Johnson, J. T., P. L. MacKeen, A. Witt, E.D. Mitchell, G. J. Stumpf, M.D. Eilts, and K.W. Thomas (1998) The storm cell identification and tracking algorithm: an enhanced WSR-88D algorithm. Weather and Forecasting, 13: 263-276.

Johnson, D., M. Smith, V. Korean, and B. Finnerty (1999) Comparing mean areal precipitation estimates from NEXRAD and rain gauge networks. Journal of Hydrologic Engineering, 4(2): 117-124.

Krajewski, W. F. and J. A. Smith (2002). Radar hydrology: rainfall estimation. Advances in Water Resources 25: 1387-1394.

Krajewski, W. F., G. J. Ciach, and E. Habib (2003) An analysis of small-scale rainfall variability in different climate regimes. Hydrologic Science Journal,48: 151-162.

Longley,P.A., M.F. Goodchild, D.J. Maquire, D.W. Rhind (2001) Geographic Information Systems and Science, John Wiley & Sons, 454p.

Maddox, R. A., J. Zhang, J.J Gourley, and K.W. Howard (2002) Weather radar coverage over the contiguous United States. Weather and Forecasting, 17: 927-934.

McCollum, J. R., W. F. Krajewski, and R.R. Ferraro (2002) Evaluation of biases of satellite rainfall estimation algorithms over the continental united states. Journal of Applied Meteorology, 41: 1065-1080.

Morin, W., W. F. Krajewski, D.C. Goodrich, X. Gao, and S. Sorooshian (2003) Estimating rainfall intensities from weather radar data: The scale-dependency problem. Journal of Hydrometeorology, 4: 782-797.

Pereira Fo, A., J., K. C. Crawford, and D.J. Stensrud (1999) Mesoscale precipitation fields. Part II: hydrometeorologic modeling. Journal of Applied Meteorology, 38: 102-125.

Reed, S. M, D. R. Maidment (1999) Coordinate transformations for using NEXRAD data in GIS-based hydrologic modeling. Journal of Hydrologic Engineering, 4(2): 174-182.

Seo, D. J., J. P. Breidenbach, and E.R. Johnson (1999) Real-time estimation of mean field bias in radar rainfall data. Journal of Hydrology, 223: 131-147.

Smith, J. A., M. L. Baeck, J.E. Morrison, and P. Sturdevant-Rees (2002) The regional hydrology of extreme floods in an urbanizing drainage basin. Journal of Hydrometeorology, 3: 267-282.

Stellman, K. M., H. E. Fuelberg, R. Garza, and M. Mullusky (2000) An examination of radar- and rain gauge-derived mean areal precipitation over Georgia Watersheds. Weather and Forecasting, 16(1): 133-144.

Vieux, B. E. and P. B. Bedient (1998) Estimation of rainfall for flood prediction from WSR-88D Reflectivity: A case study. Weather and Forecasting, 13(2): 407-415.

Vieux, B. E. and P. B. Bedient (2004) Evaluation of urban hydrologic prediction accuracy for real-time forecasting using radar. 32th International Conference on Radar Meteorology American Meteorology Society, J1.3.

Xie, H., X. Zhou, M.H. Hendrickx, E.R. Vivoni, H. Guan, Y.Q. Tian, and E. Small (2006) Evaluation of NEXRAD Stage III precipitation data over a semiarid region. Journal of the American Water Resources Association, Feb.: 237-256.

Young, C. B., B. R. Nelson, J. Smith, C. Peters-Lidard, A. Kruger, and M.L. Baeck (1999) An evaluation of NEXRAD precipitation estimates in complex terrain. Journal of Geophysical Research, 104(D16): 19691-19703.

Young, C. B., A. A. Bradley, W.F. Krajewski, A. Kruger, and M.L. Morrissey (2000) Evaluating NEXRAD multisensor precipitation estimates for operational hydrologic forecasting. Journal of Hydrometeorology, 1: 241-254.

Zhang,Y. and J. A. Smith (2003). Space-time variability of rainfall and extreme flood response in the Menomonee River Basin, Wisconsin. Journal of Hydrometeorology, 4: 506-517

Evaluating the Use of a Low-Cost Unmanned Aerial Vehicle Platform in Acquiring Digital Imagery for Emergency Response

Lewis G.

Assiniboine Community College, 1430 Victoria Avenue East Brandon, Manitoba, Canada R7A 2A9 lewisg@assiniboine.net

Abstract

This research project evaluates the utilization of a low-cost Unmanned Aerial Vehicle (UAV) digital imaging platform developed in Manitoba, Canada for emergency response situations. Such a platform allows for the timely acquisition of high resolution imagery during emergency situations by personnel with relatively limited UAV flight training.

Although military use of UAVs has been around since the First World War, their relatively high expense and operational requirements have limited their civilian use. The UAV platform, as tested in this project, is a very valuable and useful tool for disaster response situations, allowing for the easy and timely acquisition of high resolution imagery, by meeting the following criteria:

- fly fully autonomously with little pilot experience
- assembled and ready for launching in less that one half hour
- acquire high resolution imagery (greater than 25 centimetre resolution)
- transported in a midsized car
- relatively low cost ($8,000 US)

A 5.0 mega pixel digital camera enables the UAV platform to acquire colour imagery with a 22 cm spatial resolution from a height of 2100 feet. With a stable and reliable UAV platform already developed, future research will focus on the development and integration of other imaging and sampling techniques such as thermal technologies and air sampling abilities for smoke plume analysis.

1 Use of Aerial Photography in Emergency Response

The use of aerial photography as a component of emergency response is not a new concept and has been extensively studied. Most research has examined the use of conventional aircraft or satellite systems to collect such imagery. Unfortunately, both have a limited ability to provide accurate and timely imagery in emergency situations. Thus, alternative methods of acquiring imagery for emergency response need to be evaluated.

Mehrotra et al. (2004) outline the goals of an imaging system for emergency response which "aims to enhance the mitigation capabilities of first responders in the event of a crisis by dramatically transforming their ability to collect, store, analyze, interpret, share and disseminate data". The authors detail the incorporation of a variety of information technologies including aerial image processing. They conclude that the use of "...high-resolution optical imagery acquired before and after the event can provide a 'quick-look' damage assessment, distinguishing areas of catastrophic damage (ibid; p. 189)" and is extremely useful information in emergency response situations. The value of this "quick-look" imagery is largely dependent on the capability of the system to acquire imagery quickly enough to be useful to first responders in emergency situations. Kerle and Stekelenburg (2004) support this notion of the importance of real-time or at least near real-time imagery in emergency response stating "Natural and man-made disasters create a need for rapid, comprehensive and reliable information on the nature, extent and actual consequences of an event".

The ability to obtain real-time imagery can be complicated by numerous factors. Green et al. (2003), note that acquiring aerial imagery using conventional aircraft during a disaster can be difficult when there is a lack of suitable runways due either to the remoteness of the area or damage caused by the disaster itself. Satellite systems can also provide aerial imagery and have proven valuable in widespread disasters such as hurricane damage (Elsis 1997). Unfortunately, the temporal and spatial resolutions of satellite systems make them relatively ineffective for emergency response because "...the time required to schedule a satellite fly-by may delay first response efforts." (Green et al. 2003)

Jensen (2000) details the required temporal and spatial resolution for particular Urban-Suburban emergency situations. One such circumstance Jensen describes is post-emergency imagery situation, which requires a temporal resolution of 12 hours up to 2 days and a spatial resolution of 0.25 – 2 m. Although both of these criteria could be achieved with current and future satellite based image acquisition systems, atmospheric conditions such as cloud cover will always be a source of data loss (Jensen 2000).

Clearly the tremendous situational awareness that can be gained from the use of aerial imagery for emergency situations is important, but not without difficulties. The platforms that have traditionally acquired this imagery have not been able to do so with the required temporal and spatial resolutions that would make them effective in emergency response situations. Recently however, Unmanned Aerial Vehicles (UAVs) have emerged as viable platforms for the acquisition of high-resolution aerial imagery. UAVs are usually small aerial platforms that have the potential to be easily transported, assembled and quickly tasked to acquire near-real time high-resolution imagery. This allows first responders to emergency situations the ability to quickly and accurately assess the situation without putting personnel at risk. Currently "routine civil access to these various UAV assets is in an embryonic state and is only just now emerging" (Roper and Dutta 2005). The intent of this paper is to evaluate the application of these new civil UAV platforms specifically to emergency response situations.

2 Advantages of a UAV for Aerial Photography

Lillesand and Kiefer (2000) describe the components of an ideal remote sensing system. These six basic components include:

1. A uniform energy source
2. A non-interfering atmosphere
3. A series of unique energy-matter interactions at the earth's surface
4. A super sensor
5. A real-time data processing and supply system
6. Multiple data users

The use of UAVs for the acquisition of aerial imagery addresses three of the six basic components which many other commercially available platforms do not, namely: a non-interfering atmosphere; a real-time data processing and supply system; and multiple data users.

Until recently UAVs were almost exclusively the domain of militaries and not available for civilian use. This was due in part to security issues and their relatively high cost. Now a new generation of civilian UAVs are allowing others to take advantage of the flexibility of a UAV based system. These new UAVs can be purchased as complete systems relatively inexpensively (less than 10% of the cost of traditional aircraft) and without the same high maintenance costs. In addition, a special pilot's license is not required to fly a civilian UAV. Personnel can be trained as a ground control technician in as little as five days.

The ability to acquire imagery at relatively low altitudes (e.g. 122 – 640 m above the ground), afford UAVs the ability to acquire imagery below the majority of atmospheric conditions, such as cloud cover, that often plague other remote sensing systems. Atmospheric conditions such as fog and haze can have an affect on UAV based systems, but to a much lesser degree than other platforms that may have acquisition heights of 10 to 100's of kilometres above the Earth's surface.

Another great advantage of a UAV based aerial imagery system is the ability to be quickly deployed and have imagery available nearly real-time. This is due, in part, to the fact that some UAVs can be hand launched and skid landed into relatively small locations. This eliminates the need for take off and landing strips, which may be a significant distance from the emergency site in question.

Previously, the ability to acquire imagery in a timely manner for emergency response has been limited because of the costs and accessibility of the equipment and technology needed to acquire such imagery. The current generation of civilian UAVs overcome these shortcomings making them a valuable resource in emergency response.

3 Discussion of the Platform

CropCam is a civilian UAV platform built and integrated by MicroPilot Incorporated that currently sells for US$6,995.00. MicroPilot was founded in 1995 to design, manufacture and sell very small (28 grams) autopilots. Autopilots are a collection of sensors and processors which, when coupled with the global positioning system (GPS), enable users to fly various small UAV platforms fully autonomously. Although CropCam was developed for agricultural applications, its utilization of the MicroPilot system affords it the flexibility to be used for the acquisition of high resolution imagery for a variety of applications including emergency response.

3.1 Airframe of the Platform

The CropCam UAV is based on a commercially available remote controlled sailplane airframe that is modified and integrated with the MP2028[g] Autopilot. The airframe used is an Electra Pro sailplane manufactured in the Czech Republic by TopModel CZ (www.topmodelcz.com) and purchased in North America from North East Sail Planes (www.nesail.com)

(see Fig. 1). The current CropCam platform has a wingspan of 254 cm, a length of 122 centimetres, a wing area of 5058 cm^2 and a total flying weight of 2.72 kg including the batteries. The use of a standard sailplane airframe affords a number of advantages. The first advantage is that although the airframe does require a motor to perform its flights, it still maintains the ability to be hand launched and skid landed. Secondly, sailplanes still maintain very high glide rates, which result in stable and predictable flights and landings. Thirdly, the large wing surfaces of sailplanes maximize the amount of payload that it can be carried without having to be runway or bungee launched. Finally, the use of a commercially available airframe allows for the total cost of the platform to be as low as possible while still maintaining very high quality standards.

Fig. 1 Setting up the CropCam UAV in the field

The integration of the MP2028[g] autopilot (Fig. 2) allows the CropCam to function in full autonomous flight mode known as computer in control (CIC), or be controlled manually by the ground support technician, referred to as person in control mode (PIC). While in CIC mode the aircraft flies a predefined flight pattern, the specifications of which are saved in a flight file. This file is preloaded onto the autopilot's memory and allows the aircraft to fly predefined patterns including take offs and landings, with no user control. Alternatively, the platform can be utilized in PIC mode, which allows the ground support technician to manually control the aircraft by means of a standard model aircraft remote control.

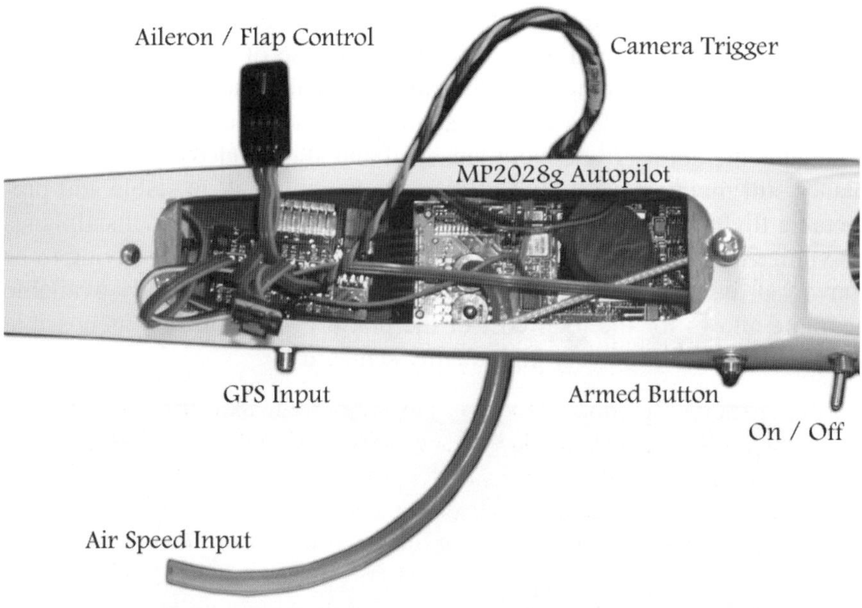

Fig. 2 Integration of the MP2028 Autopilot

3.2 Motor and Power Requirements

The current CropCam utilizes a standard R/C electric motor to achieve and maintain target flying height above the ground. The main difference between the CropCam platform and a conventional glider set-up is that the motor is mounted on the exterior of the airframe on the CropCam, to maximize the amount of room available to house the batteries. To power the electric motor, four 2100 mAh, 11.1volt, 3-cell, lithium polymer (LiPo) batteries were purchased at an approximate cost of US $279.84 (http://www.thunderpower-batteries.com). LiPo batteries have a significantly decreased weight compared to NiMH or NiCd packs, as well as the ability to provide higher voltage under load. Each of the 2100mAh, 3-cell packs weigh 142 g for a total of 568 g. Although very dependent on the wind conditions, these battery packs allow the CropCam platform to be airborne for upwards of 55 minutes. This flight time allows for the collection of aerial imagery for an area of up to 640 acres in one flight or the collection of numerous images for the same location as may be the case in some emergency response situations. After each flight, each of the

2100mAh battery packs requires a recharge. The charging time is similar to that of the flight time; that is a 55 minute flight would take approximately 55 minutes to recharge the battery pack to full capacity with the three packs being charged simultaneously. To maximize the amount of time that can be used collecting imagery of a situation, a second set of batteries can be purchased allowing for the charging of one set while the other set is flying the CropCam UAV.

3.3 Payload

The current configuration of the UAV platform allows for any number of sensors to be carried onboard. The number of sensors is limited, however, by the airframe's payload capacity. On the current platform, any payload that is heavier than six pounds would result in instability in the flight characteristics and particularly in an increase in the stall speed of the aircraft, resulting in very difficult landings. As well the added weight would increase the amount of strain on the motor batteries resulting in much shorter flight times.

As evaluated, the test platform includes a Pentax Optio S5i colour digital camera. The Optio 5si was selected for its infrared sensor, which is intended to allow users to take photographs using the optional remote control. On arriving at specific locations or coordinates, the platform is instructed to trigger a specific servo. This servo flashes an infrared sensor (Fig. 2), which has been attached in front of the cameras infrared remote control sensor and instructs the camera to capture an image. The camera is simply mounted underneath the wing of the platform in a small enclosure, using standard elastic bands. (Fig. 3)

Fig. 3 Detail of the camera mounting system

3.4 Transportation and Set up Process

One of the greatest advantages that the CropCam UAV platform has for effective use in emergency response situations is its ability to be easily transported and quickly deployed once on site. The current CropCam can be disassembled to allow for easy storage and transportation. With disassembled dimensions of only 14.6 x 32.4 x 131cm, the CropCam can easily be stored and transported in existing emergency response vehicles. If the platform is transported to the emergency site disassembled some in-field set-up is required. This includes attaching two wing tips to the main wing, one on each end, and then attaching the assembled wing to the fuselage, as well attaching the tail stabilizer on top of the tail fin. The in-field set-up process can be completed in approximately 20 minutes by one person. Alternatively, the platform could be transported fully assembled, which would eliminate the time associated with assembling the airframe. This, however, requires significantly more space: 100 x 12.75 x 51.5 in, for transportation.

3.5 Ground Control Software

The core of the CropCam UAV platform is formed by two separate components. The first is the MP2028^{g} Autopilot which, as discussed earlier, is a collection of sensors and processors that measure the aircraft position and allow for the autonomous flight of the aircraft (CIC mode). The second component is the HORIZON^{mp} ground control software. The HORIZON^{mp} software package comes with the CropCam UAV platform and was developed by MicroPilot specifically for the MP2028 series of autopilots. This software package allows the ground control technician to monitor the autopilot and airframe, change waypoints, upload new flight plans, initiate holding patterns and adjust feedback loop gains, all while the UAV is in flight.

The HORIZON^{mp} ground control software offers a user-friendly point-and-click interface (Fig. 4) which allows for the creation of flight files. Flight files are small text files that define the preloaded flight pattern and are uploaded onto the internal memory of the MP2028^{g} Autopilot before launch. In its current configuration the CropCam UAV platform flies flight patterns relative to a single starting point. This allows for the ability to create predefined flight files before the platform is needed in a time sensitive situation. Once in the field the appropriate flight file is selected and uploaded to the CropCam, based on the relative position of the target area to the starting location and the wind direction. Through the use of a 900 MHz radio modem, HORIZON^{mp} target waypoints can be modified literally "on the fly" by clicking and dragging the points representing the waypoint to the new location. Additionally, the radio modem allows for the ground support technician to monitor numerous platform sensors including, battery levels, current altitude, current speed, GPS speed and attitude to the horizon (artificial horizon).

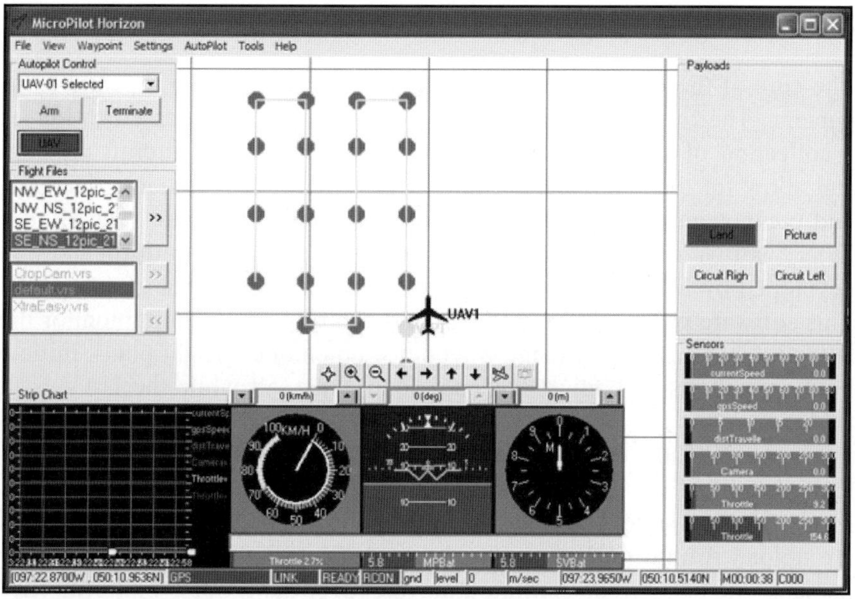

Fig. 4 Detail of the HORIZON^{mp} ground control interface

4 Flight and Image Acquisition Characteristics

Being based on a commercially available, remote control sailplane gives the CropCam a forgiving flying manner. Although the CropCam has the capability of completing flights fully autonomously, remote control airplane skills are a definite asset for the ground support technician. Having these skills allows the ground control technician to control the aircraft while in PIC mode and to be selective on the landing locations, which can significantly increase the life span of the airframe and reduce the possibility of damaging the autopilot, batteries and camera.

Currently the CropCam is rated for wind speeds of less than 30 km per hour. Although the aircraft will fly in winds greater than this, the flight stability and ability of the camera to capture accurate images becomes questionable. However, the acquisition of multiple images for a single location could be accommodated in such conditions. Another flight consideration is the space required to safely launch and land the CropCam. MicroPilot suggests a 90 m clearing for takeoff and landing if you land in PIC mode and double that if you will be requiring a CIC landing assuming 9 meter high trees on all sides (Wilde 2007).

4.1 Image Characteristics

As digital imagery is measured and stored as a series of rows and columns, the size of each cell or pixel can be calculated in terms of the amount of ground it covers. This is known as spatial resolution and is commonly expressed as a single measurement representing the ground distance one side of a pixel covers (Lillesand and Kiefer 2000). The expression of spatial resolutions allows for the assessment of the possible applications of specific imagery. In emergency response situations spatial resolutions need to be such that first responders have the ability to easily and accurately identify and locate key objects such as people, contaminant spills, and fires. As indicated earlier by Jensen (2000), this post-emergency imagery requires a spatial resolution of 0.25 – 2 m. This is the benchmark that was used in the evaluation of the CropCam system with hopes of exceeding this figure.

High resolution imagery can be achieved from small UAV platforms such as the CropCam using relatively inexpensive commercially available colour digital cameras. High resolution imagery is desirable in emergency response situations since the higher the spatial resolution (SR) the more accurately features can be interpreted and measured using image processing application or a GIS. Calculating the spatial resolution (SR) of imagery acquired from a specific digital system involves a number of factors. For purposes of this paper spatial resolution was calculated using the following equation:

$$SR_{(x)} = \frac{S_{(x)} \times H'}{nP_{(x)} \times f}$$

Where $SR_{(x)}$ represents the spatial resolution of the imagery in that specific axis, S represents the physical size of the sensor (mm) in that specific axis, which in a digital system is the Charged Coupled Device (CCD), H' represents the flying height of the platform above the ground (m), f represents the focal length of the camera used (mm) and $nP_{(x)}$ represents the number of photosites present on the CCD for that axis.

4.2 Flying Height of the Platform above Ground

The majority of the test flights were conducted at sites approximately 805 m in length and width and were flown at a height of 640 m above the ground (*H'*). A height of 640 m is used as the target image acquisition flight height for a number of reasons, but could easily be adjusted. First, Transport Canada, the governing body of UAVs in Canada, has set a ceiling of 670 m with the intent that users maintain visual contact with the airframe. (Shaw 2007) A target flying height of 640 m above the ground provides a 30 m buffer to prevent breaching this ceiling and maintains visual contact while autonomously flying patterns within a 1.6 km radius of the ground support technician. The second reason for having a 2100 foot image acquisition target height has to do with wind turbulence, which can greatly affect the ability to acquire high quality imagery from a relatively small and light platform. Test flights have determined that turbulence is often much greater closer to the ground than at 640 m where the platform is much more stable. Finally, the 640 m acquisition height was selected to maximize the amount of ground that could be covered by the platform, thereby minimizing the number of images that would need to be manipulated and analyzed for integration into an image analysis package or GIS.

4.3 Camera Specifications

The integrated Optio S5i camera utilized on the current platform captures colour imagery using a 1 / 2.5" sized Charged Couple Device (CCD), which has a physical width of 5.760 mm ($S_{(x)}$) and height of 4.290 mm ($S_{(y)}$). This CCD employs a maximum image resolution of 2560 pixels wide ($nP_{(x)}$) and 1920 pixels in height ($nP_{(y)}$) and uses a focal length of 5.8 mm (*f*). Weighing a modest 120 grams and having an approximate retail price of US$200.00, made the Pentax Optio S5i a very solid sensor choice for the CropCam platform. (Joinson 2004)

4.4 Image Resolution

Based on the previously derived equation and above specifications, the spatial resolution of the CropCam system utilizing the Pentax Optio S5i at a flying height of 640 m was calculated for both the height and width of the resulting imagery. The spatial resolution for the width of the imagery was calculated to be 24.655 cm, while the height resolution was calculated to be 24.828 cm. Although there is a slight (0.173 cm), difference between the spatial resolutions of the width and height of the pixels the spatial resolution of the system is concluded to be 24.74 cm. For the lower limit flying height of 122 m the spatial resolution was calculated to be 4.729 cm in width, while the height resolution was calculated to be 4.696 cm. At this height there is a very slight difference between these resolutions of 0.0328 cm, with the average spatial resolution being calculated to be 4.713 cm.The spatial resolution of the CropCam system has the ability to acquire imagery within the 25 cm resolution parameter set forth by Jensen (2000) for post emergency imagery. This is achieved at the highest flying height allowed and using a 5.0 mega pixel, both of which could be adjusted if higher spatial resolution is required. Even this unprocessed or "raw" imagery could enable first responders to quickly locate and identify key objects such as people, contaminant spills and fires. However, in order to be successfully implemented and analyzed using an image processing application or GIS some pre- processing may have to be completed.

5 Image Processing and Integration into a GIS

The ability to integrate imagery acquired from a UAV platform, such as the CropCam, in near real-time could be invaluable to emergency response personal. However, before accurate image analysis can take place, two post-flight image processing procedures must be completed. The first process involves stitching or mosaicing, which "is the process of combining multiple images into a single seamless composite image" (Jensen 2005). This process may not be necessary if the area of interest is small enough to be acquired with a single image. The second process is the geometric transformation process which is defined as the process of creating "consistent geometric relationships between points on the ground and their corresponding representations on the image (Campbell 2002).

5.1 Mosaicing Process

The stitching or mosaicing of multiple images acquired from the CropCam UAV platform was performed using the PTgui software package. PTgui, stands for Panorama Tools Graphical User Interface as has been developed to allow users to quickly stitch multiple photographs of a scene into one seamless image. The software automatically reads the header files (EXIF) that accompany the JPEG images and determines in what order the photographs were taken. The software then analyzes successive photographs for pixels that represent the same area in both photos, known as ground control points, however it does allow for the user to update or add new ground control points. Fig. 5 shows the PTgui interactive photo manipulation interface, which illustrates the process of compiling multiple images into one seamless image. As mentioned earlier, if the target area can be acquired in a single image this process is not necessary as multiple images would not be needed.

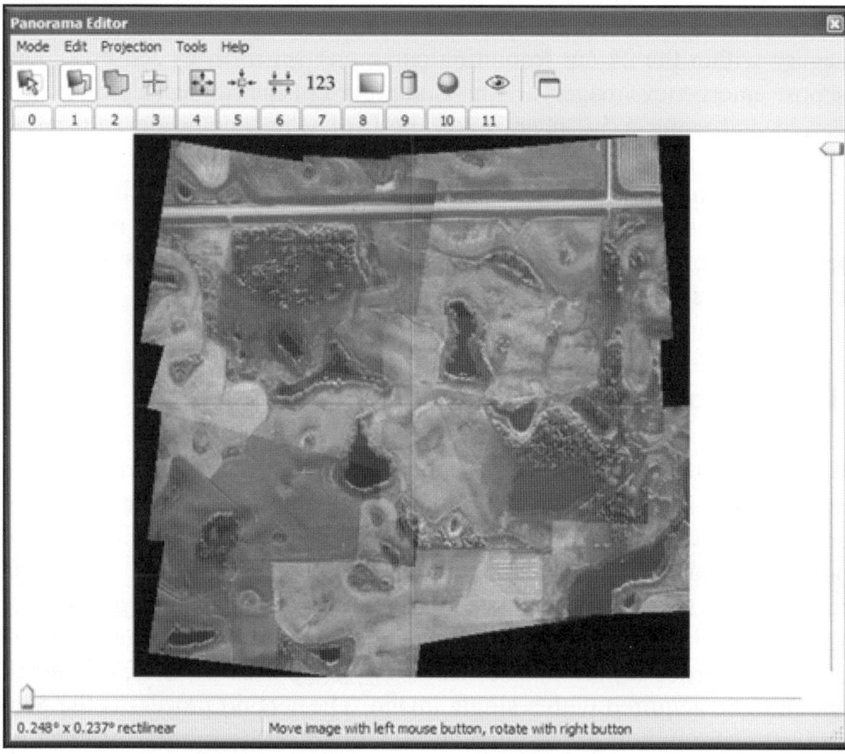

Fig. 5 Detail of the PTgui interactive photo manipulation interface.

5.2 Georeferencing Process

With multiple images stitched or mosaiced into a single image the process of georeferencing is required to allow for the integration of the imagery into a GIS. The georeferencing is the process of registering digital imagery to an established geographic coordinate system through the use of GCPs and a geometric transformation that converts file pixel coordinates (i.e. row column positions) to projected coordinates (e.g. UTM). The two most commonly used geometric correction procedures are image-to-map rectification and image-to-image rectification. (Jensen 2005) The main difference between these two methods is the source of the ground control is determined.

In the image-to-image geometric correction process the coordinates are derived from an existing image. This can be a very effective method of georeferencing however, it does require that a corrected image already exists and is easily accessible. In the image-to-map rectification process the unregistered image is corrected to ground control points either collected from an existing map or from field data collected using a GPS. This method can be much more time consuming and possibly dangerous if personal have to physically collect the ground control coordinates.

The ArcGIS® v9.1 Georeferencing tools were used to georeference the mosaiced imagery. This software offers a very easy user interface that allows the user to select ground control point that can be visible in both the unregistered and registered image. Once this is completed the transformation process can be executed resulting in a new image georeferenced to a known coordinate system, such as the Universal Transverse Mercator (UTM) system. Fig. 6 details the Georeferencing toolbar used in an image-to-image geometric correction.

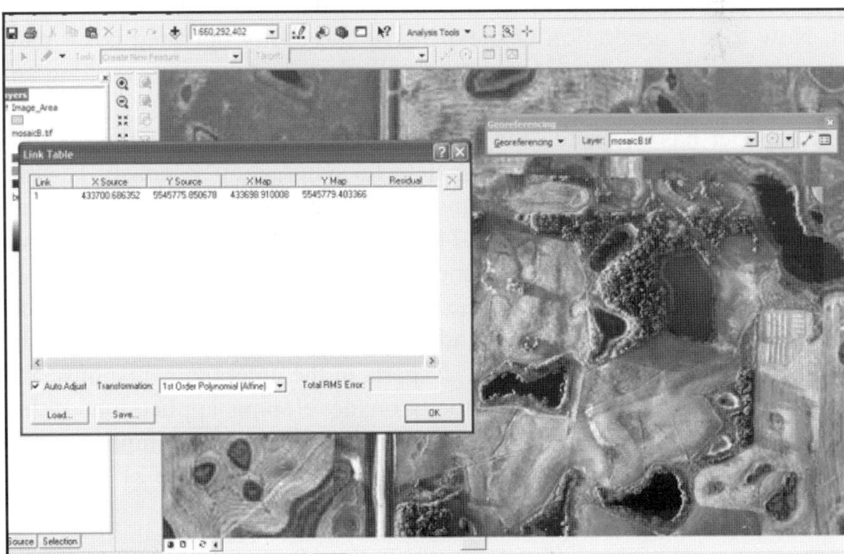

Fig. 6 Detail of the ArcGIS ®, georeference interface

5.3 Integration with GIS and GPS

Only after the imagery has been processed by mosaicing and georeferencing can it be implemented and analysed in a GIS with other geographic data. Once in the GIS, any number of feature identification processes or standard geospatial analysis tools can be employed to facilitate emergency response personal. These processes include such activities as measuring the extent of contaminant spills and fires, identification and locations of hazardous materials and the location of possible victims, all while maintaining a safe distance. In addition, this information could be uploaded to GPS units giving emergency personnel critical information to help them safely navigate the site.

6 Conclusions

The use of aerial imagery in emergency response situations has proven to be a very useful tool; however traditional methods of collection have not had the temporal or spatial resolutions required. The use of UAVs effectively addresses these issues. Unfortunately, until recently the relatively high cost and difficult operation has limited their use in the civilian market. Now a new generation of inexpensive and easily operated UAV platforms has emerged, allowing for the development of new and exciting applications of high resolution aerial imagery and sensors.

This paper has evaluated the CropCam UAV platform for the requirements of an ideal remote sensing system for emergency response. This includes the ability to be easily transported and deployed, acquiring high resolution imagery (4.71 cm - 24.74 cm), at a relatively low cost. However, more research needs to be done on new and useful sensors that could be developed and integrated into the platform. It is hoped that the commercial availability of cost effective and stable UAV platforms will foster the development of more applications. Experts in the emergency response and disasters sectors will be able to focus on developing applications rather than platforms. I am confident that with further research and improvements "only our imagination will limit the potential of UAV's in the 21st century" (Roper and Dutta 2005).

References

Campbell J. (2002) Introduction to Remote Sensing, 3rd edn. The Guilford Press, New York, NY.

Elsis R. (1997) Visualizing your landbase, proceedings 1997 ESRI International User Conference, July 8-11 1997, San Diego, CA.

Green W, Oh P. and Yoon S. (2004) An aerial image acquisition and distribution system for situational awareness, proceedings ASME International Mechanical Engineering Congress and Exposition (IMECE), November 2004,Washington, D.C., v2, pp. 1341-1346.

Jensen J. (2000) The changing face of remote sensing, report of a conference remote sensing for transportation (Appendix A), December 4-5 2000, Washington, D.C.

Jensen J. (2005) Introductory digital image processing: a remote sensing perspective, Pearson Prentice Hall, Upper Saddle River, NJ, pages 234 and 250.

Joinson S. (2004) Pentax Optio S5i review, Digital Photography Review™ (http://www.dpreview.com/reviews/pentaxs5i/)

Kerle N. and Stekelenburg R. (2004) Advanced structural disaster damage assessment based on aerial oblique video imagery and integrated auxiliary Data Sources. In: ISPRS 2004 : proceedings of the XXth ISPRS congress : Geoimagery bridging continents, 12-23 July 2004, Istanbul, Turkey. Comm. VII, TS-PG: WG VII/5. pp. 580-585.

Lilesand T., and Kiefer R. (2000) Remote sensing and image interpretation 4th edn. John Wiley & Sons, Inc., New York, NY, pages 35-37.

Mehrotra S., Butts C., Kalashnikov N., Venkatasubramanian N., Rao, R. Chockalingam G., Eguchi R., Adam B., and Huyck C. (2004) Project rescue: challenges in responding to the unexpected, Journal of the International Society for Optical Engineering (SPIE), vol 5304, pages 179-192.

Roper W., and Dutta S. (2005) Remote sensing and GIS applications for pipeline security assessment, proceedings 25th Annual ESRI International User Conference, July 25-29 2005, San Diego, CA.

Shaw L. (2007) MicroPilot Incorporated, personal correspondence.

Wilde E. (2007) MicroPilot Incorporated, personal correspondence

Automatic Classification of Collapsed Buildings Using Object and Image Space Features

Mehdi Rezaeian and Armin Gruen

Federal Institute of Technology (ETH) Zürich
Institute of Geodesy and Photogrammetry, CH-8093 Zürich
(mehdi.rezaeian,agruen)@geod.baug.ethz.ch

Abstract

We present a method based on two kinds of image-extracted features comparing stereo pairs of aerial images before and after an earthquake. The study area is a part of the city of Bam, Iran which was hit strongly by an earthquake on December 26, 2003. In order to classify damages caused by earthquakes, we have explored the use of two kinds of extracted features: volumes (defined in object space) and edges (defined in image space). For this purpose, digital surface models (DSM) were created automatically from pre- and post-earthquake aerial images. Then the volumes of the buildings were calculated. In addition, a criterion for edge existence - in accordance with pre-event building polygon lines – from post-event images is proposed. A simple clustering algorithm, based on the nearest neighbor rule was implemented using these two features simultaneously. Based on visual inspection of the stereo images, three levels of damage (total collapse, partial collapse, no damage) were considered. By comparing pre- and post-earthquake data the results have been evaluated. The overall success rate – total number of correctly classified divided by the total number of samples – was found to be 71.4%. With respect to the totally collapsed buildings we obtained a success rate of 86.5% and 90.4% in terms of producer's and user's accuracies respectively, which is quite encouraging. The results of the analysis show that using multiple features can be useful to classify damages automatically and with high success rate. This can give first very valuable hints to rescue teams.

1 Introduction

An effective disaster response planning covers three activities: 1) development of a realistic damage simulation model, 2) methods to rapidly assess actual damage, and 3) models to allocate limited rescue resources to damaged areas in an optimal way. Remote sensing techniques both by space-borne or air-borne sensors could make a very effective contribution especially to the first and second activities. The basis for these approaches is a multi-tiered procedure that allows broad regional damage assessments to be conducted using moderate-resolution data (Landsat, SPOT, ALOS/PRISM, Cartosat-1, etc.) and more detailed assessments using high-resolution information (QuickBird, Ikonos, aerial images). This concept can be considerably assisted with an automated system, using digital photogrammetry. This study outlines the principal issues for the development of image-based systems for damage and loss estimation in an emergency due to earthquake.

Effective disaster management requires assimilation and dissemination of real-time information/data to various decision makers. In a natural disaster, we should grasp the damaged areas completely and rapidly for the support of the rescue activities. Photogrammetric and remote sensing images from aerial or satellite platforms are useful because the damage situation can potentially be quickly derived from these images. However, the image information is very complex and it is for instance difficult to extract the change of shapes of roads and buildings fully automatically. Since remote sensing data observed by various platforms have different characteristics, it is necessary to consider the characteristics of each platform and sensor and the quality of data.

In this paper, first, a brief overview of damage classification by visual interpretation is presented. Then a sequence for collapse detection is given. Finally, we describe our proposed method for damage classification of buildings caused by earthquake and the evaluation of the results with regard to reference data generated through visual interpretation.

2 Visual Interpretation of Damage

Principally, the extraction of all desired information by visual interpretation is limited and depends on the specifications of the images as well as on the objects of interest. In the following we describe briefly some of these restrictions. Damages at buildings are generally described based on damage classes. This may differ from country to country. The way in which a building deforms under earthquake loading depends on the building type. Visual inspection of building damage could be conducted based on the classification of the European Macroseismic Scale (EMS 1998), shown in Figure 1. Grades 1 to 5 should ideally represent a linear increase in the strength of shaking for types of masonry buildings as well as buildings of reinforced concrete. Ogawa et al. (2000) evaluated the damage situation after the 1995 Kobe earthquake using aerial photographs. They compared the aerial photo interpretation with those of ground surveys to examine the accuracy of aerial photo-interpretation methods in detecting damages. The damage interpretation using aerial photographs in an urban area (especially for wooden building) was found to be effective for identifying "collapsed" and "severely damaged" buildings. In damage interpretation of "severely damaged" buildings the accuracy of stereoscopic photo-interpretation is higher than that of single photo interpretation. The aerial photographs allow excellent recognition of the shape and location of objects. However, since usually the aerial photographs are taken vertically to the ground surface, the recognition of minor damages and damages to side walls and columns is difficult. Hasegawa et al. (2000) used aerial high-definition television (HDTV) images, taken after the Kobe earthquake by helicopter. They investigated influences of differences in structure type, building density and human interpreter on the interpretation results. Aerial images taken at an oblique angle is effective to identify wall damages. However, in the densely built-up area with high-rise buildings even oblique images are not always suitable because of possibly substantial occlusions.

Masonry buildings	Reinforced buildings	Classification of damages
	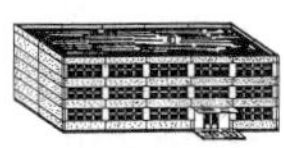	Grade 1: Negligible to slight damage (no structural damage, slight non-structural damage)
	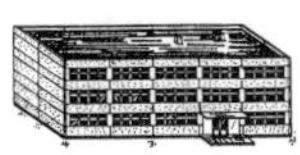	Grade 2: Moderate damage (slight structural damage, moderate non-structural damage)
	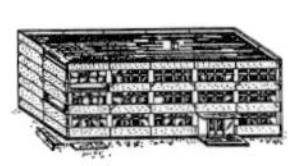	Grade 3: Substantial to heavy damage (moderate structural damage, heavy non-structural damage)
	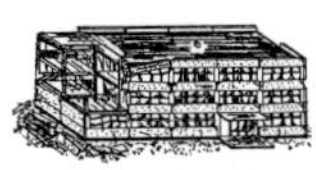	Grade 4: Very heavy damage (heavy structural damage, very heavy non-structural damage)
		Grade 5: Destruction (very heavy structural damage)

Fig. 1 Classification of damage to masonry and reinforced buildings (EMS, 1998).

resolution image data is feasible. In the following we suggest a method that tries to integrate two cues – change of volumes and edges - through an inference system in order to improve the reliability of classification.

4 Automatic Detection of Damages – Bam Earthquake

4.1 Study area

The study area is part of the city of Bam. A powerful earthquake struck southeastern Iran on December 26, 2003 and destroyed much of the city of Bam. The National Cartographic Centre (NCC) of Iran acquired pre- and post-earthquake aerial images of Bam on 1994 and 2003 (3 days after the earthquake) (Fig. 2).

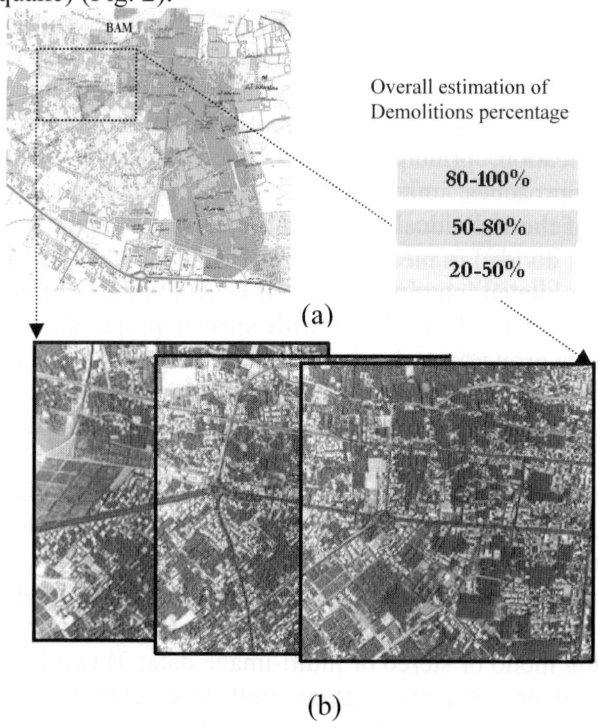

Fig. 2 The study area of Bam city, Iran, (a): Global estimation of damages, (b): Aerial photos before earthquake- 1994 (provided by NCC of Iran)

The problem here is the nine-year elapsed time between the date of pre-earthquake photo acquisition (1994) and the date of post-earthquake aerial photo acquisition (2003). The photos of both dates were visually cross-checked and those buildings that existed in both pre- and post-earthquake photos were used in the assessments. Visual comparison of the pre- and post-earthquake images provided information about the characteristics of the damage within the city. Two pairs of gray-scale stereo aerial photographs with 1:10 000 scales and 60% overlap were used for the analysis. The pre- and post-event photographs were scanned at 21-micron pixel size, giving a footprint of 21 cm on the object. The calibration reports of the cameras and ground control points were available and used as input information for the interior and exterior orientation procedures.

4.2 Cues used in image analysis

In this paper, we tried to propose a multi-feature classification method in terms of comparing stereo pairs of aerial images before and after earthquake. We use two extracted cues from object space (volumes) and image space (edges). Comparing Digital Surface Models (DSMs) of the before and after earthquake situation is one of the method to extract appropriate information. For this purpose, DSMs were created automatically from both pre- and post-earthquake aerial images at 1m spatial resolution using the DSM tool of VirtuoZo (Supresoft, v. 3.3). The DSMs generated were then assessed. The accuracy of the DSMs would directly affect the reliability of automatic detection of the damaged buildings. To evaluate the DSMs, all buildings corners, streets lines, and some single points on the roofs are manually measured which considered as three dimensional check points. The heights of these points were measured on DSMs to be assessed and compared with their corresponding heights measured manually on stereo models. The mean, max and root mean square error values for checkpoints in pre- and post-earthquake DSM differences are given in Table 1.

Table 1 Statistics of height differences in checkpoints.

	Number of points	Maximum absolute (m)	Mean (m)	RMSE (m)
Before	4944	11.64	0.64	1.88
After	4530	10.51	0.39	1.73

After automatically creating the DSMs from both pre- and post-earthquake aerial photographs, volumes were calculated before and after earthquake for each building polygon. The simple assumption made in this study is that if a building is collapsed totally or partially, then its post-earthquake volume will be less than the pre-earthquake volume. The rate of volume reduction then will be a feature to detect the collapsed buildings caused by the earthquake. As a preliminary study, we tried to find out an optimal threshold of this index for whole buildings in the study area. To determine the optimal threshold boundary, the rate of volume reduction proposed by Fung and LeDrew (1988) was used. The following accuracy indices were generated:

1. The *producer's accuracy*: is the number of correctly classified samples of a specific category divided by the total number of reference sample for that class. It is a measure of the omission error.
2. The *user's accuracy*: is the number of correctly classified samples of a specific category divided by the total number of samples being classified as that category. It measures the commission error
3. The *overall accuracy*: is the total number of correctly classified samples divided by total number of samples. It measures the accuracy of the whole data without any indication of the accuracy of individual categories
4. The *Kappa coefficient of agreement* developed by Cohen (1960): is a measure of the difference between the actual agreement between reference data and an automated classifier and the chance agreement between the reference data and a random classifier (Lillesand and Kiefer 1994). It uses all cells in the error matrix and takes into account both commission and emission errors

For our data set, the optimal threshold value is identified as 65% (rate of volume reduction with respect to the volume before the event), which provided the overall accuracy of 91% (Table 2).

Table 2. Error matrix of rate of volume reduction,

Optimum threshold: $1 - V_a/V_b = 0.65$		Reference		
Rate of Volume Reduction		Collapsed	Uncollapsed	Total
	Collapsed	240	30	270
	Uncollapsed	35	408	443
	Total	275	438	713
Producer accuracy		87.27%	93.15%	
User accuracy		88.89%	92.10%	
Overall accuracy	90.85% (Rate of volume reduction, threshold = 65%)		87.52% (DSM difference, threshold = 4 m)	
Kappa	80.62% (Rate of volume reduction, threshold = 65%)		73.21% (DSM difference, threshold = 4 m)	

Basically, it closely resembles the method proposed by Turker and Cetinkaya (2005), which uses DSM differences. Meanwhile, we have also implemented their method and achieved an overall accuracy of 87% at an optimal threshold value of 4m. The Kappa value is an index to compare the result of several classifications (Conggalton and Mead 1983). Comparing these two methods in terms of Kappa reveals a slight improvement.

The above-mentioned methods indicated the way to detect the collapsed buildings based on optimal threshold values that actually could not be considered universal. In the following, it is endeavored to employ another cue feature in order to improve collapse detection without knowing optimal thresholds. Here, we have utilized an edge-base technique to improve the performance of the detection method. First, the Canny edge detection operator was applied to the post-event aerial images in order to extract the edge pixels between the buildings and their surroundings. The one-pixel wide edges are used from both left and right images. Then, the output edge images are compared with vector lines, which were already extracted from pre-event images. In order to evaluate the degree of fit of pixel-edges to the pre-defined polygons we attempted to fit the best line to each segment of connected pixels. Then, three parameters were used: 1) the regression factor of the segmented lines (r), 2) the angle between segmented lines and actual polygon lines (θ) and 3) the length of the segmented lines (l).

To measure the degree of the match between the detected segment lines and delineated vectors of building polygons, the following formula is established:

$$0 \le f\left(\frac{\sum l_i r_i \cos\theta_i}{\sum l_i}\right) \le 1 \tag{1}$$

The function f describes the rate of fit between detected pixel-edges and actual polygon lines. The final decision concerning the damage condition of buildings is made based on average values of this function together with the rate of volume reduction (Fig. 3).

4.3 Damage classification

A three-level damage scale (totally collapse, partially collapse, no damage) appears to be adequate in representing the distribution of individual house damages in earthquakes. Accordingly, in order to examine the efficiency of the proposed method, visual inspection of building damages was conducted based on stereo-pairs of aerial photos before and after the earthquake.

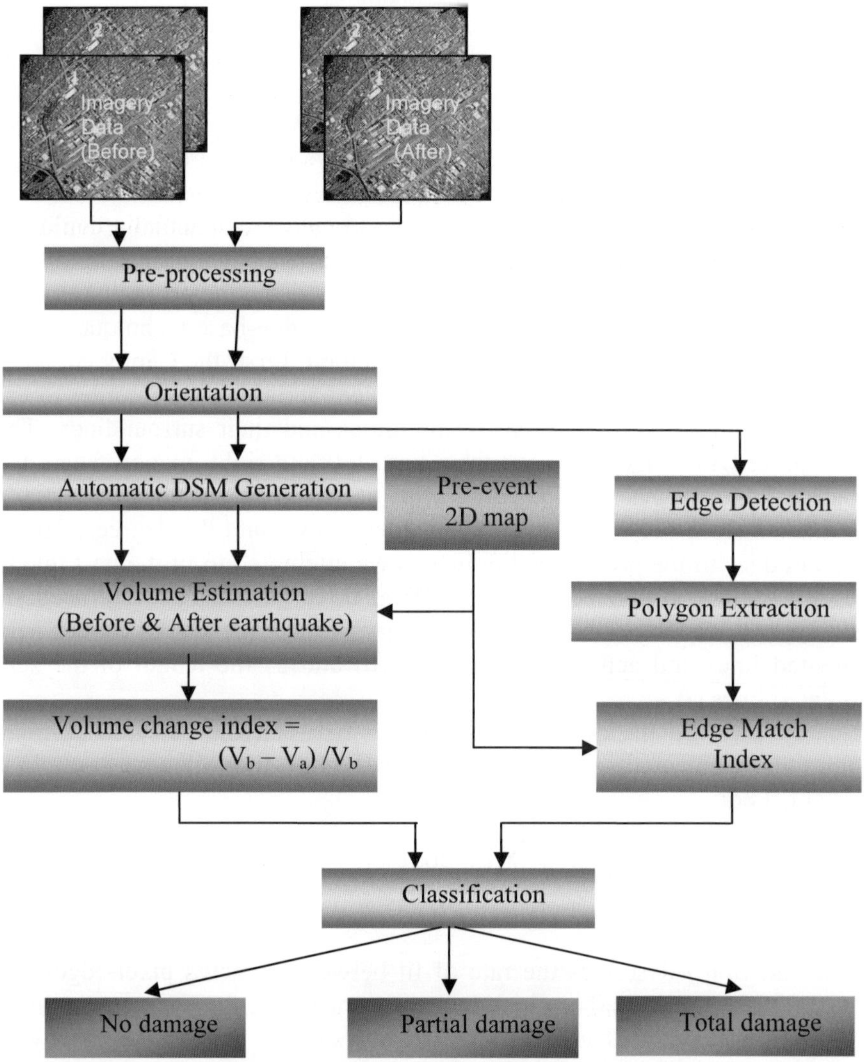

Fig.3 The processing flowchart for the damage classification

For this purpose, buildings which totally collapsed and were reduced to rubble were labeled as 'collapsed' and buildings which were partially collapsed deformed or severely leaning buildings were labeled as 'partially collapsed'. Buildings without visible damage or buildings whose damage state is difficult to identify were labeled as 'no damage'. However, the selection of suitable damage classification grades is nonstandard; the convention of the three selected grades and their qualitative descriptions may be roughly correlated to human fatalities due to demolition.

A simple clustering algorithm, based on the nearest neighbor rule was applied. The nearest neighbor rule is a simple classification technique. In the basic nearest neighbor classifier, each training sample is used as a prototype and a test sample is assigned to the closest prototype. In our study, the 'rate of volume reduction' and 'edge fit index' both constitute a feature space to classify building damages. Fig. 4 depicts the actual distribution of labeled buildings with regard to damage grades in feature space.

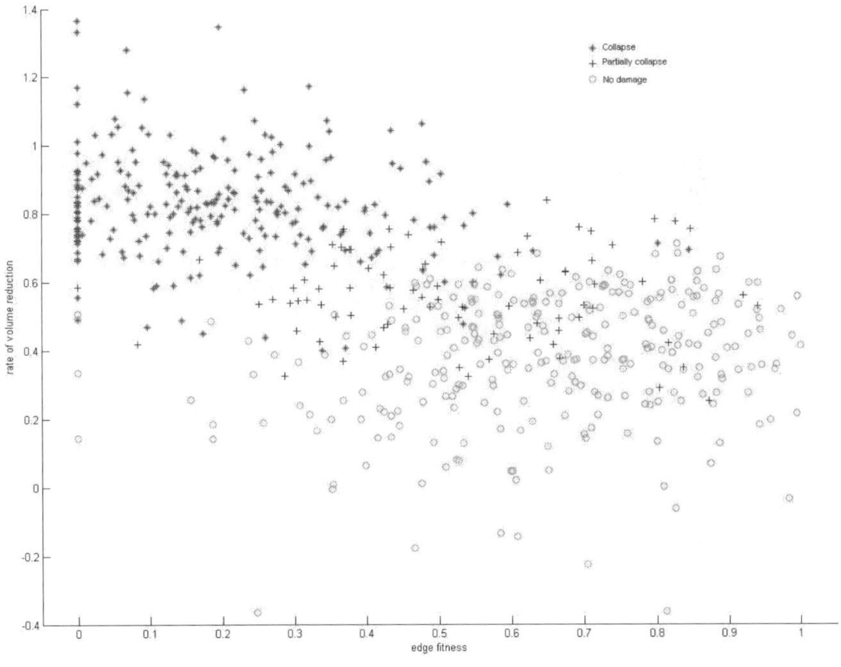

Fig. 4 Feature space and actual distribution of labeled buildings (determined by visual interpretation)

The k-nearest neighbor algorithm assumes that points close to each other in feature space are likely to belong to the same class. It bypasses the density function estimation and goes directly to the decision rule. To apply the k nearest neighbor classification method a function of Matlab 7.0.4 was used. Some sample buildings (fifteen buildings for each class) were acquired as a training set. A direct majority vote from the nearest three neighbors was employed.

5 Results

The results of the proposed damage detection approach are given in table 3. The results show that the proposed approach for detecting the collapsed buildings due to earthquake is satisfactory. Based on visual inspection of the stereo images of pre- and post-earthquake data, the results have been evaluated. A total of 619 buildings falling within the selected area of study were assessed and 251 buildings were labeled as 'collapsed', 76 were labeled as 'partially collapsed' and 292 buildings were labeled as 'no damage'. A high degree of agreement is evident between the assessment results and the reference data in the 'collapse' state. Of these buildings, 217 were correctly labeled as collapsed, providing a producer's and user's accuracy of 86.5% and 90.4% respectively. It appears that 34 buildings were omitted from the collapsed category and considered as 'partially collapsed', however, none of the collapsed building is labeled as 'no damage'. The producer's and user's accuracies of 'partially collapsed' buildings were computed as 67.1% and 26.6% respectively. Thus, the user's accuracy is significantly lower. It appears that 25 buildings were omitted from the 'partially collapsed' category. Of the total 292 buildings labeled as 'no damage', 118 were omitted from this category while 174 were labeled correctly. The producer's and user's accuracies of 'no damage' buildings were computed as 59.6% and 93.0% respectively. The erroneously categorized buildings were further investigated to find out what might have caused them to deviate from the reference data.

Table 3. Comparison between k-NN classification and visual interpretation.

	Visual interpretation			
k-NN classification	Collapsed	Partially Collapsed	No Damage	Total
Collapsed	217	12	11	240
Partially Collapsed	34	51	107	192
No Damage	0	13	174	187
Total	251	76	292	619

It appears that the main reason for 107 'no damage' buildings to be wrongly categorized as 'partially collapsed' are problems in the DSM generation due to shadows and trees. It is hard to distinguish buildings within shadows or buildings which are hidden by other objects like trees. There seem to be two main reasons that cause false classifications. The first and main reason is the absence of the features in the area where the buildings are hidden in the shadows and causes mismatching between left and right images. Therefore the volume estimation is not of sufficient accuracy in this area and leads to an overlap between 'partially collapsed' and 'no damage' in feature space (Figure 4). The second reason is the lack of edges on the boundaries to the adjacent buildings. It is therefore impossible to find line segments corresponding to the line of the building polygon.

6 Conclusions

In this paper, it has been shown that using cues from both image and object spaces of pre- and post event stereo pairs of aerial images can reveal the location of the damages – collapsed and partially collapsed – caused by an earthquake. The overall accuracy – the total number of correctly classified divided by the total number of samples – was found to be 71.4%, which is quite encouraging. In this study, we have used the k-nearest neighbor method to classify damaged buildings using training sets of buildings which were selected randomly. The results of the analysis show that using multiple cues can be useful to classify damages which may lead to particular preferences for rescue teams. The results in this test have been negatively influenced by a number of factors which in the future can be avoided by the following measures:

- Use of digital images – this improves the image quality (better radiometric depth allows to look into the shadows) and opens the possibility of faster (on-line) processing.
- Fly with larger image scale – this improves the interpretability of images (higher level of detail).
- Use multi-image approach – helps against occlusions and gives more reliability in image matching; multiple (more than stereo) overlap is easy to produce with digital images.
- Use larger focal length ("normal angle" images) – helps against occlusions.
- Use a better image matching software and image analysis procedures which are able to distinguish between trees and man-made objects

The polygon-based assessment depends on the availability of large scale and recently updated maps in the area. This could be a serious problem as well in many parts of the world.

Acknowledgments

The provision of aerial imagery, ground control information and the digital map of Bam city by the National Cartographic Center (NCC) of Iran is gratefully acknowledged.

References

Bitelli G, Camassi R, Gusella L, Mognol A (2004) Image change detection on urban areas: the earthquake case. Proceedings of the ISPRS XXth Congress, Istanbul, 35(B7), pp 692-697

Cohen J (1960) A Coefficient of Agreement for Nominal Scales. Educational and Psychological Measurement, 20(1): 37–46

Congelton RG, Mead RA (1983) A quantitative method to test for consistency and correctness in Photointerpretation. Photogrammetric Engineering and Remote Sensing, 49(1): 69-74

Fung T, Ledrew E (1988) The determination of optimal threshold levels for change detection using various accuracy indices. Photogrammetric Engineering and Remote Sensing, 54(10): 1449–1454.

Hasegawa H, Yamazaki F, Matsuoka M, Seikimoto I (2000) Determination of building damage due to earthquakes using aerial television images. Proceedings of the 12th World Conference on Earthquake Engineering, Auckland, CD-ROM, 8p

Kouchi K, Yamazaki F, Kohiyama M, Matsuaka M, Muraoka N (2004) Damage detection from Quickbird high-resolution Satellite images for the 2003 Boumerdes, Algeria Earthquake. Proceeding of the Asian Conference on Earthquake Engineering, Manila, Philippines, CD-ROM, 215-226

Lillesand TM, Kiefer RW (1994) Remote Sensing and Image Interpretation, 3rd edn, New York: John Wiley and Sons.

Ogawa N, Yamazaki F (2000) Photo-interpretation of buildings damage due to earthquakes using aerial photographs. Proceedings of the 12th World Conference on Earthquake Engineering, Auckland, CD-ROM, 8p

Turker M, Cetinkaya B, (2005) Automatic detection of earthquake-damaged buildings using DEMs created from pre- and post-earthquake stereo aerial photographs. International Journal of Remote Sensing, 26(4): 823–832

Yamazaki F, Kouchi K, Kohiyama M, Muraoka N, Matsuoka M (2004) Earthquake Damage Detection Using High-resolution Satellite Images. Proceedings of the IEEE 200 International Geoscience and Remote Sensing Symposium, IEEE, CD-ROM, 4p

Yamazaki F, Yano Y, Matsuoka M (2005) Visual damage interpretation of buildings in Bam City using QuickBird images. Earthquake Spectra, 21(S1): S329-S336

Rapidly Realizing 3D Visualisation for Urban Street Based on Multi-Source Data Integration

Zhizhong Kang[1], Zuxun Zhang[2], Jianqing Zhang[2] and Sisi Zlatanova[3]

[1]Faculty of Aerospace Engineering, Delft University of Technology, Delft, The Netherlands, z.kang@tudelft.nl
[2]School of Remote Sensing and Information Engineering, Wuhan University, China
[3]OTB, GISt, Delft University of Technology, Delft, The Netherlands, s.zlatanova@tudelft.nl

Abstract

Streets are important component of cities since they provide the best direct impression of the city. Therefore street scenes are important aspect in 3D modeling. Fast 3D City Modeling from a street level can also be quite important for emergency response by providing realistic, updated, accurate information about accessibility to and from affected areas.

This paper presents a 3D reconstruction approach for rapid 3D visualization from street level, which is based on a combination of vehicle-based image sequence and 2D vector map. The approach consists of two general steps: the rapid reconstruction of facade along the street based on side-shooting vehicle-based image sequence and 2D map, and road texture recovery using. The algorithms presented are verified by experiments on real data set.

1. Introduction

Collecting textures is one of the most critical and time consuming processes in 3D City Modelling (3DCM). Textures for terrain and roof of buildings are often acquired from aerial images and therefore the texture process can be easily automated. The textures for facades, however, are predominantly collected by digital cameras. The image processing such images has usually low efficiency and high labor intensity. Therefore, automation has been always challenging research topic addressed by many.

Many approaches are based on close-range photogrammetry. Pomaska has presented a method to reconstruct 3D model based on multiimage photogrammetry (Pomaska 1996). Recently, spherical and panoramic images are increasingly getting utilized in 3D reconstruction of building. Teller (1998) has developed a prototype system for automatic reconstruction using spherical mosaic images, in which camera's position and orientation of each spherical image is first initialized using positioning sensors, then refined through image matching. Haala and Kada (2005) have presented an automatic method based on mapping of terrestrial panoramic images to the facades of existing 3D building models.

As LIDAR (Light Detecting and Ranging) technology is progressing, 3DCM based on multiple data fusion gains increased attention (Lee and Choi 2004; Oda et al. 2004; Maas and Vosselman 1999; Schwalbe et al. 2005). However, airborne LIDAR can only capture building roofs but not the facades. This essential disadvantage prohibits their use in photo realistic walk- or drive-through scenes (Li and Chapman 2007).

Another group of researchers focus on video processing. Video's can be quite easily obtained and, consequently, extracted image sequences can be efficiently used to reconstruct textured 3D model. Wu (2005) presented an approach for textured 3D model by fusing helicopter-based video, LIDAR data and 2D vector map. Unfortunately the high cost of helicopter video as well as the dependence on weather conditions and height limit in urban area (resulting in low resolution and large distortion of facade texture), limit the utilization of this approach.

Terrestrial mobile mapping systems have been also used to acquire the complementary ground-level data and to reconstruct building facades. The resolution of textures obtained from vehicle-based sequence is much higher than those obtained from aerial or helicopter image sequence. A recent collection of the mobile mapping literature can be found in Li and Chapman (2005) and Tao and Li (2007). A common feature of mobile mapping systems is multi-sensor integration, e.g. CCD, GPS, Inertial Navigation System and dead reckoning. Although some tasks related to calibration and image data processing can be done automatically, a human operator is still needed (Li and Chapman 2005).

The approach presented in this paper, which is also based on a combination of a vehicle-based image sequence and a 2D vector map, provides high automation of the reconstruction process. This paper is subdivided into five parts. Next part presents the steps that complete the reconstruction of the 3D model for the facades. Part three explains the road texture recovery for the street surface. Experimental results are reported in part four. Part five discusses the results and concludes on further research.

2.3D reconstruction of facades

The process follows four steps, i.e. image rectification, subsection of image sequence, image mosaic and 3D facade model reconstruction in a semi-automatic way combined 2D vector map.

2.1 Image rectification

The images of facades taken with digital camera are usually very oblique with a large geometric distortion. This requires rectification of all the raw images. The challenge here, usually, is the way of computing orientation parameters (except the coordinates of projective center) using only image information. Since building facades are almost planar, there are two groups of lines in the object space parallel respectively to *X*-axis and *Y*-axis. As a result, only the focus *f* and the three angular orientation parameters φ, ω, k have be calculated applying vanishing point geometry. The facade textures are rectified employing *f* and φ, ω, k with principle point fixed at the center of image. The process of rectification is illustrated as Fig.1. *P* is the original image and P_0 is the rectified image of *P* on the plumb plane.

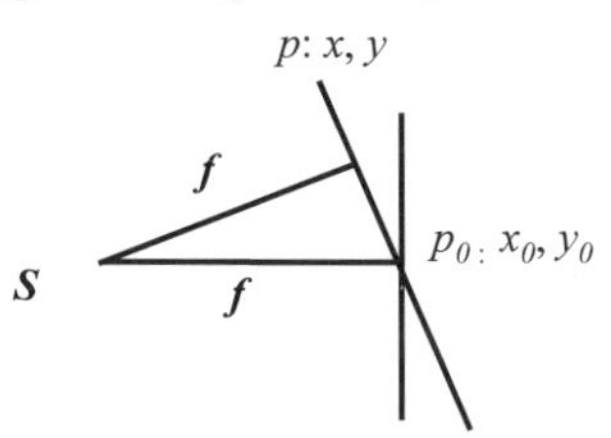

Fig.1 The process of rectification

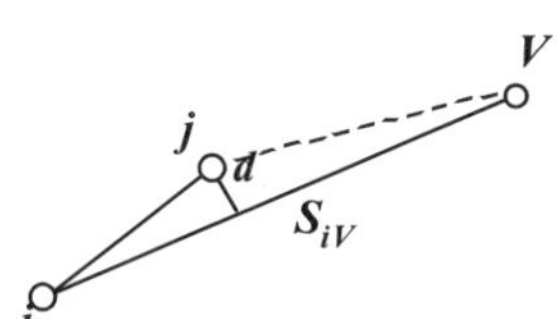

Fig.2 The constrain of Straight lines bundle

The process begins with line extraction on the image. Lines parallel respectively to *X*-axis and *Y*-axis in object space are used to calculate the vanishing points. Each straight line such as line *ij* belonging to a set of parallel lines in object space must pass through the vanishing point as shown in Fig.2. Ideally, the projections of these two group lines in the rectified image should be parallel to *X*-axis and *Y*-axis in object space. Therefore a adjustment approach is used based on constraints of straight lines bundle and known orientation of parallel lines in object space. The interior orientation parameters are fixed in the adjustment.

2.2 Image sequence subsection

To facilitate the reconstruction of the 3D facades, the image sequence is subdivided into segments corresponding to each building on the map. Accordingly, the histogram of projective difference of corresponding points is used for automatic image sequence subsection.

Fig.3 illustrates this process. The origin of the coordinate system is the projective center S_1 of the first image. The plane XY is parallel to the facade. The ideal image plane P_0 of the first image (the distance between point S_1 and P_0 is focus f) is considered as a public projective plane. As far as the corresponding points on the facade a_1 and a_1' are concerned, the projective difference between their projection points A_1 and A_1' should be very small or even zero. However, projection points A_2 and A_2' of corresponding points a_2 and a_2', which are not on the facade have a large projective difference. Accordingly the 1-D histogram along x-axis of the image coordinate system shows the projective difference (Fig.4).

Cross-roads and facades having a large range variance (Fig.5) are represented in the histogram by peak areas. Moreover, the location of range variance in image is indicated by the location of corresponding peak area in the histogram because the histogram is drawn following the moving direction of vehicle. According to the location indicated, images including cross-roads are excluded from image sequence and subsections of image sequence are divided for facades having large variance with each other.

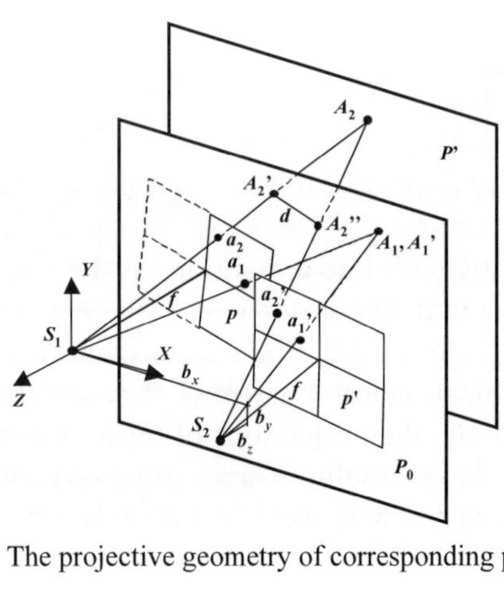

Fig.3 The projective geometry of corresponding points

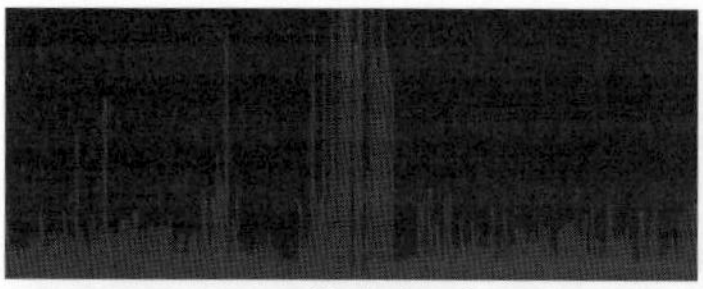

a Original

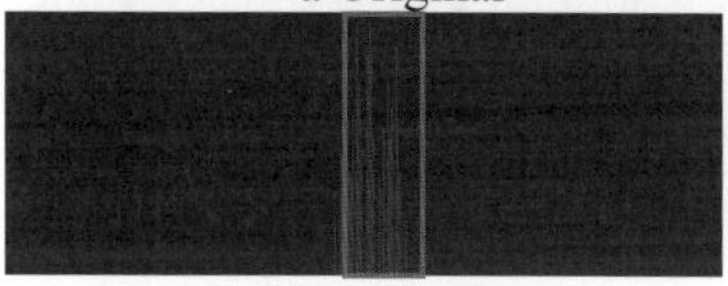

b After filtering

(1) Cross-road

a Original

b After filtering

(2) Connected buildings with large range variance

Fig.4 the histogram of projective difference of corresponding points

a. Cross-road

b. Connected buildings with large range variance

Fig.5 Segments with large range variance

2.3 Image mosaic

Usually for large buildings, facade texture is covered by a segment of image sequence as explained the previous section. To be able to complete the entire façade, images have to be combined, which is the next step. Presently, the existing image mosaic software specializes only in processing images with small geometry distortion. However, vehicle-based imagery as discussed in this article has large geometric distortions. This section presents our method for automatic image mosaic after the raw images are rectified onto a vertical plane.

As shown in Fig.3, the facade is considered as a plane and parallel to the public projective plane. Therefore, the relatively spatial relationship of the image sequence is recovered when the corresponding rays of each two images intersect on the public projective plane via adjusting the exterior orientation parameters (EOPs) of each image. Using spatial relationship of the image sequence determined above, a method similar to that of generating ortho-image is employed to generate the whole facade texture. According to the theory of optic projection, the corresponding rays of each image can intersect each other on the public projective plane via adjusting the EOPs of each image. The relative spatial relationship of the image sequence is determined by this least square adjustment.

3.4 Correspondence between facade texture and 2D vector map

After the recovery of facade texture, a 2D vector map is combined to reconstruct the 3D model of building facade. If the facade of a building is approximately planar, only two corresponding corner point pairs need to be selected to link the facade texture with the map. As a result, two parameters scale λ and rotation angle θ between facade texture and map is computed. Roof and height information which is unavailable in the map can be extracted from the facade. Accordingly, the 3D model of facade is reconstructed and geo-referenced to the 2D vector map.

However, the facade of a single building or several connected buildings possibly consists of convex or concave parts which are visible in the vector map. As shown in Fig.6, the corner points (highlighted by cross) on the convex and concave parts of vector map should be referenced to the facade texture, except for the two corner points highlighted by solid dot.

Corner point extraction

Firstly, two corresponding point pairs A1 and a1, A6 and a6 are selected in vector map and facade texture respectively as shown in Fig.6.

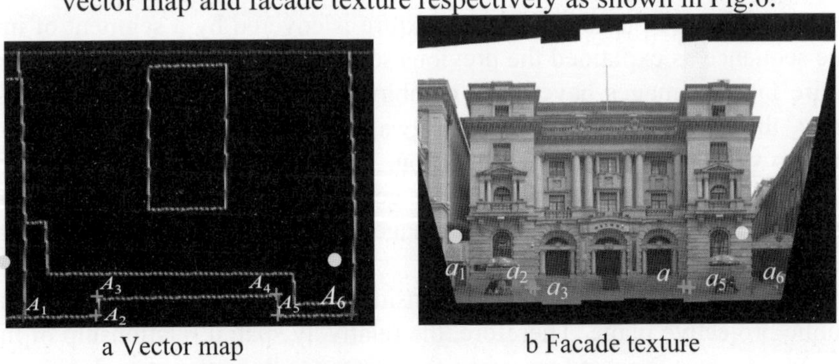

a Vector map b Facade texture

Fig.6 Building with convex or concave parts

The transformation parameters between texture coordinate system and ground coordinate system are computed with these two corresponding point pairs.

The corner points are projected onto facade texture according to λ and θ as point a_i ($i = 2,\ldots,5$) in Fig. 6b. Afterwards, those points are projected onto raw image sequence. For instance, in Fig.7, projective points in raw images of point a_5 are around the actual corners. However, corner points may not be distinct features in close-ranged images. Furthermore, due to the occlusion between objects with different range, the extraction can also fail. As it can be seen in Figs.7 and 8 that there is always a long vertical line passing the corner of wall and the corner point of interest must lie on this vertical line.

The searching for corner point is thereby limited on the vertical line. To extract the vertical line, raw images are firstly rectified onto vertical projective plane and then an image strip, which width is 60 pixels and center is the projective point, is taken out. Consequently, vertical line extraction is implemented in this image strip. As visible in Fig.8, although there might be several vertical lines extracted, the longest line is most likely on the corner of wall. As a result, the longest line highlighted in solid is picked out to search the actual corner point. As image III and IV as be concerned, the vertical lines, extracted from planes with different range, are close to each other so that only the line length is possibly not enough to determine the right line passing corner point.

Fig.7 The projective points of corner point in raw image sequence

To tackle this problem, a strategy to eliminate false accepted line is implemented based on the relative spatial relationship of image sequence acquired after image mosaic. As shown in Fig.8, vertical lines are firstly extracted from the raw image (e.g. I) in which the geometric difference between planes with variant range values is most significant since it is most likely to pick up the correct vertical line from this image. In this image, the longest vertical line is selected and projected onto other adjacent images according to the relative spatial relationship of image sequence. The projective line in image IV is highlighted in square dot and can be used as predicted position of the correct vertical line (Fig. 9). As a result, an image strip, which width is narrowed to 40 pixels and center point is the center of projective line, is taken out for vertical line extraction. Although the predicted position is close to the actual vertical line, as Fig.8, the distance is only 6 pixels between the dash line and actual solid line in the while rectangle and moreover the dash line is long as well. Under this circumstance, the dash line is possibly to be falsely accepted. Therefore, the selected line is projected into image I. As shown in Fig.10, the distance is still small between the correct projective line from image IV (highlighted in dash) and the actual vertical line in image I (highlighted in solid). The distance, however, becomes large between the false dash line and the actual solid line. Obviously, the falsely accepted line can be eliminated by this strategy.

After the extraction of vertical lines passing corner points, the line between corner points a_1□a_6 in Fig. 6 is projected into raw images and the actual corner points are the intersected points between extracted vertical lines and projective line of a_1a_6 in corresponding raw image. All extracted and selected corner points are used to calculate parameters scale λ and rotation angle θ.

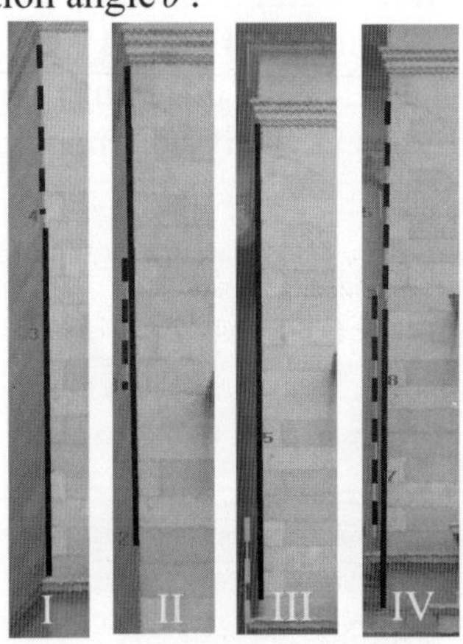

Fig.8 Vertical lines extracted

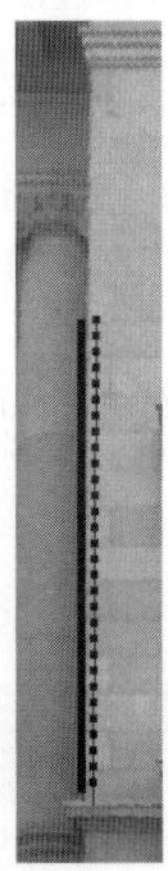

Fig.9 Line prediction

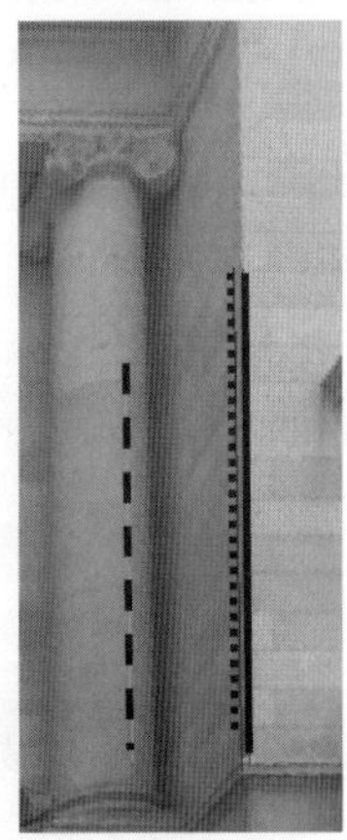

Fig.10 Error line elimination

Iterative process

Since the scale factor λ and rotation angle θ are firstly computed with the two corresponding point pairs selected, the further the cross points on the convex and concave parts on the texture away from the solid dot corner points, the larger deviation is between projective and actual points in facade texture, e.g. projective point a_4~a_7 in Fig. 11a. The iterative strategy is thereupon employed to tackle this problem.

Based on the method presented above, points a_2, a_9 closest to solid dot corner points are firstly extracted from raw images and then λ and θ are recalculated combined with extracted points and selected corner points. According to the transformation parameters recalculated, point a_3 and a_8 are extracted. This iterative process repeats till all corner points are extracted. In Fig.11b, large deviation between projective point a_4~a_7 and actual corner points in facade texture is eliminated after iterative process.

Using all of the extracted and selected corner points, combined with facade texture, the 3D model of facade is reconstructed and geo-referenced to the 2D vector map.

a Before iterative process

b. After iterative process

Fig.11 Projective points

Fig.12 Image sequence along the optic axis

3. Road texture recovery

The image sequence illustrated in Fig.12 is taken by digital camera moving along the optical axis. Since the whole road texture instead of facade texture can be covered, this kind of image sequence is a good source for road texture recovery. In this section a method is therefore presented to recover road texture using this data source. It consists of two steps: image rectification to level plane and image mosaic.

3.1 Image rectification to level plane

Similar to the facade texture recovery, raw images need to be rectified onto level plane to cover the texture close to the real road surface.

As shown in Fig.13, plane $\boldsymbol{p}$ denotes the raw image, $\boldsymbol{p_0}$ denotes the image rectified to vertical plane, $\boldsymbol{P}$ denotes the image rectified to level plane. As Fig.12, because the raw image is taken along the optical axis, the ideal image plane of raw image is the vertical plane. Therefore, the process of rectification is $\boldsymbol{p} \rightarrow \boldsymbol{p_0} \rightarrow \boldsymbol{P}$, namely the raw image is firstly rectified onto the vertical plane and then rectified onto level plane.

A least square adjustment system based on the constraint of straight lines bundle is used to compute focus f and three angular orientation parameters φ, ω, k. Then raw images are rectified onto level plane using those computed parameters. As we know, the larger is the depth of road surface, the larger geometric distortion has the rectified image. As a result, only the region with the size of 3000 pixels×2000 pixels which has smallest geometric distortion in the rectified image is used to recover road texture (Fig.14). Image mosaic is necessary to acquire the whole road texture of street with high resolution.

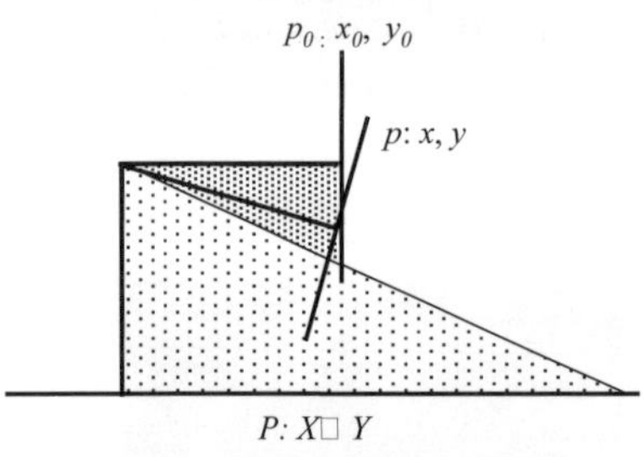

Fig.13 Rectification process

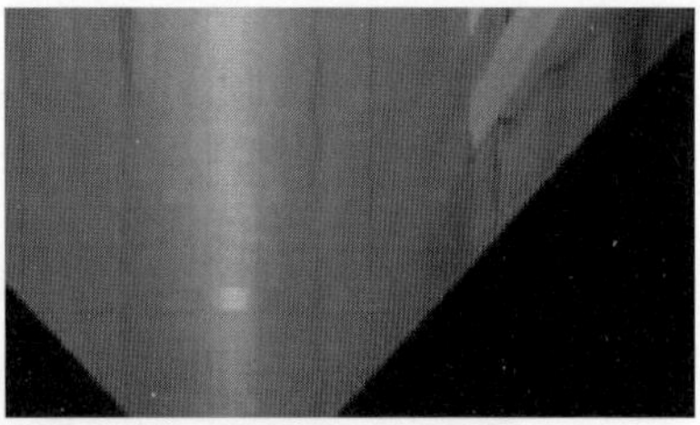

Fig.14 Rectified road texture

3.2 Mosaic for road texture

The road surface can be considered as a plane and parallel to the public projective plane. The projective geometry of corresponding points of the image sequence along the optical axis is illustrated in Fig.15. The level image plane P_0 of the first image is considered as public projective plane.

According to the theory of optic projection, the corresponding rays S_1a_1 and S_2a_2 can intersect at point A_0 on plane P_0 via adjusting the EOPs of each image. A least square adjustment is employed to calculate the EOPs by which all of coordinate difference between correspondingly projective points is least square. Using the EOPs computed, a method similar to that of generating ortho-image is employed to generate the whole road texture of street (Fig. 16).

As presented in Section 2, the 3D model of street facade was reconstructed and geo-referenced to the 2D vector map, therefore integrated with the road texture of street recovered the 3D visualization of walk-through street scene is realized.

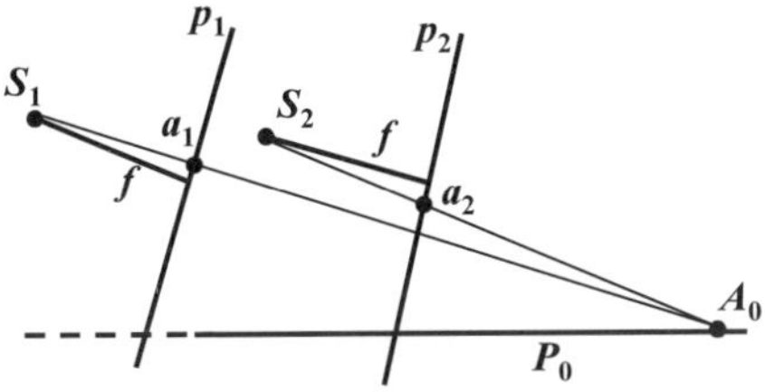

Fig.15. The projective geometry of corresponding points (along the optical axis)

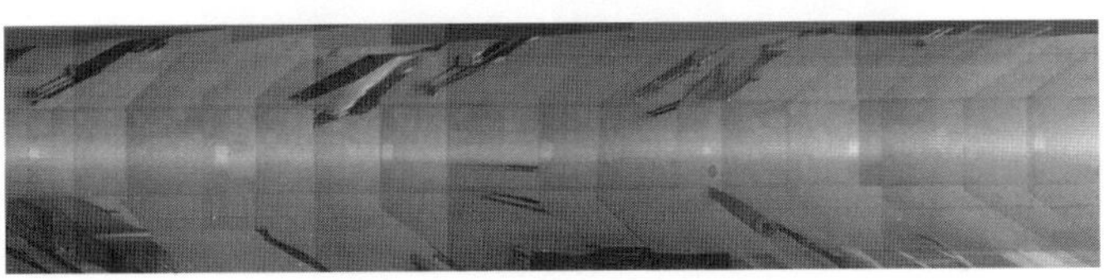

Fig.16 mosaic road texture

4. Experimental results

Experiments were implemented on two image sequences, i.e. side-shooting (Fig.17) and along the optic axis (Fig.12), taken by digital camera KODAK PROFESSIONAL DCS Pro SLR/n along the street. For side-shooting, the distance between photographic center and building is about 10m. The image size is 3000 pixels×4500 pixels. To fix the interior parameter during the process of taking photos, the focal length is set to infinitude and the function of automatic focalization is turned off. Actually, it's not necessary to process the imagery with such large image size, therefore, the raw images were compressed and image size is reduced to 1500 pixels×1000 pixels.

Fig.17 Side-shooting image sequence

Automatic image sequence subsection was implemented using the method presented in Section 2.2. After the subsection of image sequence, every building has a corresponding image segment to recover facade texture. As presented in Section 2.3, image mosaic is required to generate the whole facade texture. For comparison, facade texture is mosaic from close-ranged digital camera images respectively by the existing image mosaic software and the strip method presented in Section 2.3. The results of existing image mosaic software are shown respectively in Fig.18. Obviously, the existing software tools of image mosaic are not applicable for processing images with large obliquity of the posture of digital camera.

a. ImageSuite *b*. PhotoStitch

Fig.18 Mosaic obtained using existing software

The average time consuming for the rectification of a single image is 30 seconds. As the building facade with 82m length in Figure as be concerned, the corresponding image segment consists of 36 images. Total time of rectification is 18 minutes. For image mosaic, there are 216 unknown parameters in the adjustment system of strip method and the time consuming is 10 seconds. The whole process is fully automatic and fast. Moreover, as Fig. 19, the facade texture generated from 36 images has a very high resolution and close to the seamless mosaic.

Fig.19 Whole facade texture

Combined the facade texture and 2D vector map, the 3D model of the facade was reconstruct as shown in Fig.20 (*a*). For the facade consists of convex or concave parts, 3D model was reconstructed as Fig.20 (*b*) using the semi-automatic algorithm presented in Section 2.4.

a. Planar facade

b. Buildings with convex or concave parts

Fig.20 3D facade models

Fig.21 3D model of street

Fig.22 3D walk-through scene of street

Finally 3D models of 28 building facades along the pedestrian street were reconstructed and geo-referenced to 2D vector map (Fig.21). Using the method presented in Section 3, the whole road texture of street was recovered from 255 images along the optic axis. Integrated the road texture recovered with 3D models of building facades, the 3D model of walk-through street scene is reconstructed (Fig.22). Since terrestrial close-ranged images have high resolution and moreover are taken along the street, the walk-through street scene is close to the real impression of people on the street.

5. Conclusions

In this paper, an approach was presented that allows for fast reconstruction and 3D visualization for walk-through street scene. The image sequences (side and along the optic axis shooting) are collected by camera mounted on a vehicle. The presented approach has numerous advantages compare to other approaches:

Because the acquisition of image sequence is vehicle-based and moreover the recovery of building facade texture which is always low efficiency to existing methods realizes automation in our method, the whole process from data acquisition to 3D model reconstruction is implemented in a rapid and high efficient way.

Image sequence along the optic axis is an ideal source to recover road texture with high resolution. The texture recovery can be completed by image rectification and mosaic in an efficient way. Terrestrial close-ranged images have high resolution and since they are taken along the street, it is possible to create a walk-through 3D scene, which is very close to the reality. The approach presented in this paper is applicable also for streets with high density of buildings.

This approach is quite appropriate fast reconstruction and 3D visualization is emergency response (e.g. flood or earthquake). The process is fully automatic and very fast. Since the texture is acquired from images, the scene should have good visibility (with some exceptions, e.g. smoke). The building facades should be roughly or partly complete so that the facade model can be reconstructed from image sequence and 2D vector map. The image sequence can be acquired by a person instead of a vehicle moving along the street if the road is blocked or damaged. The developed prototype system can be further optimized to allow for near real-time reconstruction and visualization. First estimations of building damages can be then send to rescue teams via internet.

Future work will concentrate on 3D model reconstruction for more complex facade and improvement of availability for disaster management.

Acknowledgements

This research was supported by the National High Technology Research and Development Program of China (863 Program) with the serial number 2006AA12Z220.

References

Haala N, Kada M (2005) Panoramic Scenes For Texture Mapping of 3D City Models, Proceedings of the ISPRS WG V/5 'Panoramic Photogrammetry Workshop, IAPRS 34(5/W8), CD-ROM.

Lee I, Choi Y (2004) Fusion of Terrestrial Laser Scanner Data and Images for Building Reconstruction, Proc. of ISPRS [C], Vol. 35 (Part B5).

Li J and Chapman MA (eds.) (2005) Mobile Mapping Systems, Special Issue of Photogrammetric Engineering & Remote Sensing, 71(4).

Li J and Chapman MA (2007) Terrestrial mobile mapping towards real-time geospatial data collection, In Zlatanova and Li (eds.), Geospatial Information Technology for Emergency Response, ISPRS book Series (in press)

Maas H-G, and Vosselman V (1999) Two algorithms for extracting building models from raw laser altimetry data, ISPRS Journal of Photogrammetry & Remote Sensing, 54(2–3):153– 163.

Oda K, Takano T, Doihara T, Shibasaki R (2004) Automatic building extraction and 3-D city modeling from LIDAR data based on Hough transformation. Proceedings of XXth ISPRS Congress, 45 (Part B3).

Pomaska G (1996) Implementation of digital 3Dmodels in building surveys based on multiimage photogrammetry. In: International Archives of photogrammetry and Remote Sensing, Vienna, 31(Part 5): 487-492.

Schwalbe E, Maas H-G, and Seidel F (2005) 3D building model generation from airborne laser scanner data using 2D GIS data and orthogonal point cloud projections, Proceedings of ISPRS WG III/3, III/4, V/3 Workshop on Laser Scanning, Enschede, the Netherlands, September 12-14, pp 209-214.

Tao CV and Li J (eds.) (2007) Advances in Mobile Mapping Technology, ISPRS Book Series, Taylor & Francis, 175 pp.

Teller S (1998) Toward Urban Model Acquisition from Geo-Located Images, Proceedings of Pacific Graphics, pp 45-51.

Wu J (2005) Research on Rapidly Reconstructing Texture for Facades in 3D City Modeling, Acta Geodaetica et Cartographica Sinica, 34 (4): 317-323.

3D Dynamic Simulation within GIS in Support of Disaster Management

L. Hashemi Beni, M. A. Mostafavi and J. Pouliot

Department of Geomatics Sciences, Laval University, Quebec G1K 7P4, Canada
Leila.hashemi.1@ulaval.ca, Mir-Abolfazl.Mostafavi@scg.ulaval.ca,
Jacynthe.Pouliot@scg.ulaval.ca

Abstract

Modeling and simulation of dynamic phenomena helps specialists and decision makers to better understand, analyze, and predict natural disasters to reduce the related damages. Hence, several models and simulation approaches have been developed with their own strengths and limitations. One important factor which has to be taken into account when dealing with a disaster is its dynamic behavior and its geometrical and topological representation. Geographic information systems are very well adapted for spatial data organization, visualization, querying, and analysis, and may be helpful in the context of simulation and modeling of spatial phenomena (e.g. floods). However, regarding the three dimensional and dynamic nature of spatial data related to a disaster (which changes with respect to space and time), several complexities are added to the entire process of simulation using the existing GIS.

In order to use GIS as a suitable platform to implement the simulation of disaster phenomena, GIS must be improved; adding the ability to manage and model the 3D and dynamic properties of spatial data (i.e. a truly 3D GIS), as the data structures of the current GIS are static. This research work is an attempt to overcome the current GIS limitations to simulating such natural phenomenon by proposing a 3D dynamic data structure. This structure is based on 3D Delaunay tetrahedralization that deals with objects and field representation of space at the same time, and provides an on-the-fly interactive topological mesh for numerical simulation. This means that cell (or element) size and shape depend on the distribution of the data and it is possible to do local on-the-fly updating of topology after any movement or change in 3D space. In order to analyze the different capabilities of the proposed data structure, its application to flood simulation is discussed at the end of the paper.

1 Introduction

In the last decades, the damages resulting from natural and man-made disasters have increased significantly, both in amount and magnitude. The main sources responsible for such disasters are rapid population growth, global climate changes and the over-exploitation of natural resources, which lead to damages such as groundwater contamination, forest fires and floods (FIG 2006). In a recent international conference in Paris, (IPGC, 2007), the world's leading climate scientists have blamed man-made emissions of greenhouse gases for causing devastating floods, heavy rains, droughts, and an increase in hurricane and tropical storm strength. These events spark concern around the world about the future of public security; an issue which needs to be addressed carefully.

In order to effectively mitigate the hazards associated with disasters, traditional approaches are commonly used to develop solutions based on historical data. Several approaches based on simulation and modeling techniques have been increasingly used to better understand, predict, and eventually mitigate the risk of various kinds of disasters. In these models, the behavior of phenomena is sensitive to local and global spatial characteristics, as seen in the simulation of wild land fire spreading across a heterogeneous environment (Dunn and Milne 2004). Therefore, spatial modeling of these phenomena becomes very important. This is where a Geographic Information System (GIS) is a valuable tool. GIS, through its powerful capabilities to process spatial data, provides modelers with strong computing platforms for data management, visualization, querying, and analysis and thus greatly facilitate the implementation of various models (Burrough et al. 1988; Goodchild 1992; Bennet 1997; Nyerges 1993). In addition, GIS is an effective tool to integrate diverse spatial and non-spatial data types that are frequently required in disaster risk management. For example, Gunes and Kolel (2000) used GIS for emergency management operations and Rifaat and Tao (2005) developed a system which simulates earthquake damage assessment scenarios using a web-based GIS.

However, our literature review reveals that, unfortunately, GIS can not be fully exploited in the context of simulation of a 3D dynamic phenomena; e.g. flood, water contaminant, forest fire. This is because the conceptualizations of space and time embedded in the current GIS are not compatible with those in simulation models. For example, time is an explicit element in most simulation models, while the representation of time in GIS is almost non-existent and it is conceptualized as discrete slices (time step approach) (Peuquet 1994; Sui and Maggio 1999). The time step approach is incapable of handling the dynamic behavior of disaster phenomena. Moreover, GIS capabilities are not well adapted to represent and manage neither geometry nor topology of 3D spatial data (Breunig 1996; Pouliot et al. 2003; Stoter and Zlatanova 2003).Therefore, the main purpose of this paper is to provide GIS with the capability of simulating 3D dynamic phenomenon by proposing a 3D dynamic spatial data structure that can effectively deal with the characteristics of these types of phenomenon.

The remainder of this paper is organized as follows. Section 2 explains different levels of using GIS in the context of the simulation, followed by a discussion on the requirements for and the problems of simulating a 3D dynamic phenomenon within GIS. In Section 3, we present our solution to these limitations through the development of a dynamic 3D data structure. Section 4 explains how this structure contributes to the simulation of 3D dynamic phenomena such as flood. Conclusions are found in Section 4.

2 Simulation Models and GIS Capabilities

In general, modeling and simulation of physical phenomenon (e.g. flood) have been frequently applied to the analysis and understanding of phenomena, the testing of hypotheses and theories, the prediction of spatio-temporal systems behavior under various conditions and scenarios (existing and simulated, often performed to support decision making), and to new discoveries of geospatial phenomena, enabled by the unique capabilities of computer experiments.

In the past, simulations tools have generally been developed outside of GIS. However, these tools suffer from (1) insufficient spatial analytical component, (2) low performance of the spatial data visualization capability, and (3) non-conveniences in spatial data integration such as digital maps, satellite image, aerial photos (Bivand and Lucas 1997, Densham 1993, Nyerges 1993). Therefore, the integration of simulation tools and GIS was considered by many researchers and the mutual benefits of this integration were confirmed by several papers and books published on this subject (Chapman and Thornes 2003, Sui and Maggio 1999, Valavanis 2002). MODTOOLS is certainly a good example of simulation model and

GIS integration in presenting a set of computer programs for translating data of the ground-water model, MODFLOW, and the particle-tracker, MODPATH, into a GIS.

According to the level of interaction between the simulation and GIS, three integration approaches have been developed. Loose coupling approach is defined as merely passing input and output between a GIS and a process model. In this level of integration, the input into these programs is from GIS and the output of modeling is exported back to GIS for analysis and visualization. Tight coupling approach involves processing data in a GIS and a visualization model results. In this approach, although the process model is developed outside GIS, the model is configured with the interactive tools of the GIS and the data is exchanged automatically. And finally, full coupling approach or GIS-based simulation approach is defined as embedding a process model in a GIS. This approach involves implementing a process model within the GIS and taking full advantage of the built-in GIS functionalities. In addition, this allows a GIS to go beyond being a simple data management tool to offer more sophisticated analyses and simulations of natural phenomena.

2.1 Requirements of 3D GIS- based Simulation

Although there is a wide variety of implementation strategies and programming techniques in GIS, it is a difficult task to manage a 3D dynamic phenomenon (e.g. fire, flood, water contaminant) within the current GIS. For instance, for the complex process models that are frequently used in disaster simulation, GIS may require extensions to standard functions; extensions which can support multidimensional (3D and temporal) data and mesh finite element methods (Mitasova 1998), i.e. the modeling of temporal changes of objects as well as 3D spatial modeling. GIS-based simulation of a 3D dynamic phenomenon (e.g. forest fires, floods, and water contaminant) consists of two fundamental elements (Mostafavi 2001):

- The mathematical models that describe the motion of the objects in a dynamic system;
- The mesh that supports the motion (i.e. 3D data structure which is dynamic and calculates the interaction between the objects embedded in a 3D environment)

Here, the first elements are referred to as the governing equations, which describe the simulation conditions encountered at a studied site. These equations are continuous and it is very complex and nearly impossible to derive analytical solutions for them. Hence, numerical solutions are helpful in these cases. Two well-known numerical models for solving the governing equations are Eulerian and Lagrangian approaches. The solution of the governing equations is obtained by first discretizing the studied phenomenon into small particles (called elements), a regular grid (with the same shape and size) or an irregular grid (with various shapes and sizes). Next, the behavior of the phenomena for each element is approximated for each time step. By accumulating the result for each element, it is possible to approximate the phenomena being studied. In a simulation process, the discretized phenomenon is represented by a 2D or a 3D geometrical mesh which is a key link between GIS and numerical models.

In the case of spatially modeling a disaster, the mesh elements must be 3-dimensional and well-shaped to provide conformity to the complex geometries. As an example, in the case of a flood, the water land boundary is of particular interest as it is irregular and constantly changing as the flood progresses. In addition, the mesh must be flexible to modifications and editing. This ability allows the mesh to be dynamic and to be modified locally. Such a model offers the user the ability to interact with the environment and to extract information. Finally, the mesh structure must be capable of representing change over time (ex. motion). Considering these requirements for GIS-based simulation, the central question is: "do the data structures of the current GIS have the capability of supporting the three dimensional and dynamic aspects of a natural disaster?"

2.2 Complexities of 3D GIS- based Simulation

In order to answer the mentioned question, we present the main limitations of a GIS-based simulation of 3D dynamics phenomena, namely, space discretization, Time discretization, 3D geometric modeling and representation and Topology management.

Space discretization: Natural phenomena are usually continuous field, while disaster data is usually sampled by unconnected points (object). An ideal spatial data structure should ideally be able to represent objects and fields, two various ways of discretizing space. Tracking contaminants in groundwater flow is an example of this requirement. However, integrating field and object representation is still challenging in the context of GIS data structure. The existing 3D data structures in GIS, according to whether they are defined based on raster (field) or vector (object) data models, are vector- or raster- based. A vector-based data structure is a data structure used for individually representing each object with geometric primitives; usually points, lines, polygons, and solids, and the topology between objects is based on the intersection algorithms of line segments. This operation is a global process. While in a raster-based data structure, the information is represented as a grid of cells and the topology between the cells (not between the real objects) is implicit in the data structure. Raster data is best suited for continuous data representation because cells cover the whole space (Breunig 1996; Li 1994).

Time discretization: Since the data structures of current GIS are static, they can not manage the dynamics and the interaction of objects in a 3D dynamic environment. This means that integrating time-based data into GIS for simulation is complex (Mostafavi 2001). However, this complexity can be reduced by representing temporal changes in various ways. Current approaches are based on a temporal series of spatially registered snapshots which is not an efficient way to handle the dynamics of moving objects. An improved method is necessary to describe the spatio-temporal variations during simulation e.g. event-based method. In this approach, all changes are stored as a sequence of events and time is discretized based on these events.

3D geometric modeling and representation: 3D phenomena modeling needs explicit management of 3D spatial data (length, width, and height). For instance, the z dimension is required for detailed 3D data representation and quantitative assessments of disaster, while in current GIS, 3D volumes are represented as a quasi-3D representation through the use of surfaces, or as true 3D which contains three independent coordinates and can accept several z values for any given location. In quasi-3D (or 2.5D), only a single elevation (z) can be assigned to any surface at any given location, such as a digital terrain model (DTM). This model is not fully suitable to represent every type of object. For instance, in ground water risk management, several important geological structures, such as faults causing repetition of a single horizon at given locations, cannot be represented by this system. Furthermore, the space between the surfaces (volume) cannot be described. Hence, the use of a true 3D GIS would have an essential added value over 2D or 2.5D GIS. The first theoretical approaches to 3D GIS dealt particularly with topological data models and several research works attempted to solve the problem of modeling and managing 3D data (Raper 1989). The main complexities of 3D GIS are the organization of 3D data, 3D object reconstruction, visualization, navigation, and analysis in 3D environments (Stoter and Zlatanova 2003).

In addition to GIS, there are other systems, such as computer-aided design (CAD) tools, which can provide some tools for building 3D models. However, the data structures included in traditional CAD systems are suitable neither for modeling natural objects, given their complexity and enormous quantity, nor for performing subsequent spatial analysis and knowledge discovery (Pouliot et al. 2003). Another type of CAD called GOCAD (Mallet 1992), based on discrete modeling methods, has been developed to model natural objects by accounting for any series of linear constraints which may influence the data on the model. However, the database behind the software is not structured to allow 3D queries and predictive analysis.

Topology management: In a dynamic phenomena simulation, the positions of objects change with respect to space and time and we need to be able to handle them in a dynamic system. In other words, objects (unconnected points/particles) move in a 3D environment and any movement changes the relative position of objects. Therefore, a 3D topological data structure is required in order to describe their topology as well as its changes. However, current raster and vector topological data structures are static and can not be updated locally.

In the raster data model, there is complexity in simulating the motion of any specific object because space is partitioned into a regular tessellation, without any consideration for the embedded object in space (Mostafavi 2001; Gold et al. 1998). In the vector data model, it is not easy to define the region affected by a single change. Hence, a batch operation is used for topology construction. Any change in the relations between objects requires a global reconstruction of the topology. And finally, management of the interactions between moving objects in space is a non-trivial task that should be considered. Therefore, in current GIS, when simulating a dynamic phenomenon, it is not possible to do local on-the-fly updating of topology after any movement or change in 3D space. To meet the requirements and to resolve the mentioned problems of GIS-based simulation, we need a 3D dynamic structure as discussed in next section.

3. 3D Dynamic Spatial Data Structure: A proposition

Considering the simulation of disaster phenomena within GIS, as mentioned previously, the key to this objective is a dual 3D data structure that can manage the dynamic and 3D aspects of the phenomena. Hence, we propose a data structure based on 3D Delaunay tetrahedralization that can deal with discrete objects and continuous phenomena (field) at the same time, and have an interactive topological mesh for numerical simulation. Consequently, we discuss the definition, construction, and storage of this data structure.

3.1 Delaunay Tetrahedralization and 3D Voronoi Diagram

When simulating a flow (ex. flood), to represent the continuum from the discrete samples, it is necessary to make a mesh. In fact, a mesh is a partition of the space by a set of elements such that the union of all elements completely fills the continuum. A set of rules should also be defined to assign some attributes (ex. physical properties) to each node (x,y,z) in the mesh. This mesh can be constructed based on Delaunay Tetrahedral (DT), which is defined by the partitioning of the space into tetrahedrons based on the empty circumsphere test (Fig.1a). In the other words, the circumsphere of each tetrahedron does not contain any other point of the data set (Fig.1c). Thus, among all the possible tetrahedralizations of a set of points, it introduces the best one. This DT is unique for the point set, except when there are degenerate cases in the set (if five or more points are co-spherical in 3D). This definition of Delaunay triangulation generalizes to dimensions higher than three as well.

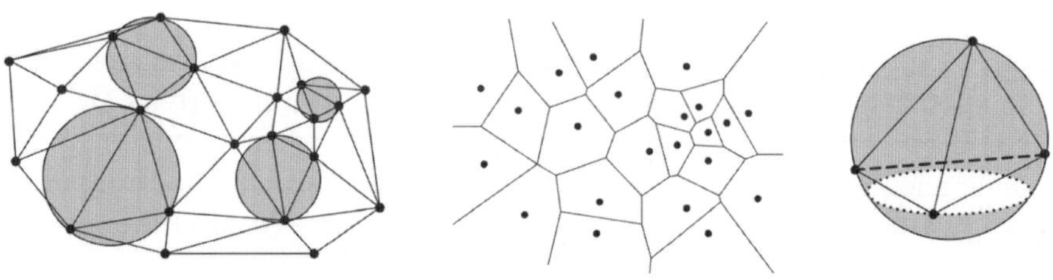

(a) (b) (c)

Fig. 1 (a) 2D Delaunay Triangulation, (b) Voronoi Diagram for the same set of points in the plane, (c) circumsphere test; circumsphere of each tetrahedron does not contain any other point of the data set.

Another interesting advantage of DT is related to its dual, i.e. Voronoi diagram (Aurenhammer 1991, Okab et al. 1992). Voronoi diagram for a given sample set (such as points, line, arcs) in N dimension space is the partitioning of the space in such a way that each point in the space is assigned to the sample closest to that point with respect to a given distance function such as the Euclidean distance (Fig.1b). In two dimensions, each cell around a sample is a convex polygon with a defined number of neighbors, and in 3D, a Voronoi cell is a convex polyhedron with convex faces. Voronoi structure is widely used to answer different types of neighborhood queries.

Duality between two structures, VD and DT, is different in 2D and 3D space (Fig. 2). In 2D, each Voronoi polygon becomes a vertex in a Delaunay triangle, each line in Voronoi (edge) becomes a line in a Delaunay triangle (edge) and each Voronoi vertex corresponds to a triangle in Delaunay triangulation. In 3D space, each Delaunay tetrahedron becomes a Voronoi vertex, a Delaunay edge becomes a Voronoi face and each Delaunay triangular face becomes an edge and vice versa (Ledoux 2006).

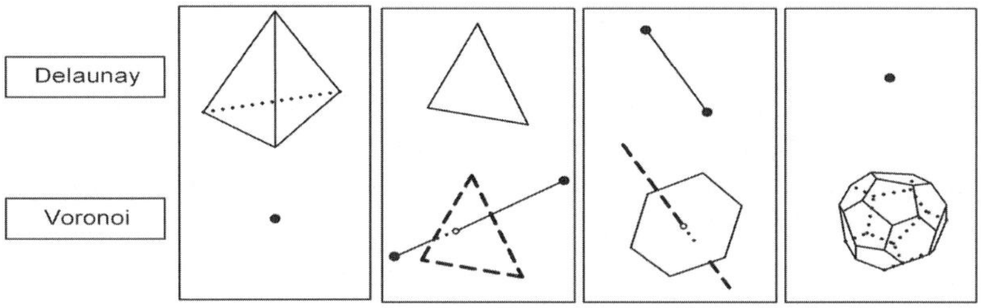

Fig. 2 Duality between two structures, VD and DT

DT and VD have simple and powerful structures and are widely applied in various applications. For instance, VD offers a high potential to automatic natural object reconstruction which allows taking into account incomplete information on the objects as seen in the work of Courrioux et al (2001). Their results demonstrate that volume reconstruction using VD is suitable for geological objects, since it gives a consistent partition of space according to the data specification. Similar results were observed using the oceanographic data in Ledoux's work (2006). Hale (2002) applied DT and VD to reservoir simulations using 3D seismic images. He demonstrates that DT and VD have the potential to answer questions concerning seismic data interpretation, fault framework building, reservoir modeling, and flow simulation.

Also, in numerical modeling, VD is adaptable to the Control Volume Finite Element (CVFE) method, since the CVFE is based on the principle of mass conservation. Thus, a volume of influence is assigned to each node (or element) and equations are defined that describe the interaction of the element with its neighbors. This interaction is expressed by means of a mass balance, a key concept which states that the difference between inflow and outflow in each 3D cell must be equal to the variation in fluid stored in the same volume. According to the steady state condition, there is no storage variation and consequently inflow exactly equals outflow (Fig. 3). This procedure is accomplished at the full domain scale and it ensures fluid conservation both locally (for each element) and globally (for the whole area discretized) (Therrien et al. 2006).

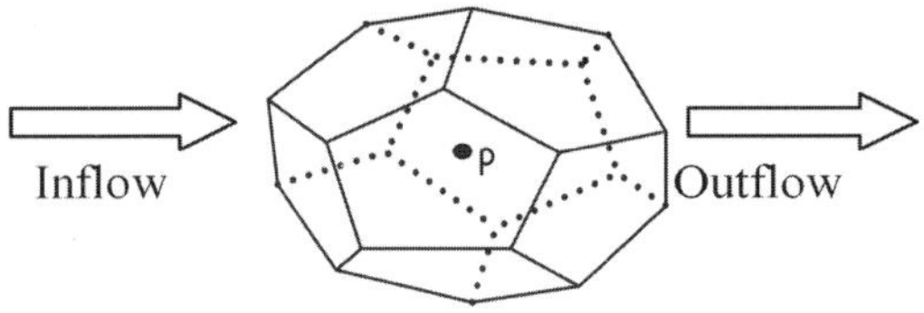

Fig. 3 A volume of influence of a node in CVFE

In this context, the properties of the Voronoi diagram make it an adaptable structure to the CVFE algorithm, with one important advantage: it simplifies the topological analysis. Considering the elegant properties of DT and VD, we believe these structures meet all mesh requirements of 3D dynamic phenomena simulation.

Construction a3D Delaunay-based Mesh

Regarding the duality between VD and DT, they represent the same concept from different points of view. Meaning, if we construct one structure, the other structure can be easily derived from the first one. Therefore, we create the target mesh based on a Delaunay tetrahedron because first, from an algorithmic point of view, it is easier to construct a tetrahedral than a polyhedral. Second, the number of vertices and neighbors of a tetrahedral is constant and manipulation (such as refinement) is easier with a tetrahedral than with a polyhedral. For the construction of Delaunay tetrahedralization, various methods are studied in the field of computational geometry, such as the incremental method, the sweep line method, and the divide and conquer method. Several algorithms have also been proposed and implemented (Aurenhammer 1991). Most algorithms for the construction of DT and VD are in 2D with just a few in 3D.

Our mesh is based on an incremental Delaunay tetrahedralization which is a dynamic method. This means, it is not necessary to know the whole data set prior to Delaunay construction, and on each time, a new object (point) can be inserted or removed in the existing data structure. On the other words, local modification is allowed after any change. While, in static Voronoi structure, the whole data set (generators) is known prior to partitioning the space according to the nearest neighbor rule. Then after inserting or deleting a generator we need to update the topology globally.

To construct and modify a dynamic DT, many basic dynamic operations in 3D are needed (Edelsbrunner and Shah 1996, Ledoux 2006, Miller and Pav 2002). The most important of these are:

- Initialization: the incremental algorithm assumes that the data set (points) is entirely contained in a big tetrahedron and then the points of data set are inserted.

- Insert: the task of this operation is the insertion of a new point in the existing structure. Insert operation includes point location and flipping operators. The task of the point location operator is finding the tetrahedron through the existing tetrahedral that contains the new (or query) point. A flipping operator is a local topological operation that modifies the configuration of the adjacent tetrahedral to satisfy the Delaunay criterion. For example, when the new point P is inserted in a tetrahedron, it may need to be split into four tetrahedrons, each having P as a vertex (Flip 14) (Fig. 4). In addition, a flip operator is used to optimize the new tetrahedral with respect to the Delaunay criterion (empty circumsphere test). This condition is tested via a determinant computation Eq. (1), when the determinant result is negative, it means that the tetrahedron is not Delaunay and a flip operation must be performed. In some cases, Flip 23 and Flip 32 are used to transform one tetrahedralization of two tetrahedrals into another one with three Delaunay tetrahedrals; and Flip32 is the inverse operation.

$$H = \begin{vmatrix} x_p & y_p & z_p & x_p^2 + y_p^2 + z_p^2 & 1 \\ x_a & y_a & z_a & x_a^2 + y_a^2 + z_a^2 & 1 \\ x_b & y_b & z_b & x_b^2 + y_b^2 + z_b^2 & 1 \\ x_c & y_c & z_c & x_c^2 + y_c^2 + z_c^2 & 1 \\ x_d & y_d & z_d & x_d^2 + y_d^2 + z_d^2 & 1 \end{vmatrix} \tag{1}$$

- Delete: the task of this operation is the Deletion of a point from the structure; similar to the Insert operation, in the Delete operation, point location and flipping operators are used to remove an object from the existing mesh. Therefore, Deletion consist of locating the point P to be deleted and restructuring the configuration of the tetrahedral adjacent to P with a sequence of flips; for example, when a point is removed the adjacent tetrahedrons may need to be merged together.

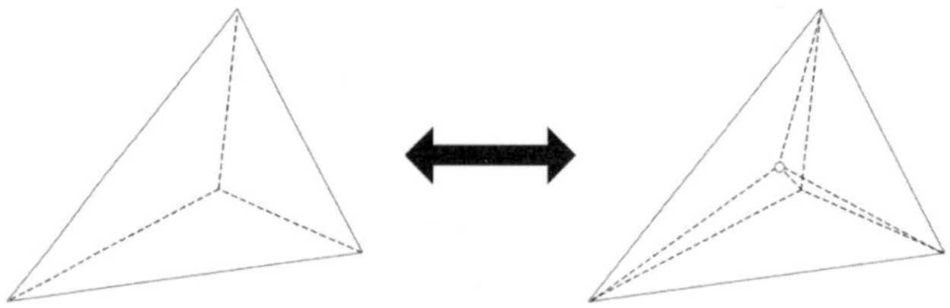

Fig.4 When a point is inserted or removed, the adjacent tetrahedrons may need to be split or merged together.

Storage Data Structure

There are at least three approaches to represent a tetrahedralization; the use of a record to represent each tetrahedron, a record to represent each face, or a record to represent each pair of faces and an edge (thus, three records per triangular face are necessary). In this research work, we use the first approach for two reasons: firstly, tetrahedral-based data structure is able to associate attributes with the mesh elements or the vertices and secondly, it has the minimum memory requirements. Thus, in our approach, the geometrical and topological information is stored based on 3D tetrahedral data structure, which includes every tetrahedron with four pointers to its vertices (nodes) and four pointers to its four adjacent tetrahedrals. However, in disaster simulation, certain analysis involves the volume and boundary of objects and requires the information on the faces and edges of the tetrahedrons. This information can be extracted form the initial data structure but computationally, it is a very costly operation which represents the main drawback of the method. This type of analysis can also be easily handled by the Voronoi diagram (dual of tetrahedron). Therefore, to solve the mentioned problem, a data structure is proposed that can manage DT and VD simultaneously. This data structure is called Augmented Quad-edge and it offers many advantages for simulation.

Augmented Quad-Edge is an extension of the popular 2D Quad-edge (Guibas and Stolfi 1985) into 3D. the Quad-Edge data structure is based on storing four directed edges for each edge of element of the mesh, two in the primal subdivision (e.g. DT) and two in the dual (e.g. VD). Every quad-edge has three pointers which allow navigation from edge-to-edge through the mesh by means of its algebraic operation. In Augmented Quad-Edge (AQE), each element (tetrahedron or Voronoi cell) is stored individually. The elements are connected together by the dual edge to the face shared by two elements. Therefore, with the aid of the AQE, any point, edge, face, and region of DT and VD can thus be retrieved.

However, in order to implement a numerical mesh based on AQE data structure; it is necessary to reformulate the traditional incremental algorithm for tetrahedral construction based on this data structure. The two important construction or modification operators that are required to manipulate elements are make-edge and split-edge. Make-edge is used for creating a new edge or face and split-edge destroys the existing edge or face. Theoretically, the AQE offers a good trade-off between storage costs and the number of topological relationships between the different elements of a subdivision. The main disadvantage of this method is high storage cost.

4 3D GIS-based Simulation of Dynamic Phenomena: Floods

As mentioned previously, for an effective natural disaster simulation based on GIS, the structure of these tools must be able to represent the 3D dynamic behavior of the phenomena. In this section, through an application example (e.g. flood simulation), we will explain how the proposed spatial data structure could contribute to his understand and modeling.

Floods are among the most common natural disasters that threaten human life (Liang et al. 2004). For example, in July 1996, exceptional rainfall over the Saguenay region (north of Quebec City in Eastern North America) caused a catastrophic flood and triggered a biophysical transformation of forests, soils, the water table, lakes and rivers, and at the same time, destroyed road systems and urban installations located near the waterways or in floodplains (Fig 5). The damages from this flood have been estimated to $1.5 billions, 16,000 people have been affected in some way and more than 7 people were dead. Although, this natural disaster cannot be avoided most of the time, the damage they cause can be reduced by its prediction where the simulation tools represent a valuable tool.

Fig. 5 The Saguenay Flood (adapted from http://en.wikipedia.org)

In order to simulate flood flow, the first step is to define a model that describes its dynamic behavior. This model is referred to as the governing equation and should be able to simulate all the conditions encountered at the studied site. The governing equation is continuous and its solution requires its discretization with respect to time and space. Then, a numerical method is used to solve the discrete equation.

Two well-known numerical methods for solving the governing equations are: the Eulerian method and the Lagrangian method (Donea et al 2004; Price 2005) (Fig. 6). The Eulerian point of view considers changes as they occur at a fixed point in the fluid. Here, the mesh is fix and, and the continuum moves with respect to the grid, and measurements are made at fixed locations (Fig. 6b). In this method, the flow equation is a nonlinear equation and numerical approximation is considered to solve the equation which leads to unreliable results. The Lagrangian point of view considers changes which occur as we follow a fluid particle (i.e. along a trajectory). In this method, the adjacency relationships between particles are defined based on the connectivity between elements. The method allows the mesh elements to move and interact locally with each other (Fig. 6c). Using this method, it is possible to easily trace individual parcels that are carried around by the flow (e.g. water pollution). According to the Lagrangian method, the flow equation does not include any nonlinear term. The result of this method is more exact than the Eulerian method and flow can be precise on the real trajectory.

In both Eulerian and Lagrangian methods, the stored connectivity between parcels (mesh elements) does not change during the simulation process and it causes problems such as mesh tangling (dam-breaking problem) (Mostafavi 2001). As a result, these methods can not easily deal with highly deforming flow such as floods. Hence, an extension of the Lagrange method, i.e. Free-Lagrange, can be used. The Free-Lagrange method uses a fully unstructured Lagrangian mesh in which the connectivity is allowed to change freely as the flow evolves, and is thereby suitable for highly deforming flows (Crowely 1985). The research work of Mostafavi (2001) is as an example of the Free-Lagrange method that was developed base on the Voronoi diagram and applied for a 2D tidal flow simulation on the globe.

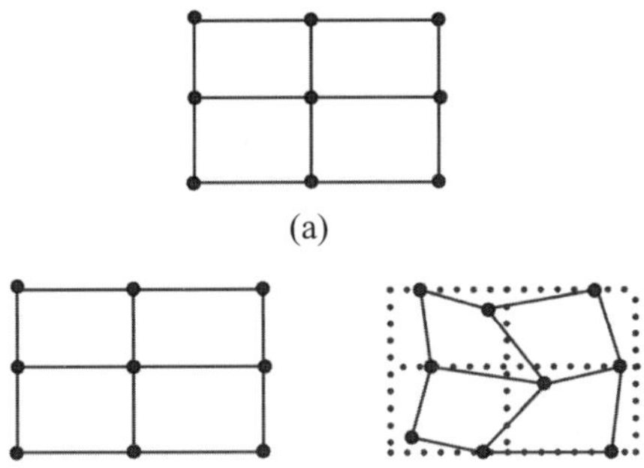

(a)

(b) (c)

Fig. 6 (a) Computational mesh for, (b) Eulerian simulation, (c) Lagrangian simulation

No matter which numerical method is used for solving the governing equation, the first step is to replace the spatial domain of interest with a mesh. Our structure is able to generate an optimal mesh. We discretize the space by a mesh whose elements are defined as Delaunay tetrahedral and the adjacency relationships (topology) are readily defined among the elements in a Voronoi diagram. In addition, the size of the mesh elements depends on the distribution of the data set. Hence, it provides a valid representation of topological and geometrical information of complex objects. Also, since the proposed structure is dynamic, it can be modified regarding resolution constraints such as, regions of high flow, particular boundaries, or unsuitable elements (such as a sliver with four vertices which is almost coplanar-with a volume of zero).This is done without having to rebuild the entire mesh.

In the case of the Lagrangian point of view, the mesh generation process is the same as in Eulerian method with the difference being that motion is represented by moving the mesh elements. This is very important in flood simulation since, in this case, the water-land boundary is of particular interest as it is irregular and constantly changes as the flood progresses. Then, the governing equation assigns a velocity vector (3D) and various physical parameters to each point (generator in VD, vertex in DT) which can be updated for each time (Fig.7). In fact, these velocity vectors may change because of the interaction between the points (particles); hence points will move in the resultant force direction and interact with each other. For this propose, it is necessary to know the neighbors of each element. This can be easily done by our topological data structure.

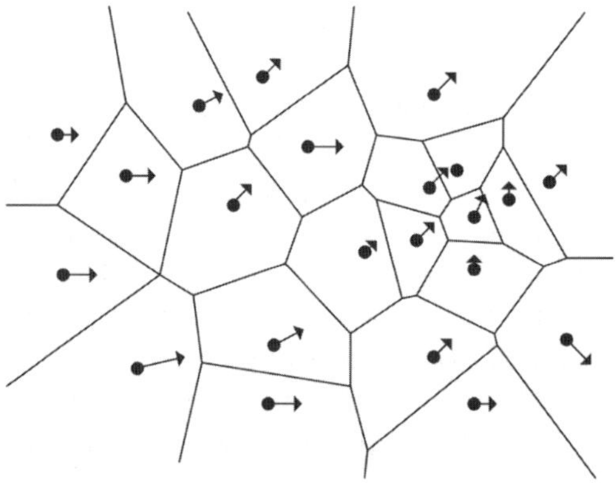

Fig. 7 Moving mesh; the particles move in the resultant force direction and interact with each other.

To update the velocity vector or physical parameters, an important question remains: "How is time discretized during the flow simulation?" Most approaches to representing change in a phenomenon are based on the snapshot model, consisting of producing a sequence of time snapshots corresponding to different time points within a given time interval (Peuquet 1994). However, it may break up the continuity of a phenomenon and may miss temporal ordering indicating the connection between events. In other words, using this approach, what has happened between two time snapshots may be lost. Moreover, in this approach, it is necessary to reconstruct the global topology after each snapshot. Advances in temporal GIS involve another approach which is referred to as an event driven method. In fact, this method detects all changes, motions, and events which occur over time (Mostafavi 2001; Mioc et al.1998; Gold 1996) and consequently, the structure is locally updated after each topological event. The data structure that supports motion is referred to as a kinetic data structure.

In order to simulate a 3D dynamic phenomenon such as a flood, it is necessary to extend the 3D dynamic DT/VD data structure proposed in this paper to a kinetic 3D data structure which supports motion in a 3D environment. The kinetic Voronoi data structure allows several objects (points or generators) to move simultaneously in the mesh with the capacity for local topology updating. In fact, continuous movement of the points in the 3D mesh are implemented by detecting all topological events on the trajectory of the points, processing them in order, and moving the points to their new positions, one by one on their trajectory, until the points arrive at their destinations. In this algorithm, a topological event occurs when a point moves in or out of the circumsphere of a tetrahedron. This must be detected to preserve the empty circumsphere criterion and then the topological modification is done locally.

Having defined a 3D dynamic topological data structure, we need to assign physical parameters to the 3D mesh for simulation purposes. These parameters are referred to as initial conditions for the simulation. Following, this step, the system is ready for the simulation of a considered dynamic phenomenon. It should be mentioned that the present work is part of ongoing research in the Center for Research in Geomatics (CRG) at Laval University. Other aspects related to 3D GIS-based simulation such as moving object data structures in a 3D environment, resolution of the mesh, boundary condition problems, and collision problems are currently under investigation by our research group.

4. Conclusion

Simulation of disasters (e.g. forest fires, floods, water pollution) within GIS is important for better understanding of observed phenomena as well as the representation and management of all steps of the process. Simulation offers the possibility of reducing social and economical losses. However, the GIS-based simulation of these 3D and dynamic phenomena has some limitations. In this paper, we discussed these limitations and, proposed as a solution, a data structure which is able to accommodate the 3D and time dimension of objects that current GIS is unable to handle. Our structure is based on 3D Delaunay with several interesting properties; it has the ability to represent both discrete objects and continuous phenomena, and to deal with static and dynamic 3D objects. Moreover, it can generate an optimal mesh for numerical simulation because it is an adaptive method; i.e. the size of the mesh elements depends on the distribution of the data points. In this mesh, after any change (event), local updates of topology are possible. Using this mesh, a fluid flow such as a flood and or a forest fire can efficiently be simulated within GIS if we use the Free-Lagrange method as the numerical model solution. In the future, several other considerations related to 3D dynamic simulation within GIS, such as the problem of moving objects in 3D enlivenment, the collision problem, and the resolution of applied mesh will be investigated in order to improve the proposed data structure discussed in this paper.

Acknowledgments

The authors would like to acknowledge the support of the GEOIDE Network under GeoTopo3D project, entitled: "Development of 3D Predictive Modeling Platform for Exploration, Assessment and Efficient Management of Mineral Petroleum and Groundwater Resources". The authors would also like to acknowledge Dr Christopher Gold from the University of Glamorgen, and Dr Hugo Ledoux from Delft University of Technology for their valuable contributions to this work.

References

Aurenhammer F (1991) Voronoi diagrams: A survey of a fundamental geometric data structure. J ACM Computing Surveys, 23(3):345–405

Bivand R, Lucas A (1997) Integrating models and geographical information systems. In: Openshaw S, Abrahart R J (eds.) Geocomputation. Taylor & Francis, London, pp. 331-364

Bennett D (1997) A framework for the integration of geo-graphical information systems and model base management. Int. J. Geographical information System, 11(4): 337 - 357

Breunig M (1996) Integration of spatial information for geo-Information systems. Springer, Berlin Heidelberg New York

Burough P A, Van Deursen W, Heuvelink G (1988) Linking Spatial Process Models and GIS: A Marriage of Convenience or a Blossoming Partnership?. In: Proceedings of GIS/ LIS'88, (Bethesda: American Congress on Surveying and Mapping), San Antonio, TX., pp. 598-607

Chapman L, Thornes JE (2003) The use of geographical information systems in climatology and meteorology. J Progress in Physical Geography, 27(3):313–330

Courrioux G, Nullans S, Guillen A, Boissonat JD, Repusseau P, Renaud X, Thibaut M (2001) Volumetric modelling of Cadomian terranes (Northern Britanny, France): an automatic method using Voronoi diagrams. J Tectonophysics, 331: 181-196

Crowely WP (1985) Free-Lagrange Methods for compressible hydrodynamics in two space dimensions. In: Fritts M J, Crowley W P, Trease H (eds.), The Free-Lagrange Method, Lecture Notes in Physics Vol.238, New York, Springer-Verlag

Densham P J (1993) Integrating GIS and spatial modelling: The role of visual interactive modelling in location selection. Int. J Geographical Systems, 1:203-220

Donea J, Huerta A, Ponthot J Ph, Rodr´ýguez-Ferran2 A (2004) Arbitrary Lagrangian–Eulerian Methods. J Encyclopedia of Computational Mechanics, 1(14): 413-437

Dunn A, Milne G J (2004) Modelling wildfire dynamics via interacting automata. Lecture Notes in Computer Science, 3305:395–404.

Edelsbrunner H, Shah N R (1996) Incremental topological flipping works for regular triangulations. J Algorithmica, 15:223–241

FIG (2006) The Contribution of the Surveying Profession to Disaster Risk Management. (A publication of FIG Working Group 8.4, International Federation of Surveyors (FIG), no.38)

Gold CM, (1996) An event-driven approach to spatio-temporal mapping. J Geomatica, 50(4):415–424

Gold C, Remmele P, Roos P R (1997) Voronoi methods in GIS. In: Nievergelt J, Roos T,Widmayer P (eds.), Algorithmic Foundations of Geographic Information Systems, vol.ume 1340 of Lecture Notes in Computer Science, pp. 21–35

Goodchild M (1992) Integrating GIS and spatial data analysis: problems and possibilities. Int. J Geographical Information Systems, 6:407- 423

Guibas LJ, Stolfi J (1985) Primitives for the Manipulation of General Subdivisions and the Computation of Voronoi Diagrams. ACM Transactions on Graphics 4(2):74–123

Gunes A E, Kolel J P (2000) Using GIS in emergency management operations. J urban planning and development 126 (3):136-149

Hale D (2002) Atomic meshes: from seismic imaging to reservoir simulation. In: Proceedings of the 8th European Conference on the Mathematics of Oil Recovery, Freiberg, Germany

IPGC (2007) Climate Change 2007: The Physical Science Basis. (Intergovernmental Panel on Climate Change (IPGC) 2007 (Report of Working Group I), Paris, France)

Lawson CL (1977) Software for C1 surface interpolation. Mathematical Software III, J. Rice ed., Academic Press, New York, pp.161-194

Ledoux H, Gold CM (2006) La modélisation de données océanographiques à l'aide du diagramme de Voronoi tridimensionnel (in French), Revue internationale de géomatique, 16(1):51–70

Ledoux H (2006) Modelling Three-dimensional Fields in Geosciences with the Voronoi Diagram and its Dual. Ph.D. thesis, University of Glamorgan

Li R (1994) Data structures and application issues in 3-D geographic information systems. J Geomatica, 3:209 -224

Liang Q, Borthwick AGL, Stelling G (2004) Simulation of Dam and Dyke-break Hydrodynamics on Dynamically Adaptive Quadtree Grids. J Numerical Methods in Fluids, 46: 127-162

Mallet J L (1992) GOCAD: a computer aided design program for geological applications. In: Keith Turner (ed.) Three-Dimensional Modeling with Geoscientific Information Systems, NATO ASI Series, pp 123-142

Miller G, Pav S E (2002) Fully Incremental 3D Delaunay Refinement Mesh Generation. In: Proceedings of the 11th International Meshing Roundtable, IMR 2002, Ithaca, New York, USA, pp. 75-86

Mitasova H (1998) Process modeling and simulations. (NCGIA GISCC Unit 130)

Mioc D, Anton F, Gold C (1998) Visualizing changes in a dynamic Voronoi data structure via time travel. In: The 6th International Conference on Computer

Graphics and Visualization in Central Europe (WSCG' 98), Vol. 2, Plzen City, Czech Republic, pp.263-269

Mostafavi M A (2001) Development of a global dynamic data structure. Ph.D. thesis, Université Laval

Nyerges T L (1993) Understanding the scope of GIS: Its relationship to environmental modeling. In: Goodchild M F, Parks B O, Steyaert T (eds.) Environmental Modeling with GIS, (New York: Oxford University Press), pp. 75-93

Okabe A, Boots B, Sugihara K (1992) Spatial Tessellations: Concepts and Applications of Voronoi Diagrams. In: Probability and Mathematical Statistics. Chichester, U.K.: John Wiley & Sons Limited, pp. 532

Peuquet D (1994). It's About Time: A conceptual framework for the representation of Temporal Dynamics in GIS, In. Annals of the Association of American Geographers, 84:441-461

Pouliot J (1999) Définition d'un cadre géosémantique pour le couplage des modèles prévisionnels de comportement et des SIG : application pour les écosystèmes forestiers, Ph.D. Thesis, EPFL Université

Pouliot J, Lachance B, Brisebois A, Rabaud O,Kirkwood D (2003) 3D geological modeling: Are GIS or CAD appropriates?. In: ISPRS Workshop, WG II/5, II/6, IV/1 and IV/2 Joint Workshop on "Spatial, Temporal and Multi-Dimensional Data Modelling and Analysis", Quebec, Canada

Price J F (2005) Lagrangian and Eulerian representations of fluid flow: Part I, kinematics and the equations of Motion. (Access date: Jan. 2007 www.whoi.edu/science/PO/people/jprice/class/ELreps.pdf)

Raper J (1989) Three Dimensional Applications in Geographic Information Systems. Taylor & Francis, London, pp. 189

Rifaat A,Tao T (2005) Integrated distributed GIS approach for earthquake disaster modeling and visualisation. In: van Oosterom P, Zlatanova S, Fendel E M (eds.) Geo-information for Disaster Management, pp. 1183-1192

Stoter J, Zlatanova S (2003) 3D GIS, where are we standing?. In: Joint Workshop on Spatial, Temporal and Multi-Dimensional Data Modeling and Analysis, Quebec City, Canada.

Sui D Z, Maggio R C (1999) Integrating GIS with hydrological modeling: practices, problems, and prospects. J Computers, Environment and Urban Systems, 23:33–51

Therrien R, McLaren R G, Sudicky E A, Panday S M (2006) HydroSphere, A three-dimensional numerical model describing fully-integrated subsurface and surface Flow and solute Transport.(Draft of Groundwater Simulations Group)

Valavanis V D (2002) Geographic information systems in oceanography and fisheries. Taylor &Francis

Ontologies for Disaster Management Response

Wei Xu and Sisi Zlatanova

GISt, OTB, Delft University of Technology, Delft, The Netherlands
Wei.xu@tudelft.nl, S.Zlatanova@tudelft.nl

Abstract

Increasing numbers of natural disasters and man-made disasters, such as earthquakes, tsunamis, floods, air crashes, etc., have posed a challenge to the public and demonstrated the importance of disaster management. The success of disaster management, amongst all, largely depends on finding and successfully integrating related information to make decisions during the response phase. This information ranges from existing data to operational data. Most of this information is geographically related and therefore when discussing integration of information for disaster management response, we often refer to the integration of geo-information. Current efforts to integrate geo-information have been restricted to keyword-based-matching Spatial Information Infrastructure (SII, may also known as Spatial Data Infrastructure). However, the semantic interoperability challenge is still underestimated. One possible way to deal with the problem is the use of ontology to reveal the implicit and hidden knowledge. This paper presents an approach for ontology development and ontology architecture, which can be used for emergency response.

1. Introduction

The five phases of disaster management, namely mitigation, prevention, preparedness, response and recovery (GRIP Animation 2006), have urged a good collaboration among various users such as the fire brigade, the police, the medical service, the municipality, Red Cross, urban planners, and other organisations or even amongst different countries. They all have to be able to work together and understand each other. These requirements are especially valid for the emergency response phase (Zlatanova et al. 2007). During this phase, we need to ensure interoperability of emergency services, and to provide appropriate information at the right place and in the right moment (van Borkulo et al. 2006).

The information ranges from existing data to operational data, of which most is geographically related. However, successfully discovering and combining geo-information in a time-critical manner for decision making is not an easy job, because the required geo-information is distributed among different organisations. Exchanging geo-information (and knowledge) by interacting on a personal/phone/fax basis is slow and might even be prone to errors. We need to allow machine automation so that we can integrate geo-information in a time-critical manner to help decision making during disaster management.

Current efforts to integrate geographic information data have been restricted to keyword-based-matching Spatial Information Infrastructure (SII, others use Spatial Data Infrastructure) (Groot and McLaughin 2000, SDI cookbook 2004). SIIs are being set up within regions, countries or even across national borders (Bernard 2002, Riecken et al. 2003), to facilitate the access to geographic information. SII supports the discovery and retrieval of distributed geospatial data sources and geographic information services by providing catalogue services and syntactic interoperability standards (Lutz and Klien 2006).

The integration of geo-information has been greatly advanced by SII. However, the semantic interoperability problem, which presents challenges for the integration of geo-information in the open and distributed environments, still exists. One possible way to deal with the problem is the use of ontology to reveal the implicit and hidden knowledge (Wache et al. 2001). Ontology is an explicit formal specification of a shared conceptualisation (Gruber 1995). Much research has been carried out on the use of ontologies for information discovery and retrieval Klien et al. 2004, Lutz and Klien 2006, Wache et al. 2001). In this paper, we suggest an approach of ontology development and ontology system architecture of geo-information to support disaster management.

The rest of the paper is divided into the following sections: Section 2 introduces the current management of a disaster in the Netherlands and explains briefly the complexity of the emergency response. By comparing different ontology perceptions and architectures in Section 3, we illustrate the potential use of ontologies to facilitate the management work for a disaster. Section 4 introduces new ontology architecture to facilitate emergency response. The last two sections discuss implementation aspects and future work.

2. Example: Discovery and Integration of Geo-Information

Let's consider the following scenario. In Fig. 1, an accident is reported to a fire station. An explosion happened at a chemical plant and toxic gas is leaking from destroyed pipes. In order to evacuate the inhabitants of the affected districts, the gas plume has to be created to plan evacuation and guide rescue teams. The forecast of the gas plume's dispersion is an essential part of this task.

For the management of such a disaster, information from different sources needs to be obtained and combined immediately. Fig. 2 gives an indication of such information: the plant's ID, the location of the plant (the exact location of the emission), real-time measured data from the field (type of released gasses and concentration), the nearest airport information (the airport code), the measurement of the wind from the weather station (the wind speed and wind direction near the place of the accident). For evacuation, we will also need other specific data (like the population of the potential area, road network).

Fig. 1 Fire at a chemical plant

One possible way to obtain the exact location of the plant is by querying databases (for example, a cadastral database or a risk map) with the name of the plant. The output, a pair of coordinates, is used to find the nearest airport and weather station from which current wind speed and direction can be obtained. The weather information and the emission rate of the gas leak are used to calculate the gas plume. A map is created, showing the plume on top of the road network and land use data of the region (Fig. 2).

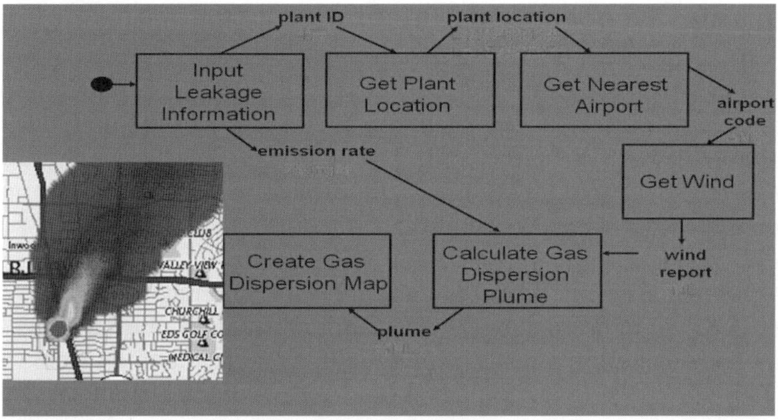

Fig. 2 Integration from various data sources

Referring to Fig. 1, if we want to find the information about the wind ('Get Wind' in Fig. 2), we will type and search the keyword 'wind' or 'prevailing wind direction' in the Disaster Management Service. We will get a large amount of results related to the keyword, among which we need to choose one (or several that are combined) that answers our question (the direction, the speed and etc.). We have to be careful about the content (semantic) of this information. Fig. 3 illustrates challenges in using the word 'prevailing direction' for wind:

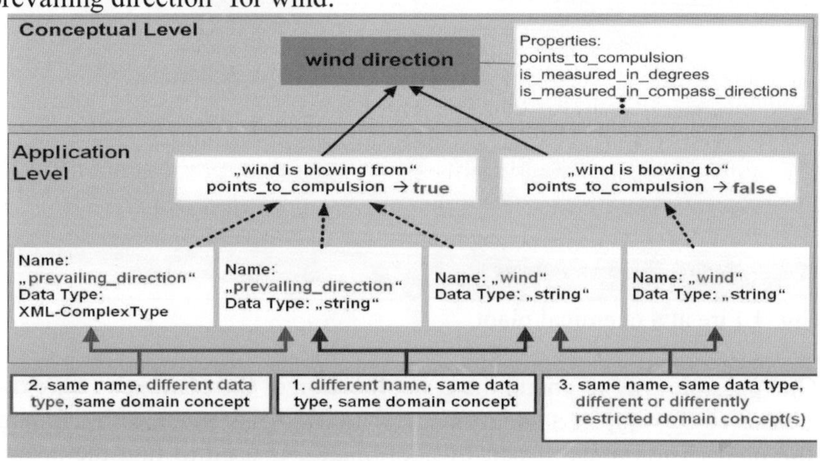

Fig. 3 Semantic heterogeneity

- 'prevailing direction' in different geo-services (or datasets) may mean the same thing, but the storage approach differs (e.g., one with XML-Complex Type and the other with String Type). This is actually the same name with the same domain concept but with different data-type.

- 'prevailing_direction' may also refer to different things — one refers to the direction which the wind is blowing from and the other refers to the one which the wind is blowing to. (Same name with same data-type but with different domain concept.)
- 'prevailing_direction' does not exist; the direction is included in the word 'wind'.

During the whole process of disaster management, we have been continuously facing such challenges. For instance, with different levels for disaster emergency, different hierarchical levels of organisation are required and therefore different combinations of datasets are needed. The content of each information source must be fully understood by and described for machines so that they can be discovered and combined automatically. It is impossible for humans to examine the information and understand content, and then discover and combine this in a short time under time pressure in disaster management.

If machines could be employed to help humans discover and combine related information, it will be possible to efficiently deal with disasters. We believe that with the help of ontologies, the content (semantic) of the information can be made explicit and thus machine process-able.

3. Disaster Management in the Netherlands

This paper is based on the research work undertaken by GDI4DM (Geospatial Data Infrastructure for Disaster Management). In the rest of this section, an overview is given of some disaster management issues in the Netherlands. The Netherlands is used as an example for better understanding the complexity of disaster management and also summarise some insights from the work being done (see also Diehl et al. 2007, van Borculo et al. 2005, Diehl and van der Heide 2005).

3.1 Organisational structure

The fire brigade, the police, the medical service and the municipality are the main actors in emergency response. Each of these organisations maintains their own data and carries out their own daily work. They have-well defined protocols how to cooperate with each other in case of a disaster. Other institutions and organisations such as Red Cross, Military forces, Ministry of Internal Affairs, etc., and more specialized organisations may also be involved in managing the disaster when needed but usually they follow direct orders from operational centres..

To specify the cooperation, and communication between the first responders, six GRIP (Common Regional Incident management Procedures) levels have been defined (GRIP 2006). On GRIP 0, the fire brigade, the police, the medical service and the municipality are doing their daily work. There is simple cooperation among them. However, with increasing the importance of incident, the need of coordination increases. For example, during GRIP 1, the mayor of the Municipality will be informed; within GRIP 2 a Regional Operational Team is formed. Depending on the magnitude and type of disaster, different organizations may become involved. For instance, emergency officers at provincial or national level are informed if the disaster affects a large section of the community; the Ministry of Internal Affairs will take the administrative lead if the disaster extends beyond the provincial; the Office of the Queen will be informed if the disaster crosses the national border (e.g. a nuclear leak). So with different levels of GRIP, different types of cooperation are established and different policies are used to form the structure to deal with disasters.

Besides the GRIP levels, each of the first responders have well-defined task described in 25-29 (depending on the province) processes. Which processes are going to be initiated depends on the type of disaster. There are processes which are related to each other (if one starts it triggers another one) and also processes, which are always active (e.g. press conferences). Examples of such processes are: performing measurements needed to compute a gas plume (completed by the fire brigade), traffic control (performed by the police), contacts with the press (a task of the municipality), etc. An initial study on the processes has revealed the following important characteristics:

- Processes exist independent of disaster types. Processes emphasise that 'core functions and the important characteristics of the disaster management are described in consistency with the other activities'.
- Processes are strictly defined and therefore very convenient to be formally described and used by system developers.
- Processes could be organised in sector clusters, i.e. the fire brigade, the police, the medical service and the municipality, which may give further indication on the information needed by a particular cluster (operating daily as one independent organisation).

A particular incident is therefore managed by combining processes with respect to the type of disaster and the GRIP levels. Since GRIP and the processes intend to outline the way working, they specify the roles of the actors and can be used as an indication for the kind of information needed for the different actors.

3.2 Data needed for emergency response

During disaster management, large amounts of data are created, maintained and integrated to deal with the disaster. These data ranges from existing data, like topographical data, to operational data, like the measurements from the field.

- Existing data are those that are created and maintained by organisations before the disaster happens. For instance, TOP10 (which is 1:10,000 topographic maps), GBKN (which is large scale basic map of Netherland 1:1,000), utility data, hydrological data, transportation data, topographic maps along rivers and roads (maintained by RWS, which is the Ministry of Transport, Public Works and Water Management), risk maps (maintained by municipalities), information about dangerous goods, statistics about population, etc. These data could help answer questions like 'which roads are available', 'what are the dangerous goods stored', 'people in the area in this moment' and so on.
- Operational data are those received from the field during disaster management. They are not available prior to a disaster. Dynamic data could be a report (description) of an incident, camera images, video clips, measurements, etc. Maintenance of operational data is usually carried out in the Commando centre but could be also by various organisations, for example, the fire brigade, the police, the medical service, the municipality, and some other specialised organisations.

Commonly the data (existing and operational) are heterogeneous in creation, maintenance and presentation:

- Data are used for multiple purposes. Users create their unique data to meet their own needs. Even when they are referring to the same problem, they might have their own interpretations. For instance, people forecasting floods might only be concerned with the height of the water, while people measuring the pollution might only have interests in the chemicals in the water.
- Data are stored in various formats. Various technologies are used to collect and create the data, and different storage systems are employed to update and maintain the data. For example, the data could be files in a relational database, an attribute in an object model, a picture or a video clip and so on.
- Users are used to their own representations. The presentations are sometimes combined in data models. For instance, the fire brigade will expect a red colour to mean danger while other people refer to red colour as any other colour, e.g., only a distinguish between green pipelines and

red pipelines (pipelines marked in green represent water pipelines and those marked in red represent fibre-lines).

3.3 Problems with disaster management — interoperability

In the management of a disaster, different people from various organisations are involved, and large amounts of heterogeneous data are required to be created, collected and integrated. So the success of disaster management could be interpreted as 'getting the right resources to the right place at the right time; to provide the right information to the right people to make the right decisions at the right level at the right time.' As we can see from previous sections, these data sources are created, and maintained individually. Interoperability problems will arise when we try to integrate these individual data sources.

With the word 'interoperability', we mean the ability of systems to provide services to and accept services from other systems and make them to operate effectively. Usually three types of interoperability are identified namely system interoperability, syntax and structure interoperability and semantic interoperability (Sheth, 1999). System interoperability refers to the ability to deal with hardware, operating systems, and communications heterogeneity, such as the instruction sets, communication protocols, different file systems, naming, file types, operation and so on. Syntax and structure (schematic) interoperability is relevant to data representation, formatting, data models, different DBMS (Data Base Management System). Semantic interoperability has more to do with the meaning of the data. Semantics refers to the aspects of meanings that are expressed in a language, code, message or any other form of representation. Sheth argues that 'semantic interoperability requires that the information system understands the semantics of the users' information request and those of information sources, and uses mediation or information brokering to satisfy the information request as well as it can' (Sheth 1999).

During disaster management, we also have problems with interoperability at system, syntax and semantic level. Sheth (1999) made a summary of the previous efforts and achievements with respect to the system and syntax interoperability. OGC (Open Geospatial Consortium) and ISO (International Standards Organisations) are current efforts to provide standards to solve syntax heterogeneity. SIIs, which are being built by different regions, countries and even across national borders (Bernard 2002, Groot and McLaughin 2000, Riecken et al. 2003), can be seen a typical example of attempts toward resolving syntax heterogeneity. SIIs support the discovery and retrieval of distributed geospatial data sources and geographic information. SIIs provide syntax interoperability only to certain extends (Lutz 2006).

Sheth (1999) concluded that a standard terminology is not prevalent within the GIScience domain and is dependent on the context of use and the user. The use of different terms and approaches causes confusion in the specification of universally accepted entities, concepts, rules, relations, and semantics as the basis of a consensual ontology. In disaster management, we also lack such a terminology. Apart from that, the semantics of the creation, maintenance and representation are not clear to the people outside the organisation. It urges that the semantics of these data understood by all the users involved in disaster management, so that they could exchange their information under high time pressure. Neches, who is one of the earliest to talk about the use of ontologies to better allow information interoperability, has pointed out we could use ontologies to represent domains and make it sharable (Neches et al. 1991).

4. Ontologies

The word, ontology, originates from philosophy, and stands for 'the study of being existence' (Ontology 2006). Computer science borrows this word from philosophy, which 'defines the basic terms and relations comprising the vocabulary of a topic area as well as the rules for combining terms and relations to define extensions to the vocabulary' (Neches et al. 1991).

There are two key points of ontology — (1) ontology is a kind of conceptualisation and (2) ontologies should be shared. According to Studer et al. (1998), conceptualisation is an abstract model of the real world phenomenon, by which the real world is identified as a set of concepts. Shared means the notions are accepted by a certain group as consensual knowledge (Studer et al. 1998, Uschold and Gruninger 1996). There are two important aspects in defining ontologies:

- Explicit: Explicit stands for the meanings of the types of concepts that are used in the conceptualisation and the constraints of their usage are defined (Studer et al. 1998, Uschold and Gruninger 1996).
- Formal: Formal refers to the fact that the ontology is defined in an artificial and well-defined language so that it is machine-readable (Uschold and Gruninger 1996).

Ontology could be defined explicit or non-explicit (implicit), with different degrees of being formal and being shared. We adopt the definition from Audi, defining ontology as 'the study of explaining reality by breaking it down into concepts, relations and rules' and share it with others

(Audi 1995). In order to solve the semantic interoperability and allow machine-automation of data integration, it is desired that the semantics of the data are defined explicitly and represented in a machine process-able way. Explicit and formal ontolgies can help us define the semantics of the data, make them sharable by different users and allow machine process-able. So when we talk about ontologies, we are referring to explicit, formal and shared ontologies.

4.1 Ontologies approaches

General speaking, there are two approaches to build ontologies, namely top-down approach and bottom-up approach. They have their own pros and cons. We will examine the two approaches and compare them in order to find our approach of building the ontologies for disaster management.

- Top-down ontologies development indicates ontology is constructed by first examining the domain of interest in general at a very abstract level and then constructing it based on top levels concepts. This is accomplished by building an abstract model of the domain of interest first (or by starting from a top level ontology) and then extending the model further to map more specific concepts from low levels. The result ontology from top-down approach is a complete model over the domain of interest. It contains all the relations between concepts within the domain of interest. When building such ontology, there are some challenging issues we need to remember. First it remains a problem for domain experts to reach an agreement over a domain, because we cannot force people to look at the problem in one single way— they might have different views over one problem. Second, it is difficult to extend the top level concepts with lower level data sources at the same time maintaining the model integrity.
- Bottom-up approach begins by looking at existing data sources from low levels (data schema, data structure labels and etc.), developing ontologies for specific individuals, and then combining them as a whole. The process starts from the low level data sources and moves towards a higher level of abstraction. The resulting ontologies from a bottom-up approach focus on the specifications of the data sources. It captures the relationship between the data sources well, like the inconsistencies, overlapping, disjoint-ness, equivalence and so on. It works well for data exchange. When building such an ontology, too many concerns are being paid to the details of the data, such as the document structure of the data sources, the implementation details of the data and so on.

In order to take the advantages of both approaches and minimise their disadvantages, we decide to use a 'hybrid approach' to build the ontologies for disaster management. We will start by examining the underlying data and at the same time will refer to more abstract concepts. So we approach from both sides: by abstracting the low level data to higher level concepts and generalising high level concepts to lower levels. The resulting ontologies will be not only general enough to exchange domain information but will also have enough information for mapping among different data.

4.2 Ontologies architecture

Visser and Stuckenshmidt (2002) and Wache et al. (2001) identified and explained three different architectures for using ontologies. A division has been made into a global (single) ontology, peer-to-peer (multiple) ontologies and hybrid ontologies.

- In a global ontology architecture, one global ontology is used to provide a shared vocabulary for specifying the semantics. All information resources have to use this shared vocabulary or a subset of it. We can have such an ontology, only when all the different parties involved in the domain of interest have a common agreement over the domain and no conflicts within the parties in a domain.
- Peer-to-peer Ontologies are characterised by demanding a distinct ontology for each service. All transformation rules from one ontology to another need to be defined manually. As a result, the relationships between two services can be identified by considering the transformation rules between their related ontologies. Peer-to-peer ontologies make sure this is a mapping between any pair of parties, but the drawback is it results in large amount of mappings.
- Hybrid Ontologies are developed to overcome the drawbacks of the other two approaches. Each service is referenced to a local ontology. In order to make them comparable, a shared vocabulary is used to build the local ontologies. The shared vocabulary contains basic terms (primitives) of a domain. The advantage of hybrid ontology is that new sources can easily be added without the need of modification.

5. Ontologies for Disaster Management

We propose to use the hybrid ontology architecture for disaster management (Fig. 4). Due to the complexity of disaster management, we believe the use of only one global ontology is unable to model the domain well.

Peer-to-peer ontologies will definitely describe the domain, but will result in relatively man ontologies since large amounts of datasets are needed to be combined and many organisations will participate. Therefore the ontologies for disaster management have to consist of data ontologies and organisational ontologies.

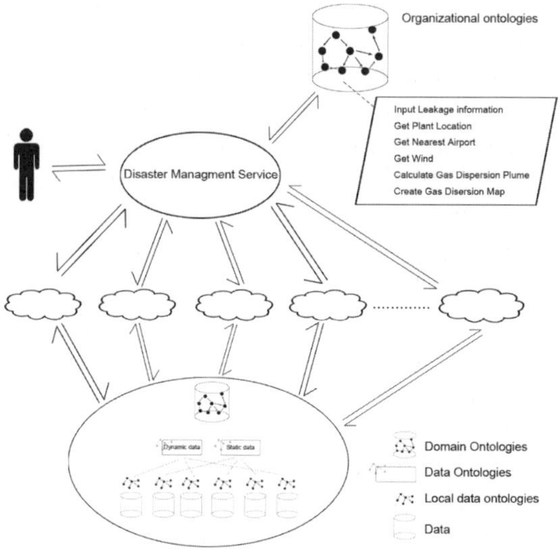

Fig. 4 Hybrid ontology architecture

Data ontologies are those that describe the datasets (e.g., topological datasets, utility datasets, cadastre datasets and so on) that are needed for disaster management. They are independent from the domain of disaster management, because all these datasets could be used for other domains, like, cadastral domain, environmental domain and so on. With data ontologies, one data set can be mapped with another (e.g., from cadastre data to utility data) or several datasets can be combined (e.g., the combination of cadastre data, utility data and topological data). In disaster management, we have plenty of data not only from the existing databases (e.g., the plant information in Fig. 3) but also from the field (e.g., the measurement of the wind in Fig. 3) that need to be processed and combined (e.g., the gasmal in Fig. 3). For each of these data, we will build a local data ontology describing the content of the data. The local data ontologies are defined together with corresponding vocabularies. In order to have grounding for all the data that are needed for disaster management, a controlled vocabulary will be made and a (possible) relation will be put on the controlled vocabulary. We give the controlled vocabulary plus the relation on it a name: upper level data ontologies. We will also build data mapping ontologies between each of the local data ontologies to the upper level data ontology. With the data mapping ontologies, each of the local data ontologies will be mapped into the upper data ontology, so that there is grounding for all the different datasets.

Organisational ontologies are those that describe the structure of organisation—how the organisation is structured, what are the responsibilities of each user within this organisation, how the users communicate with each other, how the work is carried out within the organisation, which user needs what data and so on. Organisational ontologies are dependent on the domain. For instance, the medical service will get involved in disaster management. Besides that, the medical service has its own daily task, such as sending an ambulance to do first aid rescue. As another example, let's consider the fire brigade. Apart from taking part in disaster management, the fire brigade will also do their daily work, e.g., giving advice on the storage of dangerous goods. During disaster management, people from these two organisations together with others have to cooperate with each other and work as a whole organisation (organisation of disaster management). We will examine through the process (there 25 to 29 processes in the Netherlands) and GRIP (GRIP 0 to GRIP 5) levels to work out the relations among different people in disaster management. In order to make it easier for people with different backgrounds to cooperate for disaster management, a controlled vocabulary will be worked out to help ease semantic interoperability. For instance, a vehicle might mean an ambulance to the medical service but might indicate a wagon to the fire brigade.

Generally speaking, data ontologies serve as the glue integrating different datasets. It helps disaster management to discover and combine datasets. Organisation ontologies consist of the domain knowledge of disaster management — who participates in disaster management (e.g., the fire brigade, the police, the medical service, the municipality, the Ministry, the Office of the Queen and etc.), what their task and roles are (e.g., a user in case of GRIP level 1 with process 9 needs to do measurement), what information they need (e.g., a user in case of GRIP level 1 and process 9 needs measurements of the gas from the field), how they communicate with each other (e.g., user A should send the measurement to user B after the process 9), how to cooperate to achieve disaster management (e.g., in case of a disaster at GRIP level 1, process 3 and process 4 needs to be activated and so on), and so on. The ontologies for disaster management consist of the two types of ontologies. It hides the integration of data and parts of knowledge of disaster management from the user. Any application that uses the ontologies for disaster management will maximally allow machine automation for disaster management.

As shown by the example in Fig. 2, a fire at a chemical plant has been reported to the emergency centre. The responder decides the scale of the incident according to the report and input the request in to the system. The system will make use of organisational ontologies in the database and a series of actions will be activated. The corresponding services will be invoked. With the help of data ontologies, Heterogeneous data sources will be integrated and serve as a whole database to the service. Fig. 4 illustrates the idea of the system.

6. Outlook

We have already perceived ontology architecture. Keeping this in mind, the next step will be looking at the real datasets (could be many) and building the local ontologies. We will try to analyse the relationships between the data, like super or sub relationship, equivalent, complement, disjoint and etc. The direct result of this work will be a kind a core model. At the same time, we will investigate different users and their 'language' and try to find a common vocabulary, in such a way that different terms used by different users could be mapped into the common vocabulary, i.e. we will build the domain ontology Using the definition of processes (and thus knowing the user's roles), the needs for data can be identified. Last step will be relating the users' needs and the existing data (that could be static and dynamic). Using this approach we believe we could help in automating the procedures of finding data and integrating them to answer a specific question.

Acknowledgements

The authors express their gratitude to Marian de Vries, Peter van Oosterom, Andrew Lee, and Zhuo Wang for their kind suggestions and remarks.

References

Audi R (1995) The Cambridge Dictionary of Philosophy. Cambridge: Cambridge University Press.

Bernard L (2002) Experiences from an implementation testbed to set up an international SDI. M. Ruiz, M. Gould, and J. Ramon (Eds.), 5th AGILE Conference on Geographic Information Science 2002, Palma de Mallorca, pp. 315-321.

Diehl S and van der Heide J (2005) Geo Information Breaks through sector shink, in: PJM van Oosterom, S Zlatanova & EM Fendel (Eds.), Geo-information for disaster management, Springer Verlag, Heidelberg, pp. 85-108

Diehl S, Neuvel J, Zlatanova S and Scholten H (2006) Investigation of user requirements in the emergency responce sector: the Dutch case, 2nd Gi4DM, 25-26 September, Goa, India, CD-ROM, 6p.

Groot R and McLaughin J (2000) Geospatial Data Infrastructure— Concepts, Cases, and Good Practice. Oxford University Press: Oxford, 2000.

GRIP Animation. http://www.handboekrampenbestrijding.nl/ accessed in December 2006.

Gruber TR (1995) Toward principles for the design of ontologies used for knowledge sharing. International Journal of HumanComputer Studies, 43, pp. 907928.

Klien E, Einspanier U, Lutz M, and Hbner S (2004) An architecture for ontology-based discovery and retrieval of geographic information. 7th Conference on Geographic Information Science (AGILE 2004), pp. 179188 Heraklion, Greece. Heraklion, Greece: Crete University Press.

Lutz M (2006) Ontology-Based Descriptions for Semantic Discovery and Composition of Geoprocessing Services. http://ifgi.uni-muenster.de/lutzm/ accessed in December 2006

Lutz M, Klien E (2006) Ontology-based retrieval of geographic information. International Journal of Geographical Information Science. Vol. 20, No. 3, March 2006, 233-260.

Neches R, Fikes R, Finin T, Gruber T, Senator T, and Swartout W (1991). Enabling technology for knowledge sharing. AI Magazine, 12, pp. 3656.

Ontology, Wikepedia, 2006. http://en.wikipedia.org/wiki/Ontology accessed in November, 2006.

Riecken J, Bernard L, Portele C, and Remke A (2003) North-rhine westphalia: Building a regional sdi in a cross-border environment/ad-hoc integration of SDIs: Lessons learnt. 9th EC-GI and GIS WorkshopESDI: Serving the User, A Corua: Spain, 2003.

Sheth A (1999) Changing focus on interoperability in information systems: from system, syntax, structure to semantics. In M.F. Goodchild, M. Egenhofer, R. Fegeas and C.A. Kottman (Eds), Interoperating Geographic Information Systems, pp. 530 (Dordrecht, Netherlands: Kluwer Academic).

SDI cookbook, (2004) Developing Spatial Data Infrastructure, The SDI Cookbook. http://www.gsdi.org/docs2004/Cookbook/cookbookV2.0.pdf accessed in December, 2006.

Studer R, Benjamins R,and Fensel D (1998) Knowledge engineering: principles and methods. Data and Knowledge Engineering, 25, pp. 161197.

Uschold M and Gruninger M (1996) Ontologies: principles, methods and applications. Knowledge Engineering Review, 11, pp. 93155.

van Borkulo E, Barbosa V, Dilo A, Zlatanova S, and Scholten H (2006) Services for emergency response systems in the Netherlands, accepted for Gi4DM, 25-26 September, Goa, India, CD-ROM, 6 p.

van Borkulo E, Scholten van HJ, Zlatanova S, and van den Brink A (2005) Decision making in response and relief phases, in: PJM van Oosterom, S Zlatanova & EM Fendel (Eds.), Geo-information for disaster management - late papers, pp. 47-54.

Visser U, and Stuckenshmidt H (2002) Interoperability in GISenabling technologies. Proceedings of the 5th AGILE Conference on Geographic Information Science, Palma de Mallorca, Spain.

Wache H, Vgele T, Visser U, Stuckenschmidt H, Schuster G, Neumann H, and Hbner S (2001) Ontology-based integration of information survey of existing approaches. IJCAI-01 Workshop: Ontologies and Information Sharing, pp. 108117 Seattle, WA.

Zlatanova S, Holweg D, and Stratakis M (2007) Framework for multi-risk emergency response, In: CV Tao and J Li (Eds.), Advances in Mobile Mapping Technology, ISPRS Book Series, Taylor & Francis, London, pp. 159-171.

Mapping between dynamic ontologies in support of geospatial data integration for disaster management

Mohamed Bakillah[1], Mir Abolfazl Mostafavi[2], Jean Brodeur[3] and Yvan Bédard[4]

Centre de Recherche en Géomatique, 0611 Pavillon Casault
Université Laval, Québec, Canada, G1K 7P4
[1]Mohamed.bakillah.1@ulaval.ca
[2]Mir-Abolfazl Mostafavi @scg.ulaval.ca
[3]Brodeur@nrcan.gc.ca
[4]Yvan Bédard @scg.ulaval.ca

Abstract

The effective management of disasters requires providing relevant and right information to the concerned decision makers. By its nature, disaster management involves multiple actors and organizations, potentially implying a significant volume of geospatial data coming from heterogeneous and autonomous geospatial data sources. Integration of these data sources is difficult not only because of the semantic heterogeneity of data but also because of the dynamic nature of the reality that is studied. The dynamic aspect of the reality has a direct impact in the conceptualisation of such a reality by adding different event categories to the domain ontology, thus making more complicated to apply existing methods for the mapping and integration of ontologies. In this article, we highlight some problems of heterogeneity that complicate the integration of ontologies composed of objects and events concepts; we also propose a similarity model designed to support mapping of these ontologies.

of the data. However, ontologies are also heterogeneous, since they often differ according to their level of abstraction, their terminology, their structure, the definition of concepts, etc. In this case, the semantic integration of ontologies is a necessary condition to semantic interoperability (Klein 2001). The goal of semantic interoperability is to create a semantically compatible information environment based on the similar concepts between different domains (Park and Ram 2004). The integration of ontologies, that aim at reconciliating two or more heterogeneous ontologies, can be carried out by mapping, alignment or fusion of ontologies, these processes represent increasing degrees of integration of ontologies. The mapping of ontologies consists in identifying a formal expression which describes the semantic relation between two entities belonging to different ontologies (Bouquet and al., 2005). Consequently, the mapping of ontologies is closely related to the concept of semantic similarity. Many approaches for mapping between ontologies have proposed semantic similarity models to relate concepts. Models of similarity are based on different aspects of concepts being compared; they can compare similarity between names of concepts or between their descriptions with metrics such as *edit distance* (Giunchiglia and Yatskevich 2004). The semantic similarity can also be evaluated by comparing the common and exclusive properties of concepts according to the ratio model (Rodriguez and Egenhofer 2003), or using graph-based techniques, which consists in comparing the positions of entities in their respective graphs (Rada and al. 1989; Madhavan and al. 2001) or similar relations between concepts (Maedche and Staab 2002). Methods of mapping combining several aspects of concepts, such as properties, neighbourhood and position of concepts in the graph of ontologies, are also proposed for the evolution problem to relate schemas of multidimensional geospatial databases produced at different time (Bakillah et al. 2006). Semantic relations between concepts can also be established by means of geosemantic proximity predicates (Brodeur 2004), in order to identify the nature of the relation between geo-concepts, such as overlap, inclusion, equivalence, etc., by analogy with the topological relations identified by Egenhofer (1994). Qualitative relation between concepts such as identified in this last approach is very important to established mapping relation since it convey more semantic meaning than a quantitative relation that only gives the degree of semantic similarity but does not inform on the nature of the relation. However, much of the recent research in ontology mapping presented until now are focussed on essentially static ontologies. Reality described by geospatial phenomena such as in a natural disaster evolves in time and space. Consequently, besides semantic aspects, in disaster management, spatiotemporal properties of phenomena represented in ontologies are of major importance since it is recognized that geospatial data, which are described by thematic, spatial and temporal properties, are necessary in order to achieve efficient management of disasters (Cutter at al., 2003).

2.2 Spatiotemporal Properties of Geospatial Phenomena

If analysis of spatial data is often based on quantitative measurements, it is often needed to reason on semantic of geospatial data to help in geographical knowledge discovery. Semantic of geospatial data includes topological relations among geospatial objects, which can be displayed by the 9-intersection model (Egenhofer, 1994). Semantic of geospatial data also includes qualitative proximity relations referring to distance among geospatial objects. Reality described by geospatial phenomena constantly evolves in time. In order to capture this evolution, until now, succession of states of the reality in time was usually represented in GIS by snapshots of this reality (Peuquet, 2001). For example, temporal changes in a city were described as a collection of maps representing the same city at different times. Thus, the underlying temporal representation was a discrete one, and the main entity to be represented was the geospatial object. This discrete representation of different states of the reality (the set of geospatial objects) makes it difficult to represent changes between the snapshots. Thus, the contemporary research focuses on more realistic representation of changes that could allow dynamic analysis of spatial data. So one of the issues is how to develop appropriate spatiotemporal data models that could be able to represent those changes (Tryfona and Jensen 1999). A key idea is to understand that entities that compose reality differ according to their temporal mode of existence (Grenon and Smith, 2004). In the first temporal mode of existence, entities have a continuous existence. These entities preserve their identity over time even though they encounter various changes. Examples of these entities are geographic object such as countries, mountains, buildings, etc. In the second temporal mode of existence, entities are occurrents and they unfold trough a period of time. Such entities represent changes and are more often termed as processes or events: passage of a storm, crossing of a border, destruction of a building, etc. This vision of the reality results in extended kinds of ontologies including events. Object-Event models such as GEM, Geospatial Event Model proposed by Worboys and Hornsby (2004) are based on this underlying ontological theory where entities of the world may be seen either as object or as events. The Geospatial Event Model extends the traditional object-based geospatial models by representing classes of geo-objects and geo-events, each of which may have attributes, relations to other geo-objects or geo-events. For example, the *plane_take_off* event may have attribute *depart_time* and the relation *take_place_on* with the object *runway*. Now modeling events also brings the question of what type of relation may involve events and objects. In the Geospatial Event Model, based on the terminology of Grenon and Smith, the following classes of relations for events and objects are described: *creation*, *sustaining in being*, *reinforcement/degradation*, *destruction*, and *splitting/merging*, so each of these classes may include different verbs describing possible relations. Finally, just as geospatial objects, events have thematic, spatial and temporal features. Events are related to a spatiotemporal setting: they may be described on a point, a line or a region of space, they may be instantaneous or they may extend on a finite interval of time. All these possibilities need to be represented in the ontology of a dynamic domain.

Objects-event models provide a foundation for modeling dynamic geospatial domains and enable to fulfill tasks such as event detection. The utility of event-based models as already been proven in domain such as representation of events in dynamic video scenes (Kojima et al. 2002; Olivier et al. 2004; Xin and Tan 2005). Semantic analysis of events in dynamic scenes, which aim at helping in visual surveillance, has essential task to detect and track moving objects, label and classify them, ascertain and represent special relationships among them or with the environment, and finally analyze and express behaviors or events in this dynamic scene (Shah 2002).

3 An integrated Approach for Reasoning with Geospatial Object and Events

In this article, we will propose an approach for reasoning with geospatial objects and events, in order to facilitate the semantic mapping between ontologies that represents geospatial object and event concepts. Reasoning with relations between geospatial object and event concepts of different ontologies can help to discover knowledge that is relevant for disaster management. The first goal of our approach is to describe by mean of examples what are the heterogeneities that may arise between ontologies of geospatial object and event concepts, which are obstacles to integration of these ontologies. The second goal is to propose an hybrid model, that is, a quantitative and qualitative similarity model, that will help to achieve mapping of concepts between these ontologies. The hybrid model and the resolution rules we will define can be used to reason with relation between object and event concepts of two ontologies.

3.1 Representation of object and event concepts

In a dynamic reality such as in natural disaster, representation of events is important, as the data sources that capture this reality such as in data captured by a video surveillance, contains objects and events at the same time. However, as geospatial objects, events may have multiple and heterogeneous representations. For example, it is well-known that geospatial objects may be represented at different levels of spatial granularity, thus giving the problem of multi-representation. The same fact may arise for events, which may be seen in different scales depending on applications. Moreover, events and objects may represent a same reality, and just as there are different methods to cope with the problem of reconciliating ontologies that represents similar geospatial objects, we need methods to help in the reconciliation of ontologies that represent geospatial objects together with geospatial events. In this section, we will describe how geospatial objects and events can be represented in our approach and we will provide a working example. Later in the next section we will present some of the heterogeneity conflict we have identified that can affect ontologies describing objects and events.

Concepts are mental constructs used to understand and communicate different ideas that help to structure the domain of discourse (Gahegan and Pike, 2006). In our approaches, concepts can be geospatial objects or geospatial events. Any concept has an associate spatiality, temporality and set of thematic features. Spatiality of a concept denotes its position in space, according to a reference, and its geometry. Types of geometries may be point, lines, polygons and volumes. Temporality of concepts is the duration of concepts in reality. Geospatial objects have a duration that can be described by a valid time interval $[t_s, t_e]$, where t_s is the time of appearance of the object in reality, and t_e is the time when the object disappears from the reality. However, geospatial event does not necessarily have a temporality that can be described by an interval, since some events may be considered as instantaneous. Consequently, types of temporality for an event can be a time point, t_{event}, describing an instantaneous event, or a time interval $[t_s, t_e]$, describing a spanning event. For example, the event *vehicle crossing a country border* may be considered as instantaneous, while the event *construction of a bridge* may have a duration of one year. Finally, semantic of concepts (objects or events) is described by their thematic features, such as color, function, material, etc. In order to facilitate discussion, we adopt the representation of concept given in Figure 1. In this figure, OT, OS, OR and OF refer respectively to object temporality, object spatiality, object relations and object features. ET, ES, ER and EF refer respectively to event temporality, event spatiality, event relations and event features.

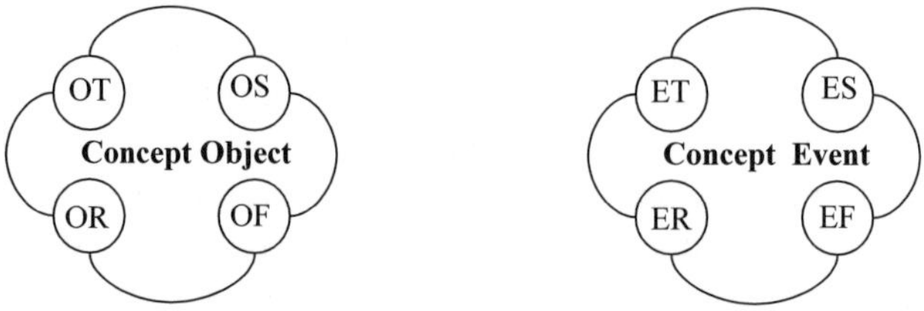

Fig.1. Geospatial Object and Event Representation

In addition, in an ontology each concept has relations with other concepts of the ontology, so the ontology can be represented as a graph with labelled nodes, which are concepts, and labelled arcs, which are relations between concepts. Relations between concepts depend on the type of concepts they involve. We distinguished four categories of relations, that is: relations between geospatial objects (OO_Rel), relations from a geospatial object to a geospatial event (OE_Rel), relations from a geospatial event to a geospatial object (EO_Rel) and relation between events (EE_Rel). These categories of relations are shown in Figure 2.

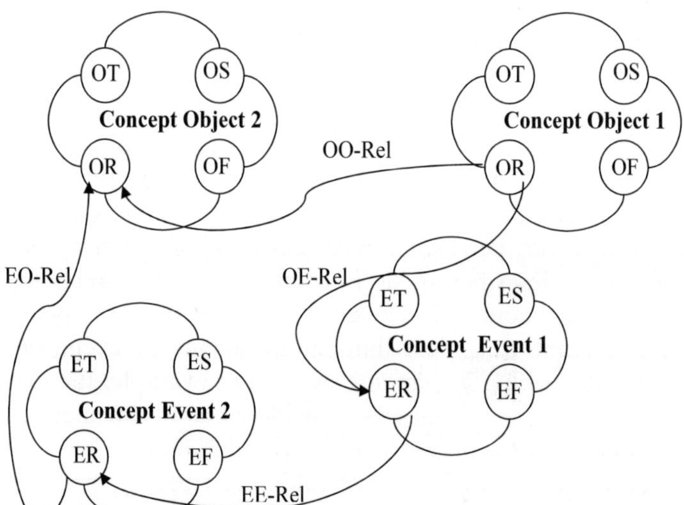

Fig.2. Categories of Geospatial Object and Event Relations in object-event ontology

In the following, we will show an example in order to present problems encountered in disaster management when having to deal with ontologies representing object and event concepts. Suppose on the one hand we have an ontology that represent the risk of fire for different places and buildings of a city. Each instance of concept (places and buildings) of this ontology have different attributes that are relevant to assess risk of fire, for example, the number of floors, the last time the building was inspected by the fire department, etc. This ontology is evolving because of the apparition of new types of buildings, new attributes that are relevant to assess risk of fire are considered (for example the attribute emergency exit), some buildings are destroyed (for example Building B), some new roads are been constructed, etc. On the other hand, consider an ontology that have represented and classified all kinds of incidents (events) that may be related to fire incidents, with their causes, the place and time where they happen, the severity of damage, etc. On Figures 3 and 4 we present a small part of the city that we will take as our working example. In this example, we also suppose that for prevention of fires, analysts in disaster management want to evaluate if the risk of fire associated to different geospatial objects such as forest cover and buildings was correctly assessed. So they need to perform a comparative analysis between geospatial objects, their attributes, and the incidents that really happened. For example, they may ask the question: does frequent inspections of buildings are related to reduced fire incidents?

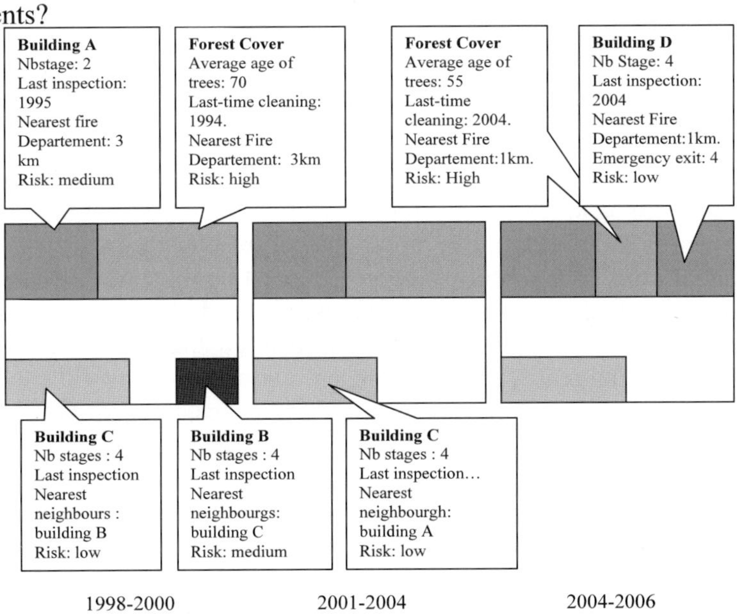

Fig.3. Working example: snapshots of reality city at different times

The problem in this scene is that ontology representing the city and the ontology representing the fire incidents use different representations of time (snapshots and events) so answering the question requires to relate both ontologies. In the following section, using this working example, we will define some heterogeneities that are obstacle for the semantic reconciliation of these ontologies.

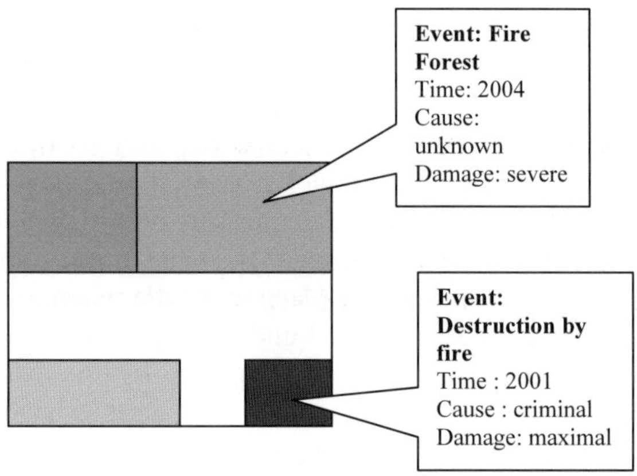

Fig.4. Working Example: events related to fire incidents for the same reality

3.2 Heterogeneity conflicts in evolving geospatial objects and events ontologies

We have seen that an event may be defined as an obvious change of important features which we are interested in (Howarth and Buxton 2000). This suggests that an event may be represented with a different manner in different ontologies. In this section we will describe some of the heterogeneity conflicts that may arise when comparing ontologies which contain geospatial object and event concepts. The first conflict we describe is the conflict we called heterogeneity of temporal paradigm. This heterogeneity results from the different temporal point of view that can be taken to describe a changing reality. On one side, entities of the real world may be abstracted as object concept, for example the concept *speed*. On the other side, the same reality depicted by the concept *speed* may be described by the event concept *displacement of an object*. In our framework, we will call this kind of divergence between these ways of describing temporality as *heterogeneity of temporal paradigm*. This heterogeneity is specific to ontologies containing event concepts since in ontologies where concepts can only be object concepts, there is no need to choose between the different temporal paradigms.

Conflicts caused by heterogeneity of temporal paradigms

These conflicts arise when a same concept is represented using different temporal paradigms in distinct ontologies:

1. A concept is represented as a geospatial object concept.
2. The same concept is represented as a geospatial event concept.
3. The concept is represented as a set of geospatial object concepts succeeding in time.

An example of this heterogeneity is that the destruction of the building B is represented by a succession of scenes of 1998-2000 to 2000-2004 in the ontology of geospatial objects and it is represented by the event: destruction by fire of building B in the incident ontology. So this example show that an event can be represented explicitely by a geospatial event concept, or that it can be represented implicitely by a serie of states for the same object at different times.

Conflicts caused by temporal heterogeneity.

Another type of conflict is the conflict caused by temporal heterogeneity. Temporal heterogeneity arises when a same event concept has different temporal types, i.e. a time interval and a time point in different ontologies. For example, in airport ontology, the event of plane take-off may be considered as an instantaneous event (by referring to the moment to plane get off the runway) or as a spanning event (if length of time between the moment the plane starts to roll and the moment where it takes off is important to consider). In our fire incident example, temporal heterogeneity between events may be that a same event *destruction by fire* is described by a temporal point (2001) or more precisely by a time interval [from 2001/12/20 13:40; to 2001/12/20 23:12].

Conflicts caused by spatial heterogeneity

Spatial heterogeneity between multiple representations of geospatial objects is already a recognized problem, and an obvious example is that at different scales a city may be seen as a point on a country map or as a polygon on a regional map. In the same way, events may be seen spatially at different scales, for example in meteorology the passage of a depression may be seen as lines (pressure iso) or as a region affected by this depression. In our fire incident example, spatial heterogeneity between events may be that the event *destruction by fire* can be associated to a building or, in an ontology requiring more precision in the representation of this event, to only a certain part of the building that was affected.

Conflicts caused by semantic heterogeneity

Conflicts of semantic heterogeneity affect both object concepts and event concepts. They include different names for the same concept (object or event), or different definitions to describe the same concept. Different thematic features may be used to describe a same concept, for example, the event *destruction by fire* can have attributes *time of event*, *cause*, and *degree of damage*, while in another ontology the same event destruction by fire can have attributes *time of event*, *nearest fire department*, *intensity of fire*. A concept can have different relations with surrounding concepts in different ontologies. This includes concepts having relations with different concepts or concepts having relation with the same concept but where the nature of the relation is different.

In order to cope with the different conflicts identified in this section, we propose in the next section a hybrid model for similarity mapping that will help to relate object and event concepts of different ontologies.

4 A Hybrid Similarity Mapping Model for relating object and events concepts

The conflicts due to heterogeneities described in the previous section need to be addressed in order to make it possible to share knowledge between different ontologies. One way of achieving this task is to identify the relation between concepts of distinct ontologies with a model of similarity. It can be seen that a model being designed to relate concepts of ontologies describing geospatial object concepts and geospatial event concepts must consider the following cases:

- an object concept of a first ontology can be related to an object concept of a second ontology
- an object concept of a first ontology can be related to an event concept of a second ontology
- an event concept of a first ontology can be related to an object concept of a second ontology
- an event concept of a first ontology can be related to an event concept of a second ontology.

To obtain relevant information in order to map geospatial object or event concepts, we argue that both qualitative and quantitative aspects of similarity between concepts are relevant and, therefore, have to be considered. Qualitative aspect gives information about the nature of the relation between concepts, for example if the concept is more general than the other one or if they are only overlapping. Quantitative aspect gives information about the degree of similarity between the concepts. Both approaches are necessary in order to identify the most coherent relation between concepts with respect to reality and with respect to the user need.

Consequently, in our approach we propose to use a model that gives both qualitative and quantitative output. We also consider that a complete model should consider all the aspects of definition of concepts, that is, semantic similarity through thematic features, spatial features and temporal features and finally relations between concepts. In the following, we will first give definition for our framework and define what constitute the hybrid model of similarity mapping, then we will show how it can be used with resolution rules and mapping strategies to map between event-object ontologies.

SpatioTemporal Object-Event Ontology

A spatiotemporal object-event ontology O^i is defined by $O^i=(C, R)$ where $C=\{c_i, i=1,2,...\}$ is the set of spatiotemporal concepts of the ontology, and R is the set of relations $r=(c_i, c_j, type\_rel)$ between concepts of the ontology.

Concepts of SpatioTemporal Object-Event Ontology

Set of concepts of the SpatioTemporal Object-Event Ontology can be of two categories: Object concepts and Event concepts, in order to represent classes of geospatial objects and geospatial events of the real world. Object concepts are labelled OC and Event concepts are labelled EC. Each concept c is defined by a set of properties which are further refined in the following categories: thematic properties F, temporal properties T, spatial properties S and set of relations R with other concepts of the ontology (Figure 5 and 6).

Relations between concepts of SpatioTemporal Object-Event Ontology Relations between concepts may be of four different categories, depending on categories of concepts being involved: relations from object concept to object concept (OO_Rel), relations from object concept to event concept (OE_Rel) (Figure 5), relations from event concept to object concept (EO_Rel), relations from event concept to event concept (EE_Rel) (Figure 6). Those relations proposed in our framework are general and in further works they could be further refined in more specific relation. However, as we would consider more specific relations between event and object concepts, reasoning on these relations would be hard unless we introduce an ontology of event that could classify these relation in order to allow inferencing between them.

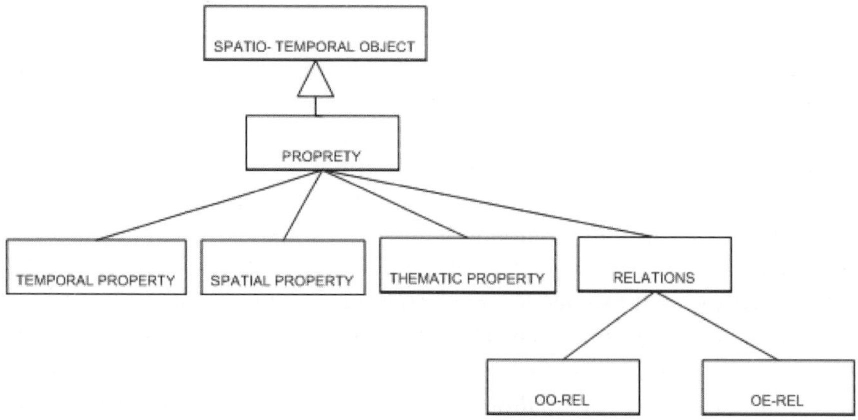

Fig.5. Description of object concepts of SpatioTemporal Object-Event Ontology

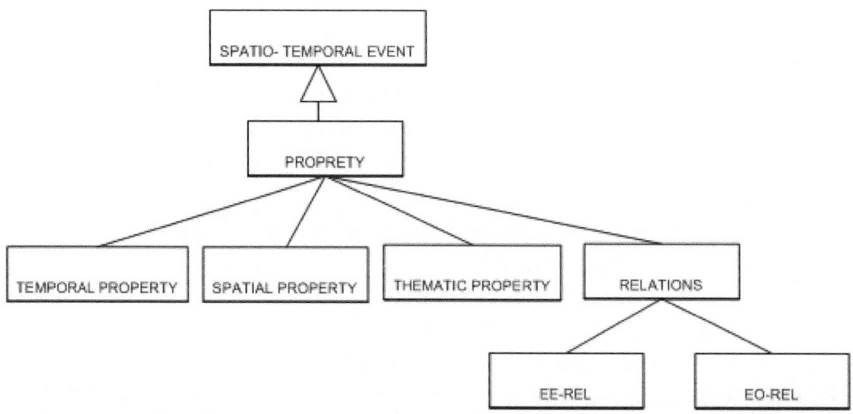

Fig.6. Description of event concepts of SpatioTemporal Object-Event Ontology

Semantic Similarity between concepts of SpatioTemporal Object-Event Ontologies

Semantic similarity between concepts of the ontologies depends on the type of concept being compared, either object concepts or event concept, since object concepts don't share the same relation with object concepts than with event concept, and reciprocally. For the development of our semantic similarity model between object concepts and event concepts, we adopted a similar representation of concepts proposed in the geosemantic proximity model (Brodeur 2004), which described concept as set of intrinsic and extrinsic properties, and adapted it to describe similarity between event concepts. This leads us to identifying new relations that can be established between event concepts, or object and event concepts. In the model we propose in this article, semantic similarity is a function of similarity between thematic properties and relationships properties of concepts that are compared. By considering that thematic properties of a concept are analogous to internal part of concept and relationship properties are analogous to boundary between the concepts and the surrounding concepts of ontology, semantic relation can be analogous to topological relation between regions as described by Egenhofer (1994). The set of thematic properties is labelled OF for object concept and EF for event concept, while the set of relationship properties is labelled OR for object concept and ER for event concept. The subscript i for example in OF_i indicate set of thematic properties for object concept c_i.

Event-to-Event semantic similarity

The model of semantic similarity between two event concepts E1 and E2 is a matrix which compares their EF and ER sets of properties:

$$S(E1, E2) = \begin{pmatrix} EF_1 \cap EF_2 & EF_1 \cap ER_2 \\ ER_1 \cap EF_2 & ER_1 \cap ER_2 \end{pmatrix} \tag{1}$$

Semantic similarity between events is a qualitative relation that depends on the state of this matrix, that is on whether the four intersection sets it contains are empty or not. We denote by $\varnothing$ an empty intersection, by $\neg\varnothing$ a non-empty intersection and by $\equiv$ an intersection that is equivalent to union of the set of properties. An empty intersection $\varnothing$ express that there is no common properties between the compared sets and a non empty intersection $\neg\varnothing$ express that there is at least one common property between the compared sets. An intersection that is equivalent to union $\equiv$ indicates that all properties of set being compared are common, in other word, sets are equals. Thus, we identified that semantic similarity between events can verify one of the relations on Figure 7.

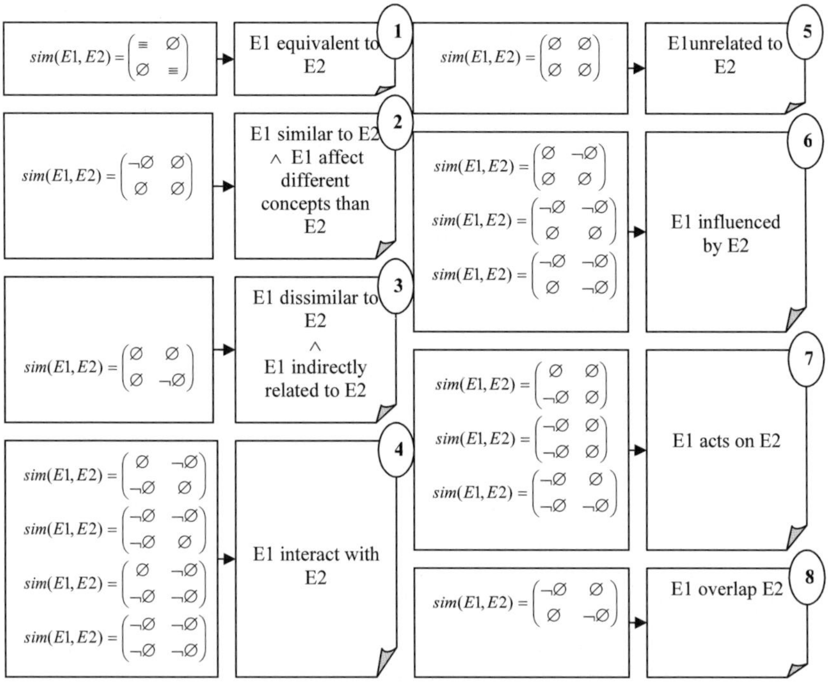

Fig.7. Relations of semantic similarity between two events E1 and E2

The semantic similarity relations expressed on Figure 7 allow relating events from different ontologies. In the case 1, E1 and E2 have the same sets of thematic and relationship properties so they are equivalent. In the second case, E1 and E2 share some thematic properties but they affect different objects and they may be related to different events. The anti-symmetrical case is the third one where event have no thematic properties in common but they share relationship so they may affect the same objects concepts or be related to similar events. Consequently in the third case both events E1 and E2 may be indirectly related by mean of a third concept. In the case 5, E1 and E2 share no thematic properties and no relationship so they are not related. In the case 6, thematic properties of E1 are common with relationship of E2, meaning that E1 is partly determined by one ore more relationship of E2. Consequently, E1 is influenced by E2. The case 7 is the opposite of case 6 where relationships of event E1 are common with thematic properties of event E2, then E1 acts on E2. Case 4 combines cases where E1 is influenced by E2 and E1 acts on E2, so we define this relation as E1 interact with E2. Finally, in the case 8, we give the relation where event concepts E1 and E2 may overlap because they share some thematic and relationship properties to a certain degree.

Event-to-Object semantic similarity

The model of semantic similarity between event concept E1 and object concept O1 is a matrix which compares their EF, OF, ER and OR sets of properties:

$$S(E1, O1) = \begin{pmatrix} EF_1 \cap OF_1 & EF_1 \cap OR_1 \\ ER_1 \cap OF_1 & ER_1 \cap OR_1 \end{pmatrix} \tag{2}$$

$$S(O1, E1) = \begin{pmatrix} OF_1 \cap EF_1 & OF_1 \cap ER_1 \\ OR_1 \cap EF_1 & OR_1 \cap ER_1 \end{pmatrix} \tag{3}$$

In the same way as for semantic similarity between event concepts, semantic similarity between event concept and object concept depends on the state of this matrix, and gives relations shown on Figure 7 (replacing E2 with O1), since object and event concepts may also be related by relations Acts on, Influenced by, Interact with and may overlap if an object concept is represented by an event concept in the other ontology, or inversely.

Object-to-Object semantic similarity

The model of semantic similarity between event concept O1 and object concept O1 is a matrix which compares their OF and OR sets of properties; this matrix verifies one of the cases presented in Figure 7:

$$S(O1, O2) = \begin{pmatrix} OF_1 \cap OF_2 & OF_1 \cap OR_2 \\ OR_1 \cap OF_2 & OR_1 \cap OR_{21} \end{pmatrix} \tag{4}$$

Now we proposed that the overlap sets contained in the matrix of equations (1), to (4) can be evaluated quantitatively in order to give the degree of semantic similarity between concepts being compared. Quantitative degree of similarity is required in order to determine if the semantic similarity relation that has being identified by semantic similarity matrix from equation (1) to (4) are significant. For example, if degree of overlap between properties is very low, it would be wrong to consider that two concepts, for example, interact with each other. Degree of overlap, denoted by $D=D(c_1, c_2)$, compared the number of common (thematic or relationship) properties to the number of (thematic or relationship properties) of the first concept c_1; it always gives a value between 0 for disjointness and 1 for equivalence relations, and is of the general form:

$$D(f, semantic, c_1, c_2) = \frac{f(c_1) \cap f(c_2)}{f(c_1)} \quad (5)$$

where *f* is a set of (thematic or relationship) properties. Degree of overlap is computed for each intersection set of the matrix (four intersection sets), and to obtain a global semantic similarity D(semantic, c_1, c_2) which range between 0 and 1, we compute the normalized sum of these four overlaps. Semantic similarity value is also asymmetrical since qualitative relations are asymmetrical. Now we also consider degree of similarity between temporal and spatial properties by considering the overlap of temporal attributes and spatial attributes of concepts, D(temporal, c_1, c_2) and D(spatial, c_1, c_2). The degree of spatial and temporal overlap can be computed for instances of concepts, which have values for spatial and temporal attribute. Degree of spatial and temporal overlap can be computed by adapting the general form of overlap given by eq. (5). For concepts that have no values for spatial and temporal attributes, spatial and temporal primitives can be considered as single properties. In order to give a global value of similarity, that is global similarity between two concepts, we define a three dimensional similarity space composed of the semantic similarity, temporal similarity and spatial similarity axis. In this similarity space we can define the vector of similarity V(c_1, c_2) :

$$V(c_1, c_2) = \left(D(semantic) \quad D(temporal) \quad D(spatial)\right) \quad (6)$$

Global similarity between concepts is related to the length of this vector in the similarity space:

$$Global\_Sim(c_1, c_2) = \sqrt{\frac{1}{3}\sum_{i=1}^{3}(D(f_i, c_1, c_2))^2} \quad (7)$$

Results of assessing similarity in a qualitative and quantitative way are the first steps toward mapping the object-event ontologies. For reasoning on the obtained results, we need to define a way to analyse this result in order to decide if two concept are to be mapped or not and which relation is going to hold between them. For this purpose we propose to define some example of conflict resolution rules to resolve some of the conflicts given in section 3.2 and example of mapping strategies to handle the mapping process. Resolution rules are logics to follow for a given state of the relation between concepts. These resolution rules and strategies are shown on table 1.

Table 1: Resolution rules and mapping strategies

Identified relation	Conflict to resolve	Resolution Rules	Mapping strategy
global_sim (O1, E1)=1	Temporal paradigm heterogeneity : object and event may refer to same reality	O1 and E1 are same entity	Establish a synonymy relation between O1 and E1
global sim (O1, E1) between 0 and 1 exclusively	Temporal paradigm heterogeneity: object and event may refer to a similar reality but O1 and E1 shows either semantic,spatial and/or temporal heterogeneity	1) If O1 and E1 are semantically unrelated O1 cannot be similar to event E1 even if they share spatial or temporal properties 2) If O1 and E1 are semantically related , but temporally or spatially disjoint, O1 cannot be equal to E1 because they don't overlap in space and time 3) If O1 and E1 are semantically related, and temporally /spatially non disjoint, O1 can be related to E1	1)Establish no direct relation between O1 and E1 2) Establish no direct relation between O1 and E1 3)Establish semantic, temporal and spatial relation between O1 and E1 4) Establish no relation between O1 and E1
global sim (O1, E1)=0	O1 and E1 represent different reality	O1 and E1 are not related	Establish no direct relation between O1 and E1
global sim(O1, O2)=1 ∧ r=(O2, E2, OE_Rel)	Finding matching objects in both ontologies to find object-event relation	Since O1 and O2 are equivalent, relation r between (O2, E2) also hold for between O1 and E2	Establish relation r between O1 and E2
global sim(O1, O2) between 0 and 1 exclusively and there is relation r between (O2, E2)	Finding matching objects in both ontologies to find object-event relation	Since O1 and O2 are related, relation r between (O2, E2) is used to find relation between O1 and E2. For example if O1 and O2 are related by O1 is specialisation of O2 and O2 participate to event E2, we can deduce O1 participate to E1.	Deduce relation between (O1, E1) from relation between O2 and E1.
global sim(O1, O1)=0	O1 and O1 represent different reality	O1 and O2 are not related	Establish no direct relation between O1 and O2

With these resolution rules and mapping strategies, we underline some general rules that can be used to map event and object concepts of ontologies. In application such as disaster management, this can help for example to discover if some geospatial objects were implied in an event represented in another ontology. The following example illustrates the results of the proposed similarity model for mapping event and object concepts from two ontologies, the first one containing *Geographical concepts* and the second containing *Incidents* (event concept).

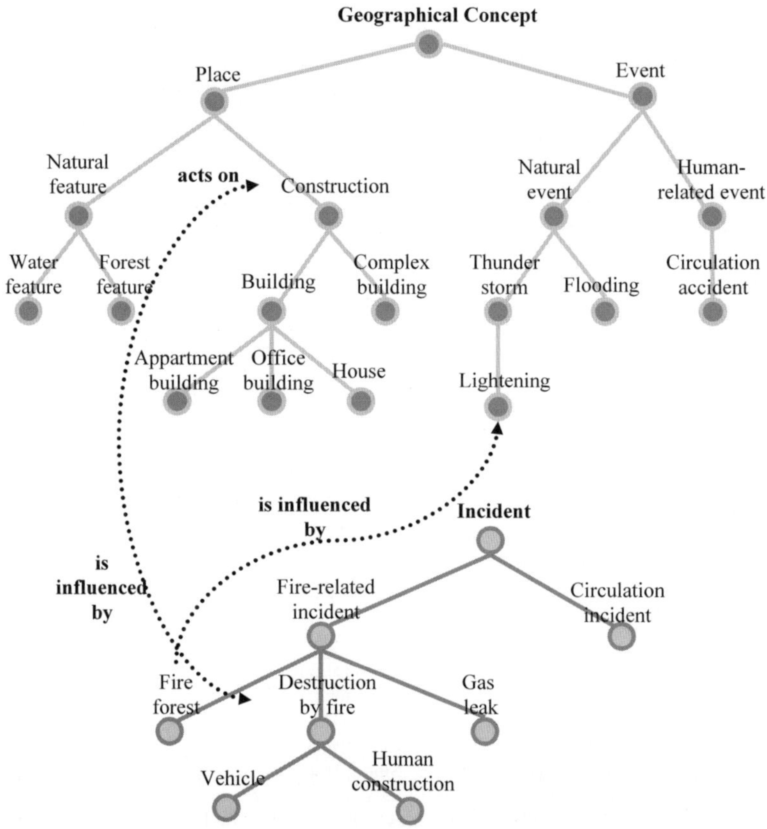

Fig.8. Example of semantic mapping between object and event concepts

In this example, we consider first the comparison of two event concepts *Fire forest* and *Lightening*. The event concept *Fire forest* has thematic property *cause = {human negligence, thunder storm}*, temporal property *time interval* and spatial property *region* and *EE_Rel={is_a Fire-related incident}*. The event concept *Lightening* has thematic property *force = {strong, weak}*, temporal property *time interval*, spatial property *region* and *EE_Rel={is_a ThunderStorm}*. Comparing these event concepts give a semantic similarity and a global similarity:

$$S(Fire\_forest, Lightening) = \begin{pmatrix} \varnothing & \neg\varnothing \\ \varnothing & \varnothing \end{pmatrix}; \text{global sim= 0,76} \quad (8)$$

Thus, according to Figure 7, *Fire forest* event concept *is influenced by* the event concept *Lightening*. Global similarity is an important feature of similarity because it gives a weight to the relation *is influenced by*. For a very weak similarity, we may not consider that a mapping relation should be established between concepts even if the semantic similarity give the qualitative relation *is influenced by*, or any other relation.

As a second example, we consider the comparison of event concept *Destruction by fire* and object concept *Construction*. The object concept *Construction* has thematic property *{Nb_of_stages, Date_of_Last_ inspection, Nearest fire Departement, Risk}*, temporal property *time interval* and spatial property *region* and *OO_Rel={is_a Place, generalises Building, generalises Complex building}*. The event concept *Destruction by fire* has thematic property {*Cause, Damage, Nearest fire Departement}*, temporal property *time point* and spatial property *region* and *EE_Rel={is_a Fire-related Incident}* and EO_Rel={*act_on Vehicle, act_on Human construction*}. Comparing these event and object concepts give a semantic similarity and a global similarity:

$$S(Construction, Destruction\_by\_fire) = \begin{pmatrix} \neg\varnothing & \neg\varnothing \\ \varnothing & \varnothing \end{pmatrix}; \text{global sim= 0,57} \quad (9)$$

Thus, according to Figure 7, *Construction* object concept *is influenced by* the event concept *Destruction by fire*. Conversely, if we evaluate the inverse relation we have:

$$S(Destruction\_by\_fire, Construction) = \begin{pmatrix} \neg\varnothing & \varnothing \\ \neg\varnothing & \varnothing \end{pmatrix}; \text{global sim= 0,61} \quad (10)$$

That is, we find the relation destruction by fire acts on construction, which is the reciprocal relation of Construction is influenced by Destruction by fire. Global similarities are different because the assessment of overlap is asymmetric, and the concept *Destruction by fire* has less distinctive properties than the concept *Construction.*

In our working example, with this similarity model we could find which buildings and place were implied in fire incident and thus analyse attributes that were not necessarily represented in the fire incident ontology. For example, if we ask the following query: what are the constructions having low risk of fire that were implicated in the event destruction by fire? The mapping relation established between construction from the ontology of geographical concepts and destruction by fire from the ontology of incidents can be used to answer this type of query.

The rules we have underline are general but we can fix explicitly a whole set of resolution rules in a given domain with a known set or possible relations from objects to events in order to automate as much as possible the mapping process. Results may be even more concluding using a knowledge base to reason with different categories of relations that can be defined depending on context. In future work, we will define further this mapping model with the development of a function that considers interaction between thematic, spatial and temporal properties in order to consider more relations between object and event concepts.

5 Conclusion and Future Works

In this paper, we discussed the problems of heterogeneity that can be encountered when we want to share knowledge between ontologies representing geospatial objects and geospatial events in disaster management applications. We have proposed an hybrid similarity model that considers both qualitative and quantitative approaches for the similarity measurements between object and event concepts in different ontologies, and that can help to resolve some of the described heterogeneities, and help in discovering new relations between event and object concepts. Based on the results of this model, we have proposed some resolutions rules and mapping strategies to help in discovering relations between objects and events. This helps us to better analyse a dynamic phenomena such as in natural disaster management. This work is a part of ongoing PhD project and research will be conducted to improve the proposed approach by considering types of relations between objects and events in a real world case. In future works, we attempt to extend this work to help in achieving semantic interoperability in a dynamic network of data sources. This will help to a more effective decision making in disaster management by providing solutions for sharing relevant geospatial data between multiples data sources.

Acknowledgment

This research was made possible by an operating grant from Natural Sciences and Engineering Research Council of Canada (NSERC).

References

Arpinar B, Sheth A, Ramakrishnan C, Usery EL, Azami M, Kwan MP (2006) Geospatial Ontology Development and Semantic Analysis. Transaction in GIS, 10(4): 551-575.

Bakillah M, Mostafavi M A, Bédard Y (2006). A Semantic Similarity Model for Mapping between evolving Geospatial Data Cubes. OTM Workshops 2006, LNCS 4278, pp.1658 – 1669.

Brodeur J (2004). Interopérabilité des Données Géospatiales: Élaboration du Concept de Proximité Géosémantique. Thèse de doctorat, Université Laval.

Chen H, Perish F, Finin T, Joshi A (2004) SOUPA: Standard Ontology for Ubiquitous and Pervasive Applications. The First Annual International Conference on Mobiles and Ubiquitous Systems: Networking and Services, pp.258-267.

Cutter SL, Richardson DB, Wilbanks, TJ (2003) The Geographic Dimension of Terrorism. New York and London: Toutledge.

Deen S M, Ponnamperuma, K (2006). Dynamic Ontology Integration in a Multi-Agents Environment. Proceedings of the 20th International Conference on Advanced Information Networking and Applications (AINA'06).

Egenhofer M, Mark D M, Herring J R (1994). The 9-Intersection: Formalism and Its Use for Natural-Language Spatial Predicates. Santa Barbara, CA, University of California, National Center for Geographic Information and Analysis Technical Report 94-1.

Flouris G, Plexousakis D, Antoniou G (2005). Evolving Ontology Evolution.

Fonseca F T, Egenhofer M J, Agouris P (2002). Using ontologies for integrated Geographic Information Systems. Transactions in GIS 6: 231–57.

Grenon P, Smith B (2004) SNAP and SPAN: Toward Dynamic Spatial Ontology. Spatial Cognition and Computation 4(1), pp.69-103.

Giunchiglia, F., Yatskevich, M., 2004. Element Level Semantic Matching. In *Proceedings of Meaning Coordination and Negotiation Workshop at International Semantic Web Conference.*

Gahegan M, Pike W (2006). A Situated Knowledge Representation of Geographical Information. Transactions in GIS 10 (5), pp. 727–749.

Helfin J, Hendler J (2000) Searching the Web with Shoe. In Artificial Intelligence for Web Search, Paper from the AAAI Workshop, pp. 35-40.

Howarth RJ, Buxton H (2000) Conceptual Descriptions from Monitoring and Watching Images Sequences. Image and Vision Computing 18: 105-135.

Kashyap V, Sheth A (2000) Information Brokering Across Heterogeneous Digital Data: A Metadata-Based Approach. The Kluwer International Series on Advances in Database Systems.

Klein M (2001) XML, RDF, and Relatives. IEEE Intelligent Systems, 15 (2): 26-28.

Kojima A, Tamura T, Fukunaga K (2002) Natural Language Description of Human Activities from Video Images Based on Concept Hierarchy of Actions. International Journal of Computer Vision, 50 (2), pp. 171-184.

Madhavan, J., Bernstein, P., Rahm, E., 2001. Generic Schema Matching with Cupid. In *Proceedings of Very Large Data Bases Conference*, pp. 49-58.

Maedche, A., Stabb, S., 2002. Measuring Similarity between Ontologies. In *Proceedings of International Conference on Knowledge Engineering and Knowledge Management*, pp.251-263.

Oliver N, Garg A, Horvitz E (2004) Layered Representations for Learning and Inferring Office Activity from Multiple Sensory Channels. Computer Vision and Image Understanding, Special Issue on Event Detection in Video, 96 (2), pp. 163-180.

Park J, Ram S (2004) Information Systems Interoperability: What Lies Beneath? ACM Transactions on Information Systems.

Peuquet DJ (2001) Making Space for Time: Issues in Space-Time Data Representation. Geoinformatica 5(1), pp.11-32.

Rada, R., Mili, H., Bicknell, E., Blettner, M., 1989. Development and Application of a Metric on Semantic Nets. *IEEE Transactions on Systems, Man and Cybernetics*, 19(1), pp.17-30.

Rodriguez, M.A., Egenhofer, M.J., 2003. Determining Semantic Similarity Among Entity Classes from Different Ontologies. *IEEE Transactions on Knowledge and Data Engineering*, 15(2), pp. 442-456.

Salton G, McGill M (1983). Introduction to Modern Information Retrieval. McGraw-Hill, New York.

Shah M (2002) Guest Introduction: The Changing Shape of Computer Vision in the Twenty-First Century. International Journal of Computer Vision, 50 (2), pp. 103-110.

Sheth A, Thacker S, Patel S (2003). Complex Relationships and Knowledge Discovery Support in the InfoQuilt System. VLDB Journal 12(1):2-27.

Tryfona N, Jensen CS (1999) Conceptual Data Modeling for Spatio-Temporal Applications. Geoinformatica 3(3), pp. 245-268.

Worboys M, Hornsby K (2004) From Object to Events: GEM, the Geospatial Event Model. GIS Science 2004: 327-344.

Xin L, Tan T (2005) Ontology-based Hierarchical Conceptual Model for Semantic Representation of Events in Dynamic Scenes. Proceedings of the 2nd Joint IEEE International Workshop on VS-PETS, Beijing, pp. 57-64.

Winter S (2000). Uncertain Topological Relation between Imprecise Regions. International Journal of Geographical Information Science, 14(5), pp. 411-430

An Open GeoSpatial Standards-Enabled Google Earth Application to Support Crisis Management

Scott Pezanowski, Brian Tomaszewski and Alan M. MacEachren

The GeoVISTA Center, Department of Geography, The Pennsylvania State University, 302 Walker Building, University Park, PA 16802 USA Ph: 814-865-3433 Fax: 814-863-7943
Email: {spezanowski, bmt139, maceachren}@psu.edu

Abstract

Google Earth (GE) and related open geospatial technologies have changed both the accessibility of and audience for geospatial information dramatically. Through data rich applications with easy to use interfaces, these technologies bring personalized geospatial information directly to the non-specialist. When coupled with open geospatial data standards, such as Web Map Services (WMS), Web Features Services (WFS), and GeoRSS, the resulting web-based technologies have the potential to assimilate heterogeneous data from distributed sources rapidly enough to support time-critical activities such as crisis response. Although the ability to view and interact with data in these environments is important, this functionality alone is not sufficient for the demands of crisis response activity. For example, GE's standard version currently lacks geoanalysis capabilities such as geographic buffering and topology functions. In this paper, we present development of the "Google Earth Dashboard" (GED), a web-based interface powered by open geospatial standards and designed for supplementing and enhancing the geospatial capabilities of GE. The GED allows users to create custom maps through WMS layer addition to GE and perform traditional GIS analysis functions. Utility of the GED is presented in a use-case scenario where GIS operations implemented to work with GE are applied to support crisis management activities. The GED represents an important first step towards combining the ubiquity of GE and geospatial standards into an easy-to-use, data rich, geo-analytically powerful environment that can support crisis management activity.

1 Introduction

Geographical information and technologies are essential to crisis management activities (National Research Council, 2007). Easy to use, data rich environments such as Google™ Earth (GE) are making geospatial technologies accessible and customizable to non-specialist end users for crisis management applications (Nourbakhsh, 2006). Furthermore, the proliferation of data availability through open geospatial standards such as Web Map Services (WMS), and Web Features Services (WFS) creates the potential for rapid assimilation of heterogeneous data sources to support disaster response (Open Geospatial Consortium, 2005). In this paper, we present our preliminary results on the "Google Earth Dashboard" (GE-Dashboard), a web-based interface to Google™ Earth powered by open GIS standards and the Adobe Flex open-source software environment. The GE-Dashboard has been designed to supplement and enhance the existing geospatial capabilities of Google™ Earth (GE) with functionality to incorporate WMS layers into GE and to perform traditional GIS spatial analysis functions. We discuss the technical approach used to create our solution, provide a use-case scenario of the GE-Dashboard, and offer avenues for further research.

2 Problem Domain

Although the ability to view data in GE is important, simple viewing functionality alone is not sufficient for the demands of time-pressured disaster response. For example, GE currently lacks geo-analysis capabilities such as geographic buffering and topology functions that may be used to support situation assessments during a disaster, such as to derive casualty counts in a region or track the population impacted by the spread of chemical plume.

Google™ Earth (GE) demonstrated its effectiveness as a geospatial disaster response tool and platform during the Hurricane Katrina crisis of 2005 (Google Earth Blog, 2005; Google, 2007) and continues to be used for a wide variety of crisis management activities. It is an easy-to-use virtual globe environment with an eye-catching design that allows users to efficiently zoom and pan around the globe and add data layers from both

Google and other groups through the Keyhole Markup Language (KML)[1]. Recognizing the effectiveness and value of the free, public version of GE to support crisis management, we sought to expand the geospatial capabilities of GE without comprising the ease of use that GE affords.

3 Solution

The solution we developed to create the GE-Dashboard consists of seven high level architectural and conceptual components. These include the Open Geospatial Consortium (OGC) Web Map Services (WMS), Web Feature Services (WFS) and Stylized Layer Descriptors (SLD) standards, the Adobe Flex web development platform, the open source Geospatial database postGIS, the open source geospatial web server GeoServer, and networking components of GE itself. In this section, we provide brief sketches of each of these components to provide context for how each of these in turn were used to create the GE-Dashboard.

3.1 OGC Web Map Services (WMS), Web Feature Services (WFS) and Styled Layer Descriptors (SLD)

The Open Geospatial Consortium, Inc. (OGC)[2] is a non-profit, international, voluntary consensus standards organization that is leading the development of standards for geospatial and location based services. Standards maintained by the OGC allow for open and extensible software application programming interfaces to be used in geographic information applications.

The Web Map Service (WMS) standard[3] allows maps of spatial data to be produced as images, typically in PNG or GIF formats. End users can view WMS maps using a simple client such as a web browser. Data that are served up as images provide the advantage that the underlying data can not be altered or stolen. However this also limits user interaction with data. The Web Feature Service (WFS) standard[4] typically produces maps of spatial data in an XML format known as the Geographic Markup Language

[1] KML is Google Earth's proprietary markup language used for adding symbols to the GE map, and performing other functions

[2] http://www.opengeospatial.org/

[3] http://www.opengeospatial.org/standards/wms

[4] http://www.opengeospatial.org/standards/wfs

(GML)[5]. Because data are being retuned in an XML format, a thicker client is needed to render the geography encoded in GML. An advantage with WFS is that users can query and style data at will and transactional WFS allow users to save data to a server. However, large volumes of WFS data can overwhelm limited bandwidth. Styled Layer Descriptor (SLD)[6] encoding is an XML-based specification for implementing user defined descriptions of how geographic data is rendered in terms of colors and symbols.

3.2 The Adobe Flex web development platform

Adobe® Flex[7] is a rich-internet application development framework. It is based on Adobe Flash®. Flex differs from Flash in that it has been designed to develop full, web-client applications and has native support for web-service consumption and database access. This differs from Flash, which is primarily used for creating animations and other vector-based graphic applications. Elegant and pleasing web applications can be quickly developed with Flex. Although maintained by a commercial software company, a free Flex SDK is available for download, thus making source code written in Flex available for open-source community use.

One drawback of using Adobe Flex is that the user needs to have the Flash Player version 9 installed. This may cause limitations such as the Adobe Flash Player not being supported on the Linux OS (GE is also not available for the Linux OS). However, the Flash player is available on 96% of all internet-capable computers, and thus makes a good choice for browser plug-in based applications (Adobe Systems, 2007).

3.3 PostGIS

PostGIS[8] is an open-source software component that "spatially" enables the object-relational postgreSQL database. For example, with postGIS, geographic objects such as lines, points, and polygons can be added to standard database tables. postGIS also provides functionality for spatially enabled SQL queries. postGIS follows the OGC's Simple Feature Specfications for SQL standard[9].

---

[5] http://www.opengis.net/gml/

[6] http://www.opengeospatial.org/standards/sld

[7] http://www.adobe.com/products/flex/

[8] http://postgis.refractions.net/

[9] http://www.opengeospatial.org/standards/sfs

3.4 GeoServer

GeoServer[10] is an open source web mapping server that allows geospatial data to be served over the web. GeoServer uses the WMS and WFS protocols for serving data in image and XML formats from a variety of sources that include shapefiles and postGIS-enabled databases.

3.5 The Google Earth Network Link Object

The Google Earth Network Link object[11] allows KML data to be accessed in GE from an outside URL or other remote location. It is a critical component to custom GE applications that need to add external data in real-time.

4 GE-Dashboard

4.1 Overview

The GE-Dashboard is a web-based interface, developed in Flex and designed to accompany GE. Since it is a web-based interface, the GE-Dashboard is loaded into the built-in web browser control of GE to create a seamless interface. The GE-Dashboard allows users to add their own custom WMS Layers and allows users to perform GIS functionality on those layers in GE. The overall architecture of the GE-Dashboard can be seen in Figure 1.

[10] http://docs.codehaus.org/display/GEOS/Home
[11] http://earth.google.com/kml/kml_tags_21.html#networklink

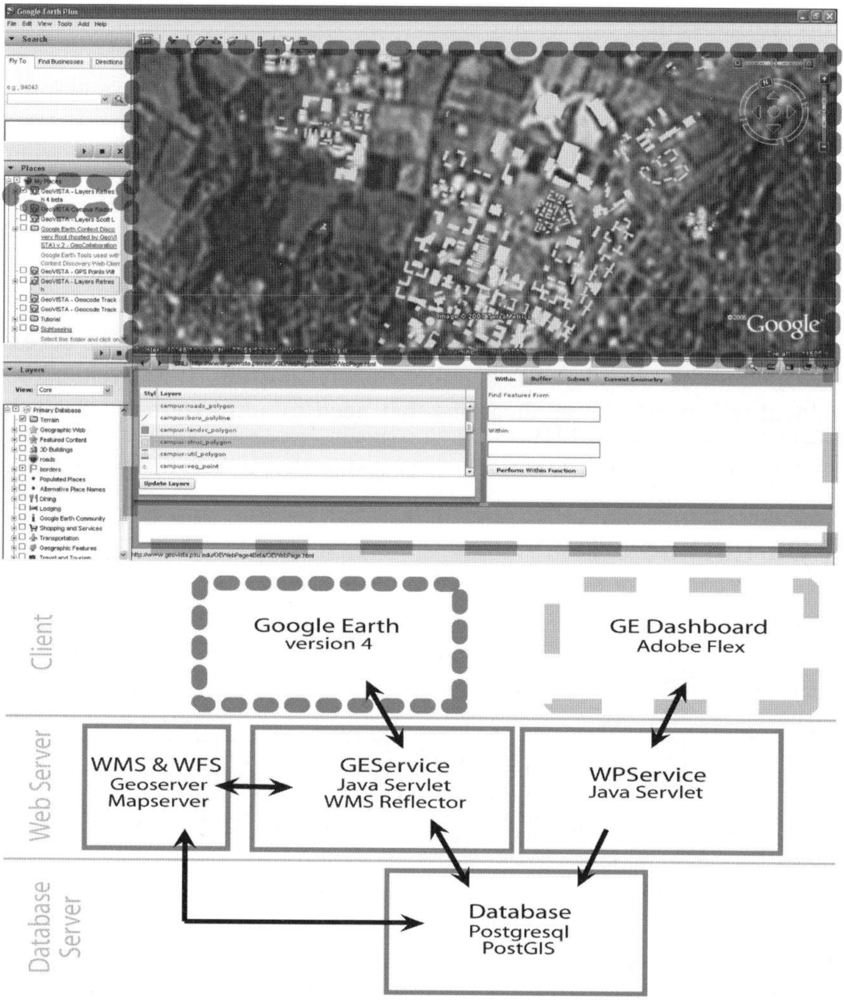

Fig. 1 – The overall architecture of the GE-Dashboard. Dotted boxes in bottom schematic diagram represent actual interface components outlined in the application screenshot above.

In this figure, two important components are shown that allow the GE-Dashboard to function. First is the custom WMS Reflector for GE. Second is a database which serves as both a back-end and middle-ware component that GE and the GE-Dashboard communicate with.

In its basic form, a WMS reflector is a server application that is accessed by a client such as GE, and communicates with a WMS-enabled server like Geoserver. In our implementation a request from GE comes through the network link in to the WMS Reflector. The WMS Reflector then requests a WMS map image from Geoserver. Geoserver then returns this image to GE where the image is overlaid onto the main GE globe.

The GE-Dashboard creates an interface that makes it easy to add WMS layers, turn WMS layers on and off and also allow users to visually see results of a GIS function on the GE globe through the dynamic styling of the WMS layer (such as highlighting features returned by a spatial query).

The back-end database of the GE-Dashboard records user action. For example, if the user decides to view two particular WMS layers, the layer names and their SLD styles are recorded in the database along with some ancillary data such as the IP address of the users' computer. This function ensures that the correct layers are drawn for the correct user, as GE can not distinguish between different users accessing the back-end database.

4.2 GIS functionality

A key aspect of the GE-Dashboard is the users' ability to perform GIS analysis functionality with the data present. One function supported currently in the GE-Dashboard is a within function. A drag-drop graphical interface was implemented in Flex to perform this function. A detailed description of how a user of the GE-Dashboard can use the GIS functionality is provided below (in section titled "GE-Dashboard Process Flow").

GIS functionality of the GE-Dashboard is provided by sending custom WFS requests to Geoserver, which are then passed on to the backend postGIS database for processing, thus load-balancing all of the computationally intensive work onto the server. This allows the results of the GIS function to be returned quickly to the client.

Geoserver provides an eloquent way to access the back-end database through a web client, and serves as a middle-ware between the client GIS functionality request and back-end processing of those requests. In particular, a WFS within function request is similar to that shown in Sample Code 1, below.

```
<?xml version="1.0" encoding="utf-8"?>
<wfs:GetFeature service="WFS" version="1.0.0"
  outputFormat="GML2"
  xmlns:campus="http://www.geovista.psu.edu/campus"
  xmlns:wfs="http://www.opengis.net/wfs"
  xmlns:ogc="http://www.opengis.net/ogc"
  xmlns:gml="http://www.opengis.net/gml"
  xmlns:xsi="http://www.w3.org/2001/XMLSchema-instance"
  xsi:schemaLocation="http://www.opengis.net/wfs
```

```
http://schemas.opengis.net/wfs/1.0.0/WFS-basic.xsd">
  <wfs:Query typeName="campus:structures">
    <ogc:Filter>
      <ogc:Within>
        <ogc:PropertyName>the_geom</ogc:PropertyName>
          <gml:MultiPolygon...{rest of Geometry here}
      </ogc:Within>
    </ogc:Filter>
  </wfs:Query>
</wfs:GetFeature>
```

Sample Code. 1 - Sample WFS Query Request.

In this case, the campus:structures layer is queried for all of the features within the Geographic Markup Language (GML) geometry. This xml is posted to Geoserver's WFS, where the request is translated into SQL to query the postGIS database. Geoserver also translates the records returned from the query back into GML that is then returned to the client.

In the WFS query process, the GE-Dashboard constructs the query from a GML template and adds user input. Adobe Flex makes it programmatically easy to post this request to Geoserver and parse the results through Flex's native HTTPService Object. An HTTP post in Adobe Flex looks much like Sample Code 2.

```
<mx:HTTPService id="filterFunctionRequest" showBusyCursor="true" url="http://www.geovista.psu.edu/geoserver/wfs" result="showResults(event)" method="POST" contentType="application/xml" resultFormat="e4x" useProxy="false" >
        <mx:Request>
        WFS GML here.
        </mx:Request>
</mx:HTTPService>
```

Sample Code 2 - Sample Adobe Flex HttpService tag POST request.

4.3 SLD Service

Another important aspect of the GE-Dashboard is the SLD Service. This service provides dynamic cartographic styles and symbols to the WMS layers appearing in GE. The purpose of the SLD Service is to accept SLD xml which is posted to this service from a client application, and to then write this SLD xml to a file on the server which can be accessed from the web and Geoserver. Finally, it returns the url of this SLD file to the client application.

With this approach, the GE-Dashboard can pass Geoserver the SLD file url and dynamically style any WMS layer. The url for dynamically styling layers in Geoserver looks something like that of Sample Code 3.

Sample Code 3 – Sample GetMap request with SLD file XML.

```
http://www.geovista.psu.edu/geoserver/wms?request=GetMap&s
rs=EPSG:4326&width=400&height=400&SLD=http://www.geovist
a.psu.edu/GEService/SLDFiles/highlightStructures.sld

    highlightStructures.sld

    <StyledLayerDescriptor version="1.0.0">
      <UserLayer>
        <Name>campusstructures</Name>
        <UserStyle>
          <FeatureTypeStyle>
            <Rule>
          <Name>Rule 1</Name>
          <PolygonSymbolizer>
            <Fill>
              <CssParameter
name="fill">#FFFF00</CssParameter>
              <CssParameter                        name="fill-
opacity">0.5</CssParameter>
              </Fill>
              <Stroke>
                <CssParameter
         name="stroke">#FFFF00</CssParameter>
                <CssParameter                    name="stroke-
         width">1</CssParameter>
         </Stroke>
          </PolygonSymbolizer>
            </Rule>
            </FeatureTypeStyle>
        </UserStyle>
      </UserLayer>
</StyledLayerDescriptor>
```

Geoserver does have the ability to accept the actual SLD xml as a url parameter. This method was not used because Microsoft Internet Explorer truncates urls at a specific character length. This often causes problems when adding a lengthy xml body into the url. Therefore, the SLD Service was necessary. The SLD Service allows for a simple url to the SLD file to replace the actual SLD body xml.

***Scenario**: There is a chemical leak on a University Campus. This chemical leak is airborne and could potentially affect thousands of students, faculty and staff. The disaster manager brings up GE and adds the necessary layers to the visualization to get an overview of the situation. The manager next drags the building footprint layer into the* within *functions' first text box. Then the plume model polygon layer depicting the chemical leak is dragged into the second text box of the* within *function. This allows the user to find all building footprints or structures within the plume model. The buildings identified are highlighted on GE and the attributes of the buildings are shown in the GE-Dashboard. The* building footprints *layer is joined to a table that has the Registrar's estimate for population in each building at the current time. This allows the manager to develop a plan for evacuation of potentially affected buildings.*

6 Future Research

We foresee several directions that developments with the GE-Dashboard can take.

The GE-Dashboard is a web-based interface designed to be used with GE's built in web browser. This allows the GE-Dashboard to work effectively as a seamless application with GE. However, because the GE-Dashboard is a web application, it could easily be opened in an external browser and its functionality made available to any other client mapping application that can display OGC-standards compliant data. Furthermore, the GE-Dashboard could also be completely embedded as a component in any other Adobe Flex-based client that needs mapping support. We are currently experimenting with these variants on the general model.

Future research also involves extending the GE-Dashboard to include additional GIS functionality. Both postGIS and Geoserver support many other GIS Functions that could be connected to the GE-Dashboard (e.g, buffering, intersection). One challenge of implementing additional GIS functionality is the design of easy to use, intuitive interfaces to access different types of GIS functions through the GE-Dashboard.

In addition to extensions to functionality of the system, there is potential to use the GE-Dashboard to study use of web mapping tools in crisis management activities. The current GE-Dashboard stores records of user activity in a database. This record can support study of how users interact with the interface when carrying out crisis-related tasks (Robinson & Weaver,

2006). This could help improve the interface as well as aid in training during crisis management exercises.

7 Conclusion

The GE-Dashboard represents an important first step towards combining the ubiquity of GE with open geospatial standards into an easy-to-use, data rich, geo-analytically powerful environment that can support disaster and risk management activity. Further research developments using this approach can lead to improved crisis management systems that can overcome challenges in information availability and interoperability across heterogeneous software environments, development platforms, and sources of data.

Acknowledgements

The research reported here has been supported by the National Science Foundation under Grant EIA-0306845. This work is also supported by the National Visualization and Analytics Center, a U.S. Department of Homeland Security program operated by the Pacific Northwest National Laboratory (PNNL). PNNL is a U.S. Department of Energy Office of Science laboratory.

References

Adobe Systems. 2007, Flash Player Statistics. Available: http://www.adobe.com/products/player_census/flashplayer/2 March 2007].

Fuhrmann, S. & Pike, W. 2004, User-centered Design of Collaborative Geovisualization Tools. In J. Dykes & A. MacEachren & M. J. Kraak (Eds.), Exploring Geovisualization (pp. 591-609): Elsevier.

Google. 2007, Hurricane Katrina Imagery. Available: http://earth.google.com/katrina.html2 March 2007].

Google Earth Blog. 2005, Hurricane Katrina Archives. Available: http://www.gearthblog.com/blog/archives/hurricane_katrina/2 March 2007].

National Research Council. 2007, Successful Response Starts With a Map: Improving Geospatial Support for Disaster Management. Washington, D.C.: National Academies Press.

Nourbakhsh, I. 2006, Mapping disaster zones. Nature, 439.

Open Geospatial Consortium. 2005, OGC announces Risk and Crisis Management Working Group. Available: http://www.opengeospatial.org/pressroom/pressreleases/4223 March 2007].

Robinson, A. C. & Weaver, C. 2006, Re-Visualization: Interactive Visualization of the Process of Visual Analysis. Workshop on Visualization, Analytics & Spatial Decision Support at the GIScience conference, Münster, Germany, pp.

Tomaszewski, B. 2003, Emergency Response and Planning Application Performs Plume Modeling. ArcUser, 6, 10-12.

Web Service Orchestration of OGC Web Services for Disaster Management

Albrecht Weiser and Alexander Zipf

i3mainz, Institute for Spatial Information and Surveying Technology
University of Applied Sciences FH Mainz, Germany
zipf@geoinform.fh-mainz.de

Abstract

Flexibility and reusability are major goals when developing complex applications based on OGC Web Services (OWS). Within the project OK-GIS (www.ok-gis.de) we evaluate the suitability of the Web Service Orchestration (WSO) technology as possible solution for disaster management scenarios. We present an example of a part of an evacuation scenario after a bomb has been found. This scenario includes in particular the need for emergency route planning. We evaluate how the actions to be performed by the system supporting the rescue workers can be mapped onto a service chain of basic OWS. The service chain is represented as a BPEL document and can be executed in a web service orchestration engine, such as Oracle BPEL-engine. BPEL is a standard for service orchestration and means Business Process Execution Language.

1 Free and open GIS for disaster management - the project OK-GIS

"OK-GIS" is the acronym for Open Disaster Management using free GIS (http://www.ok-gis.de). One of the main goals of the OK-GIS project is the development of a flexible toolbox of software components and services based on a Spatial Data Infrastructure (SDI) that uses open standards (e.g. by OGC, ISO, W3C etc.) This framework should be applicable for various scenarios and requirements in case of a disaster situation. The approach should be independent from specific scenarios. This shall be achieved by a highly configurable system. Technically this freedom can be gained by using profile dependent configurations on the one hand side and through the aggregation of high-class functionality by using more fine-grained base services, in particular standardised OGC Web Services (OWS) on the other hand side. The latter will be combined to specific workflow chains according to the demands of a corresponding profile. Research was carried out in order to test the possibility for OWS to use the technique of Web Service Orchestration (WSO).

We do present the idea of Web Service Orchestration of OWS and the practical experiences we have made implementing this. We discuss how OWS and in particular an Emergency Route Service (ERS) as well as an OpenLS Route Service (RS) can be used within a service chain orchestrated through BPEL scripts within a scenario where a bomb has been found and people need to be evacuated. The work of the firemen and rescue team that plans this evacuation shall be supported through this. The basic workflow as defined in the rules of the firemen is shown in figure 1.

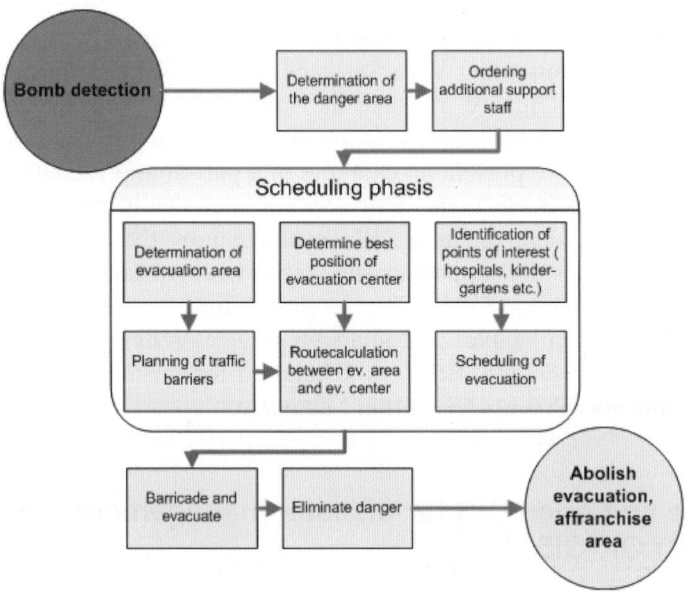

Fig. 1: Simplified evacuation scenario in case of finding a bomb

The above figure shows the workflow of a bomb finding and evacuation scenario. A part of that scenario will be mapped onto a BPEL use case in a later paragraph.

2 Orchestrating OGC Web Services with BPEL

The aim of OK-GIS project is to develop a modular architecture which ensures flexible reaction to a range of demands in emergency management. Complex functions and workflows of a certain granularity will be achieved through the skilled aggregation of standardized base services (OWS). These new aggregated functionalities can then also be used as web services on their own. So far the middleware performing the aggregation usually needs to be developed using conventional programming languages. This way all the advantages of object oriented languages can be used. The price to pay, however, is that every modification and compilation has to be made within the programming platform. The promise of the alternative WSO is to provide an easy and flexible way to link services in given patterns, especially through the configuration. This is done using so called orchestration scripts. By using WSO (Peltz 2003), it should be possible to map workflows as service chains through XML based languages such as BPEL (Business Process Execution Language), which can be configured with graphical tools instead of hard-coded programming. These BPEL scripts will be executed in corresponding WSO engines. Similar ideas and methods have been discussed by Kiehle et al. (2006) and Lemmens et al. (2006). The suitability of this concept will be reviewed based on real experiences and tests.

In OK-GIS, OWS offer the base for all further functionalities, as OK-GIS focuses on the spatial aspect of disaster management - i.e. finding, handling, managing, visualizing and analysing spatial data. With respect to WSO OWS unfortunately are subject to certain constraints: One of the main problems is the missing SOAP support (Simple Object Access Protocol) of OWS (OGC 03-014). Instead the services communicate through the http-protocol. However, the WSO engines usually use the SOAP-protocol. As a result we need to answer the question whether or not the orchestration of OGC web services is possible at all with BPEL. To summarize the result in short we can say, that our tests have shown that it is generally possible to orchestrate OWS using BPEL. Some technical problems were encountered because OGC Services (OWS) are not SOAP Web Services but XML-RPC[1] via HTTP-Get/Post. These findings will be discussed in the following in more detail.

[1] RPC: Remote Procedure Call - protocol for calling operations and procedures of a service/application from a remote point. Uses mostly http.

3 BPEL and what else? - Dependencies and Severities

BPEL and its extension for Web Services BPEL4WS is based on technologies like WSDL (Web Service Description Language), XML Schema and XPath. The highest influence on BPEL has WSDL, because the BPEL process model is built upon the WSDL 1.1 Specification. How do all these parts fit together?

BPEL itself defines the syntax for an XML based programming language (Juric et al. 2004). Although with some limitations, it is nevertheless an adequate programming language because of some properties like:

- loops (while)
- algorithm branch (switch)
- definition of variables

The result of a complete process definition is a BPEL script that will be interpreted by an orchestration engine like Oracle BPEL Process Manager or Active BPEL (Active Endpoints). The engine can be seen as a runtime environment that interprets the BPEL script. But a complete process definition does not only consist of the BPEL-based program sequence, but of the so-called “endpoint” references that reference the services used within the service chain. This is the part that WSDL plays in that concert: WSDL is the interface definition to the involved services. It keeps both- the information of the abstract operational structure of the services and the information of the concrete binding information like service-URL, servlet mapping path and binding protocol. Because WSDL defines the base for defining a BPEL-process, it has also the most critical potential of danger to its breakdown.

Due to the differences of OGC web services (OWS) to “normal” web services there are some incompatibilities between the W3C-built WSDL and OWS. One belongs to the special Key Value Pair encoding (KVP) of the WMS GetMap and the WCS GetCoverage requests; the other belongs to a missing raw binary format in XML Schema definitions of WSDL that would be necessary to accept a returned WMS GetMap- or WCS GetCoverage- binary (OGC 04-060r1). These limitations and some more issues are part of our evaluation and proof of concept tests.

4 Orchestration exemplified through the use of Emergency Routing during a bomb discovery

A possible emergency scenario could look like the one we want to discuss now.

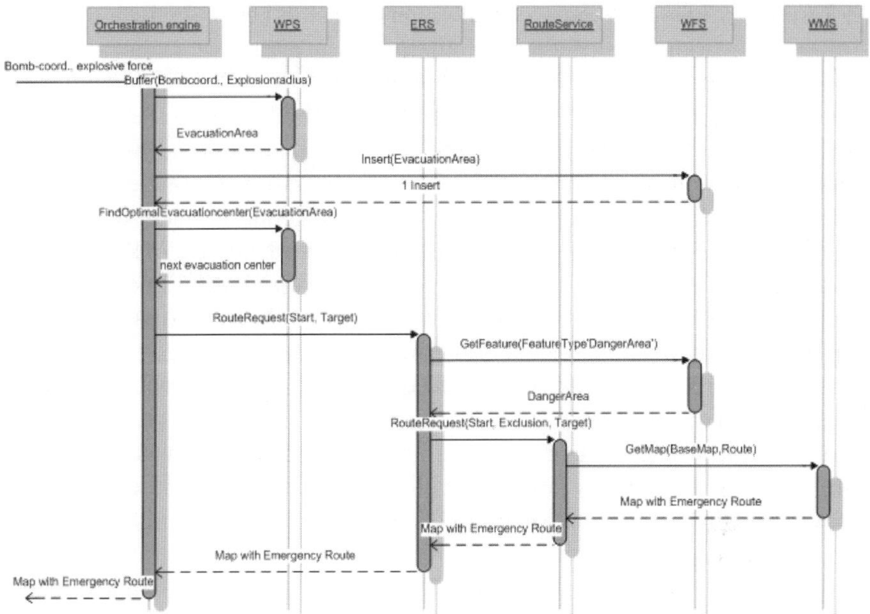

Fig. 2: UML Interaction diagram of involved OWS in the evacuation scenario (simplified)

The following scenario is given: A bomb was found at some place. Now it is necessary to evacuate the people living within a given distance. A part of the scenario includes that it is necessary to calculate an emergency route between the staging area, where all the people from the danger zones are being concentrated and the shelter. The emergency route needs to bypass the danger zones. This will be used as an example to explain how to orchestrate the OWS base services within a service chain, using a WSO engine, such as the Oracle BPEL orchestration engine.

The BPEL script executed by the orchestration engine (OE) integrates a WPS (OGC Web Processing Service), a WFS (OGC Web Feature Service) and an (Emergency Route Service) ERS using an OpenLS Route Service (Neis 2006). The OE acts as a web service on its own that is wrapping the complete process chain. This aggregated ("virtual") service needs as input parameters from the user merely the coordinates of the bomb, as well as its explosive force. The evacuation area can then be determined in several different ways. One option is that the explosive force (available within an OE internal XML look up table) can be mapped to a certain evacuation radius. With the bombs coordinates and the radius, the WPS can identify the evacuation area by using a simple buffer function (cp. Heier and Kiehle

2006). Of course also more complex versions can be implemented such as information about the affected population in the area.

In the next step the OE starts a WPS process called „*FindOptimalEvacuationCenter*" which defines the nearest shelter from the evacuation area. That WPS process implements typical GIS functionality like buffer and intersection (Stollberg 2006). Using the acquired geometries of the danger zones and the shelter locations a RouteRequest is sent to the ERS. The ERS gets all the dangerous and closed areas from the WFS using the *GetFeature* function. These are integrated into the request and sent to the RouteService as *AvoidAreas*, along with start- and destination points. The OpenLS RS then calculates the route including the *RouteGeometry*. This is then transferred through a *GetMap*-Request by the route service as a *User-Layer-SLD* (which is possible with the version 1.1.0 of the SLD standard) along with the parameter for a base map. The client receives a map with the specified emergency route as well as further relevant information regarding this route as resulting output.

5 Feasibility

The question that arises when discussing the scenario that was explained above: "Is it actually possible to build a service chain like postulated above"? What are possible limitations of OWS using such a technology like WSO?

In order to answer the questions we performed a range of test. The answer in short is that we successfully orchestrated several OGC Web Feature Services (WFS) and OGC Web Catalogue Services (CS-W). These services can be taken as examples for the feasibility of all http-POST based OWS.

In particular the OGC WMS GetMap request with the Oracle BPEL Process Manager (OPM) 10.1.2 and 10.1.3 was tested. The promise from Oracle was to also support transportation of binary raster data through the engine. The previous version supported only base64Binary, a lexical encoded XML binary type based on the 64 numeration. The latter version also supports hexBinary which bases on the hexadecimal numeration. But our tests showed, that none of them grant success. A raw, uncoded binary type would be necessary to accept a GetMap response, returning maps as binary raster data.

A possible solution would be a proxy server acting as binary transcoding service. It would transcode the uncoded binary bitmap to a base64Binary. That would of course increase the amount of traffic on the wire and the necessary computing power by the factor 3, because the bit-

map would have been decoded between WMS and BPEL process and also recoded between BPEL process and client.
A qualified question may be: is piping of large binary data like a WMS-map or a WCS-coverage through orchestration engines sensible at all? It is very likely that it would slow down the response time of the whole system. But displaying spatial data is essential in disaster management. Street maps have to be shown, directions have to be displayed, hazard zones to be mapped.

There is another solution to that problem: Instead of maps being exchanged, references to the maps can be communicated:

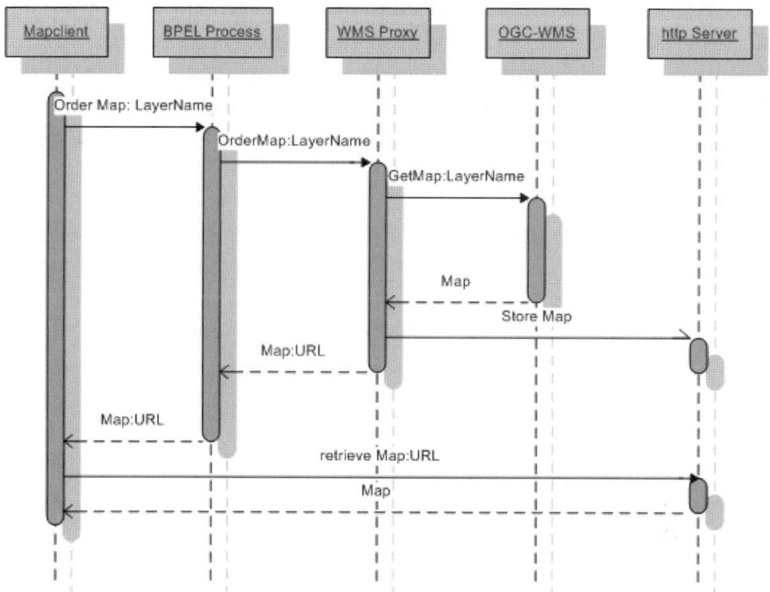

Fig. 3: UML sequence diagram of an alternative WMS orchestration using WMS-proxy and a map client

The above figure shows a possibility to embed an OGC WMS (or alternatively WCS) in a BPEL process chain.
The map-generation was initiated by the map client. The map client is not completely OGC compliant, as it handles the response slightly differently as explained here: First the client requests a map in an OGC compliant way by passing the desired LayerName (and all other required GetMap parameter) to the BPEL process (instead of the WMS directly). The process hands the parameters over to the WMS proxy-server. The proxy launches an OGC compliant GetMap request to the WMS and receives the GetMap response from the WMS. Subsequent the proxy stores the physical part of

the map on a simple http-server directory and returns the URL of the stored map back to the BPEL process. The process gives the information back to client. The map client then collects the stored map from the specified URL.

The main difference to an OGC compliant map retrieval of a WMS client is, that the WMS-map generation is a synchronous operation. In this example calling the map and displaying it was divided in two (partly asynchronous) operations.

The OpenLS Route Service (RS) described above does this alike. It works as a WMS proxy server that stores the map instance on an http-server and gives back the RouteMap which contains the route summary, route instructions, URL to the map-instance etc. (Neis 2006).

Another possibility totally different to the one above is the usage of an SLD inline Feature. It is an SLD 1.1 concept where the SLD contains a collection of simple features "inline" its body. Fig. 4 shows the service chain:

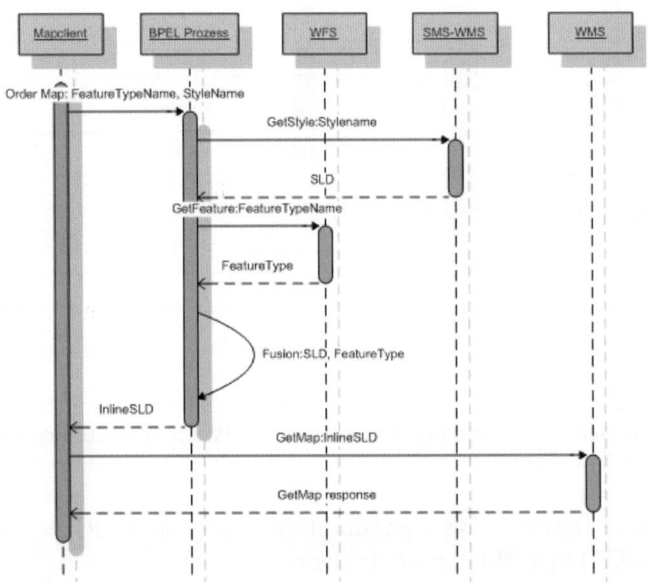

Fig. 4: UML sequence diagram of an alternative WMS orchestration using WFS, SMS-WMS and a map client

A map client is the base of operations. It passes a request to the BPEL process. The process requests the current style of a special Style Management Service-WMS (SMS-WMS) (OGC 04-040). The SMS-WMS responds an SLD to the process. Subsequent the process requests a Feature

Type from WFS. The BPEL process merges the responded *FeatureCollection* and the SLD to an *InlineFeature*-SLD and passes it to the calling map client. The map client sends the inline-SLD to a compliant WMS and displays the resulting map.
There are currently only a few WMS implementations able to compute an inline-SLD. One is the geoserver WMS 1.4.0. It supports WMS 1.2.0 and SLD 1.1.0.

These two examples show how a WMS can be integrated into an orchestration process in spite of the current limitations. This means that the handling of maps can be incorporated in a larger orchestration process like presented in Fig. 2.

6 Web service allocation security - BPEL ensured find-bind

The availability of all necessary services is one of the most critical points of view for service oriented architecture (SOA) in an emergency response situation. Particularly in a supraregional natural disaster the electrical and electronic infrastructure may suffer and therefore support of a disaster management system might fail partially when it is really indispensable.

Therefore an effective emergency support system (ESS) needs a high level redundance in its participating services.
If the SOA has information about several redundant services on infrastructural independent locations, it can choose between these services.
This means that it is necessary to implement an algorithm that enables the system to choose automatically the necessary service that is currently up and running.

Fig. 5 shows the whole “secure Web Service Allocation” BPEL-process.

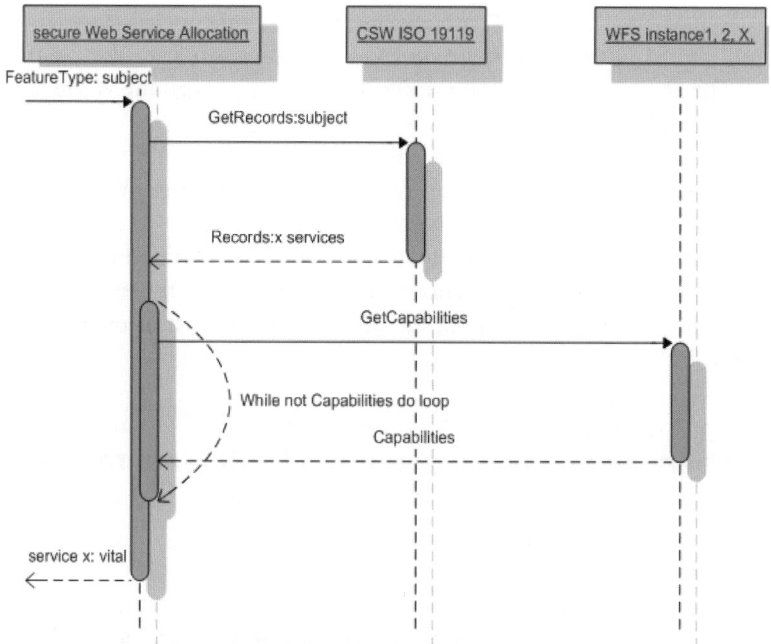

Fig. 5: UML sequence diagram of the secure Web Service Allocation

As a simple example the process starts with the request of a special *FeatureType*. That may be a client-application where this feature was selected in a dropdown list. The BPEL process launches a semantically tolerant GetRecords request to an ISO 19119 catalogue service with the specified subject. All records with according ISO *subject* elements will be returned to the process. These records represent different instances of redundant Web Feature Services. They will be tested in a case-expression for the ISO element *MD_DataIdentification.SupplementalInformation*. If the service appellation of this element correlates with a service appellation of a service defined in the BPEL process, it will try to invoke the WFS. For these purposes the WFS GetCapabilities operation is being requested from that service.

If it answers with a capabilities response, the process exits the loop, because a vital WFS instance that serves the desired subject was found. The calling client was sent the information "service X is up" so that the client in the next step can request the desired feature from the service. Figure 6 shows the process in Business Process Model Notation – BPMN which has been derived from UML.

The aim of the process is a necessary and integral part of a successful working disaster management system and hence can become a future module of the OK-GIS platform. The only constraint in this way of service allocation is, that the services cannot be defined during runtime. It is not necessarily the case that all available services are found in a GetRecords request, but all possible services need to be previously defined in the process to be available if necessary.

That means for the above example that if the GetRecords response returns WFS_1 and WFS_2 they can be invoked. But it's not possible to expect that, if the CSW finds a WFS_4, the service can be invoked at runtime by arranging the specific service information to a WSDL and a BPEL definition. All endpoints must be well known in the run-up to the original definition.

The above BPEL process can be combined with other processes like the FeatureType Retrieval we already implemented:

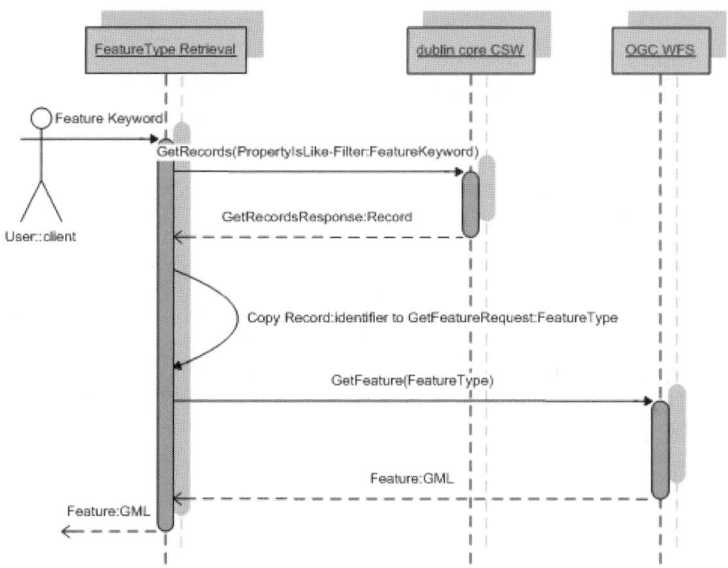

Fig. 6: FeatureType Retrieval in the OK-GIS SDI

For the OK-GIS GDI it was necessary to join the metadata with the relating geodata. To manage this there are (at least) two possibilities:

One was to use the ISO 19115 metadata schema. There are 2 possible elements to contain the relation to the geodata:

- *MD_Metadata.identificationInfo/MD_(Data)Identification.citation/CI_Citation*, which contains the ID of the geodataset

- *MD_Metadata.distributionInfo/MD_Distribution.transferOptions/MD_DigitalTransferOptions.onLine*, where the link to the http-GET request can be placed

In this way a "normal" web service could send a request to the WFS after obtaining the record from the previous *GetRecords* request.

The other possibility to join metadata with geodata was to use a Dublin Core (DC) CSW and BPEL, because in conjunction with BPEL the metadata information of an http-link is not really necessary. All the necessary information (like the reference to the relating geodata-service) can be found in the corresponding WSDL.

As a BPEL process for a valid *GetFeature* request only the *FeatureTypeName* is necessary. All further information like service, etc. can be initiated in the BPEL script. The DC element *Identifier*[2] was used for containing the *FeatureTypeName*. The *FeatureTypeName* (plus Namespace-prefix) establishes the relationship to the concerning dataset, hence the usage is compliant to the Dublin Core (DC) metadata standard on the one hand and on the other hand it allows the usage in a *GetFeature* request.
This means that the process flow works very similar to that described in the paragraph on Web Service Allocation Security.
A difference to the example above is that the catalogue used is a Dublin Core compliant CS-W. It returns the records found in exchange to a requested *Feature* keyword. This keyword can be any name of the *FeatureType* the user is looking for. If there is an according name in the DC *description* element, the record will be found.
Another difference concerns the handling of the records. The requestor wants one special *FeatureType*. For this purpose the response is tested on its element *SearchResults* that contain the attribute *numberOfRecordMatched.* If the attribute contains more than one hit, the result is ambiguous and will be terminated after informing the requestor about this condition. If the attribute contains no hit, the name of the *FeatureType* could not be resolved. If its exactly one hit, a *GetFeature* request will be send to the corresponding WFS asking for the FeatureType with that special FeatureTypeName. In a final step the *Feature* will be returned to the requestor.

We made some test series to get an average response time. The medial for the *FeatureType Retrieval*-process was 3 seconds on a not too new desktop computer. Before every single test the browser cache and history was cleared. Most measured values ranged from approximately 2 to 3 seconds (if all endpoint services were up and unemployed). Always at the first test after startup of the engine we got some overlong process durations

[2] According to the Dublin Core Metadata initiative the purpose of the element *identifier* is: "An unambiguous reference to the resource within a given context"

(from approximately 8 until 22 seconds).
So we infer that the engine is caching internal connection parameters.

7 Current Work

Orchestration of OGC web services (OWS) is not a brand-new topic although it is still highly interesting as only recently real experiences using technologies like BPEL have been reported. Within OGC discussions on service oriented architectures with OWS have started at least since the OWS2 common architecture discussion paper (OGC 04-060r1) from 2004. Although there were some earlier considerations to adjust the architecture towards compatibility to common web services (OGC 03-014) the OWS2 paper offered the first helpful suggestions in this direction.
Hence there are some first trials to provide "real" SOAP-based web services technology for OWS in contrast to the earlier XML-RPC. For example some WSDL-files can be found at the official OGC schema sites for Web Feature Service (…/wfs/1.1.0/wsdl). The WSDL are readily configured for http-GET-, –POST- and SOAP-binding.
When testing the WSDL with an Oracle BPEL designer 10.1.2, the WSDL were accepted from the engine internal XML processor, but no WSDL structure was displayed and hence it was not possible to be used within the orchestration engine. Some results of our investigations about the reasons of the failure follow in the next section:

On the other hand the XML schema files for *GetFeature* request and response provided by OGC worked well inside a self-made WSDL. That's a great relief, because the challenge in creating WSDL isn't the WSDL itself, but the XML schema type definitions of the very complex OWS messages.

Recently, within the internal OGC discussion paper "OWS4 Workflow IPR" (OGC 06-187) has been written that presents some results from the OWS4 initiative. It contains some BPEL use cases for/related to the OGC Geo Processing Workflow. Within the defined service chains there are used WFS and Web Processing Services (WPS).
There are also examples of BPEL workflows used in complex ESA and NASA Spatial Data Infrastructures (SDIs). The NASA use case uses the grid operation of a WPS, triggered by a Web Coverage Service. The ESA use case is a bit more complicated and much more interesting: There a Sensor Planning Service (SPS) of the OGC Sensor Web Enablement (SWE), a Web Coordinate Transformation Service (WCTS) and a Web Coverage Service are being aggregated to a service chain. The SPS catches

raw level 1a (not geocoded) SPOT satellite images and transforms them via WCTS to a geocoded ortho image. The BPEL process passes the reference to that ortho image back to the calling client. Unfortunately within the paper there is no explanation of the interaction between client and process, but the use of the WCS is similar like the use of WMS described in the paragraph "feasibility of Orchestration through the use of Emergency Routing during a bomb discovery"

Some initiatives react on the increasing demand of W3C compliant web service interfaces. So did 52°North, Münster, Germany for their recently released Sensor Observation Service (SOS) and Web Notification Service (WNS). They offer a few already configured WSDL for the current version of their service (the pre standards of SOS and WNS are still heavily under construction) (Claussen 2007).

8 Conclusion

BPEL and Web Service Orchestration attract more and more interest within the GI community due to their increasing popularity in main-stream computer science. The proof-of-concept evaluation presented here shows that it is possible to create an added-value by combining and aggregating OGC Web services. Of course the use of BPEL must be reasonable and purposeful. It only makes sense to use orchestration where a continuous service chain without human intervention is possible. But in particular in many cases within disaster management there are a lot of unforesseable factors influencing the service chain in a dynamic way. This means that it is on the one hand necessary to find relativeley stable standard chains (small parts of a larger workflow, that can act as modular building blocks) and on the other side ist is necessary to research how such BPEL scripts can be assembled in a even more dynamic way even for non-technical users. This needs work on relevant user interface concepts in order to ease the highly dynamic ochestration of OWS on the fly.

Because of the big benefit when using standards to develop a platform for disaster management is that this results in an highly adjustable toolkit .

In OK-GIS we develop a spatial data infrastructure and the necessary server and client applications to support some typical use cases for the fire department. But the fire brigade is here only a typical example representing for the varienty of disaster task forces. The aim of this work is to make it possible to add new use cases to the overall system, without the need for difficult changes within the code or even theW program structure. Therefore we decided to test new technology, use standards like BPEL and standards-based software: With that middleware technology it is possible to create loosely coupled service networks with a high reliability. These can use and integrate other (OGC-) web technologies like the OGC Sensor Web Enablement (SWE). For the use case of a flood scenario we test the integration of SWE services into our OK-GIS SDI (Claussen 2007) in order to implement a web based flood early warning system.

References

Biermann J, Gervens T, Henke S (2005): Analyse der Nutzungszenarien und Aktivitäten der Feuerwehr . Internal Project Report. Projekt OK-GIS.

Botts M, Robin A, Davidson J, Simonis I (2006): OpenGIS® Sensor Web Enablement Architecture Document. V. 1.0. OGC Ref. Nr. 06-021r1. D. P.

Chen L, Wassermann B, Emmerich W, Foster H (2006): Web Service Orchestration with BPEL. Dept. of Computer Science, University College London

Claussen K (2007 in work): Umsetzung von OGC Sensor Web Standards im Katastrophenmanagement. (OGC Sensor Web for disaster management). Master Thesis. University of Applied Sciencens FH Mainz. (working title).

Fitzke J, Greve K; Müller M and Poth A (2004): Building SDIs with Free Software - the deegree Project. In: Proceedings of GSDI- 7, Bangalore, India.

Friis-Christensen A, Bernard L, Kanellopoulos I, Nogueras-Iso J, Peedell S, Schade S, Thorne C (2006): Building service oriented applications on top of a spatial data infrastructure – a forest fire assessment example. AGILE 2006.

Juric M, Mathew B, Sarang P (2004). Business Process Execution Language for Web Services. Packt publishing Ltd., Birmingham, 2006

Kiehle C; Greve K & Heier C (2006): Standardized Geoprocessing – Taking Spatial Data Infrastructures one step further. Proceedings of the 9th AGILE International Conference on Geographic Information Science. Visegrád, Hungary.

Lemmens R, Granell C, Wytzisk A, de By R, Gould M, van Oosterom P (2006): Semantic and syntactic service descriptions at work in geo-service chaining. Proc. of the 9th AGILE Int. Conference on Geographic Information Science. Visegrád, Hungary

Neis P. (2006): "Routenplaner für einen Emergency Route Service auf Basis der OpenLS Spezifikation". Dipl. Thesis. University of Applied Sciences, Mainz.

OASIS. http://www.oasis-open.org/apps/org/workgroup/wsbpel/

Open Geospatial Consortium Inc. OWS2 Common Architecture. Hrsg. OGC. RefNum. OGC 04-060r1; Vers. 1.0.0; Status: OGC Discussion Paper.

Open Geospatial Consortium Inc. OWS 1.2 SOAP Experiment Report. Hrsg. OGC. RefNum. OGC 03-014; Vers. 0.8; Status: OGC Discussion Paper.

Open Geospatial Consortium Inc. OWS-4 Workflow IPR. Hrsg. OGC. RefNum OGC 06-187; Vers. 1.0.0; 2007-03-11 Status: internal OGC Discussion Paper.

Open Geospatial Consortium Inc. Style Management Services for Emergency Mapping Symbology. Hrsg. OGC. RefNum OGC 04-040 Vers. 1.0. Status: OGC Discussion Paper

Peltz C. (2003); Web services orchestration and choreography. IEEE Computer.

Stollberg B (2006): Geoprocessing in Spatial Data Infrastructures - Design and Implementation of a Service for Aggregating Spatial Data. Diploma Thesis. University of Applied Sciences FH Main

Agent-Based Simulation of Spatial Cognition and Wayfinding in Building Fire Emergency Evacuation

L. Hajibabai[1], M. R. Delavar[1], M. R. Malek[1] and A. U. Frank[1, 2]

[1] Center of Excellence in Geomatics Engineering and Disaster Management, Dept. of Surveying and Geomatics Engineering, Engineering Faculty, University of Tehran, Tehran, Iran, lbabai@geomatics.ut.ac.ir, mdelavar@ut.ac.ir, malek@ncc.neda.net.ir
[2] Institute of Geoinformation and Cartography, Technical University of Vienna, Gußhausstr. 27-29 E127, A - 1040 Vienna, Austria
frank@geoinfo.tuwien.ac.at

Abstract

There is a need to understand how people and environment react in a fire building emergency. Sometimes in the wayfinding process decision errors may occur mainly based on topological errors of the signage. A situation is critical if a decision about which path to take cannot be made with certainty, especially in a crisis situation. An agent-based simulation of human's behavior in escaping from the fire with due attention to the building's signage and dynamic nature of fire propagation affecting the wayfinding task is outlined in this paper. The hypothesis of the paper is that successful navigation is possible if the agent is able to make the correct decision through well-defined cues in critical cases, so the design of the building signage is evaluated through the agent-based simulation. Construction of mental representations of spatial environment and exploring models in the agent-based simulation have been proposed and a computational model successfully tested in an indoor complex hospital environment in different situations and the evacuation time from the building is computed. The most appropriate signage design resulted in the shortest evacuation time in various situations.

1 Introduction

One of the most disastrous forms of collective human behavior is the kind of crowd stampede induced by panic, often leading to fatalities as people are crushed or trampled. Sometimes this behavior is triggered in life-threatening situations such as fires in crowded buildings. Although engineers are finding ways to alleviate the scale of such disasters, their frequency seems to be increasing with the number and size of mass events. An example of such topics which has been considered in this paper, is fire building evacuation modeling. The ability to enable efficient circulation of people in heavily populated enclosures is important to the day to day operation of large complex buildings. More importantly, it is an essential design feature in the event of emergency situations. The aim of this research is to support people's wayfinding and escaping process using optimum quality and placement of building signage which helps them finding the exits in a safer way.

As architects continue to implement novel concepts in building design, they are increasingly finding that the fixed criteria of the traditional methods of prescriptive building codes are too restrictive. This is due in part to their almost total reliance on configurational considerations such as travel-distance and exit width. Furthermore, as these traditional prescriptive methods are insensitive to human behavior or likely fire scenarios, it is unclear if they indeed offer the optimal solution in terms of evacuation efficiency.

The emergence of performance based building codes together with computer based evacuation models offer the potential of overcoming these shortfalls and addressing the needs not only of the designers but also the legislators. However, if such models are to make a useful contribution they must address the configurational, environmental, behavioral and procedural aspects of the evacuation process. The result is that the fire safety engineer is empowered to realize the full value of mathematical models to help in the prediction of human behavior during evacuation, and to determine the consequences of fire under a variety of conditions. This in turn enables them to design and implement safety measures which can potentially control, or at least reduce the impact of fire.

Intelligent approaches for modeling fire safety evacuations is a useful new method to evaluate capability of such methods. In this paper we report of an agent-based simulation of escaping from the fire building and computed the total evacuation time for various building signage design and compared the results and found the optimum placement and design of the building informational cues. This agent-based modeling uses concepts from cognitive science and wayfinding research. Cognitive research will lead to improved systems that take advantage of an understanding of human's geo-spatial perception and conception, including that of geo-spatial experts. A geo-spatial information technology that is more responsive to human factors in its design will greatly improve the effectiveness and efficiency of spatial decision-making. This includes both general concerns about the limited geo-spatial knowledge in the population and more general concerns, such as environmental change.

A common task of humans is to navigate from one place to another, often in an unknown or critical area. Typically, maps are used for navigation. Maps may not be available or their use may not be possible due to the critical situations. Recent research on human wayfinding has a balanced focus on mental representations and processes of wayfinding (Raubal and Worboys 1999). The goal-driven reasoning chain that leads to action begins with incomplete and imprecise knowledge derived from imperfect observations of space. Actions result in further observations, derived knowledge and further actions, until the goal is achieved or the wayfinder gives up. A special case of wayfinding in a complex building, that is finding exit ways in a fire emergency, is the application of the model.

The goal of the previous work of the agent-based simulation of human behavior was to develop a computational theory of perceptual wayfinding (Raubal 2001). That theory uses an agent-based approach and can just explain people's wayfinding behavior in unfamiliar buildings. It is different from previous computational models for wayfinding, which were built to investigate how mental representations are created, stored and used. These models assume that people become familiar with their environments over time and, therefore, acquire cognitive maps (Garling et al. 1998; Moeser 1988).

Artificial intelligence researchers in computer science and other disciplines have developed simulations of spatial intelligence, in some cases as part of the design of mobile robots. In this research we have mainly focused on agent-based simulations for optimum designing of the building signage system in order to simplify the task of wayfinding in an unknown complex building during fire emergency. To represent and simulate people's processes of spatial cognition and wayfinding, it is necessary to understand how people immediately make sense of spatial situations while performing a wayfinding task which will occur in a building during fire emergencies. Some of the assumptions for rescue operation in these situations may fail and so we should use intelligent approaches in these contexts.

2.1 Spatial Cognition

Research in the cognition of geo-spatial information deals with human perception, memory, reasoning and communication involving the spatio-temporal objects and events, both in the real world and in their digital representations. Basic research in geo-spatial cognition is relevant to many issues involving geo-spatial information such as data collection and storage, graphic representation and interface design, spatial analysis, interoperability, decision-making and the social context of GIS. We believe that many aspects of geo-spatial information system usability, efficiency and profitability can be improved by greater attention to cognitive research. It is claimed that our experience and interaction with the environment is based on the use of recursive and imaginative patterns.

Spatial cognition is composed of several elements including landmarks, route maps and survey maps. Landmarks are identifiable environmental markers associated with specific locations (Chown 1995). Route maps are sequences of instructions, often involving landmarks, which describe at a personal level how to get from one location to another (Appleyard 1969). Survey maps are similar to topological maps and describe the spatial layout of the environment as opposed to reflecting a specific navigational task. Spatial cognition concerns the study of knowledge and spatial properties of objects and events in the world. Cognition is about knowledge: its acquisition, storage and retrieval, manipulation and use by humans and intelligent machines. Generally speaking, cognitive systems include sensing and perception, thinking, imagery, memory, learning, language, reasoning and problem solving. Spatial properties include location, size, distance, direction, separation and connection, shape, pattern and movement. Human spatial cognition is an interdisciplinary research area in cognitive science (Raubal 2001).

A widely accepted model of spatial learning suggests that spatial knowledge of places is developed in a sequence of three stages or elements. First is landmark knowledge, unique features or views that identify a place. Second is route knowledge, based on travel routines that connect ordered sequences of landmarks. The third stage is survey knowledge, knowledge of two-dimensional layouts that includes the simultaneous interrelations of locations.

2.2 Wayfinding

Wayfinding, getting from some origin to a destination, is one of the everyday problems that humans encounter (Richter and Klippel 2004). It is a purposive, directed and motivated activity (Golledge 1999). Human wayfinding research investigates how people find their ways in the physical world, what they need to find it, how they communicate directional information and how people's verbal and visual abilities influence wayfinding (Raubal et al. 1997). According to Lynch (1960) wayfinding is based on "a consistent use and organization of definite sensory cues from the external environment". Wayfinding is a complex human activity involving moving along while evaluating alternatives and making decisions. It is defined as a spatial problem solving process with the three sub-processes including decision-making, decision execution and information processing (Timpf 2005).

Wayfinding typically requires planning and the ability to stay oriented while moving. Navigation is this coordinated and goal-directed travel through space. It consists of two components, locomotion and wayfinding. Locomotion refers to the guidance of oneself through space in response to local sensorimotor information in the immediate surrounding and includes such tasks as identifying surfaces of support, avoiding obstacles and moving toward visible landmarks. Locomotion generally occurs without the need for an internal model or cognitive map of the environment. Wayfinding refers to the planning and decision-making that allows one to reach a destination that is not in the immediate sensory field and includes such tasks as choosing efficient routes, scheduling destination sequences orienting to non-local features and interpreting verbal route directions.

Two fundamental processes are involved in orientation during navigation. The first process is the recognition of external features or landmarks. In some cases, the recognized landmark is the goal, but more commonly the landmark aids orientation by serving as a key to know the spatial relations stored in an internal cognitive or external cartographic map. The second process is updating a sense of orientation by integrating information about movement speed, direction and/or acceleration without reference to recognized features. There are four classes of environmental variables that influence wayfinding performance within built environments: visual access, architectural differentiation, signs and room numbers to provide identification or directional information and plan configuration (Weisman 1981). Spatial knowledge is assumed to consist of landmark, route and survey (configurational) knowledge (Siegel and White 1975). The following guidelines for good wayfinding signage are developed (Arthur and Passini 1992):

- Wayfinding information should be given at the point that a decision needs to be made.

- Legibility is important. The information on the sign should be perceived with ease. Make sure signs are not hidden from view by bushes, trees, etc.
- Readability is the ease with which information can be understood. People glance for messages rather than read them word for word. Keep the messages short. Don't expect long messages to be read from moving cars.
- Reliability relates to the question of whether or not the signs give correct information. Signs with false information can lead to stress, frustration and mistrust.
- Accessibility, meaning that persons with disabilities can read and use the signs.

3 Methodology

The navigating human is viewed as an agent with state and performs actions such as perceiving information from the real world and moving through the environment. Given a sequence of landmarks between the current position and a desired destination, the agent executes the appropriate steps necessary to reach the destination point. Most of the wayfinding problems are due to the poor quality of the building signage or lack of them in some critical decision points. Therefore, designing signage is an important task which is done and analyzed successfully in this research in order to gain the most optimum design and placement of the building cues (Hajibabai 2006). Fig. 1 shows the agent's state at a decision point. As part of the agent's state, the actual planned path is marked as a directed graph with arrows.

Fig. 1 Decision taken at every decision-point (Hajibabai 2006).

An action is modelled as a physical displacement of the wayfinding agent. The agent moves in the building environment and changes its position in it.

The wayfinding agent observes the situation in the real world and learns about the building environment. Learning occurs when its observations do not confirm its belief about the exit ways and their observable attributes. The wayfinding agent is an active learner, which means that it uses the learned information in its planning algorithm when it plans the traveling path. On the contrary, a passive learner simply tries to learn the utility of being in different states without actively using the learned information. Learning is called to be incremental if the agent tries to update its old hypothesis whenever a new example or information arrives (Krek 2002). Planning happens at every decision point where the agent has to select the next way to continue wayfinding via the optimal path to the destination. The agent's decision is based on the plan; it takes the path suggested by the planning program as the first step on the planned path to the destination.

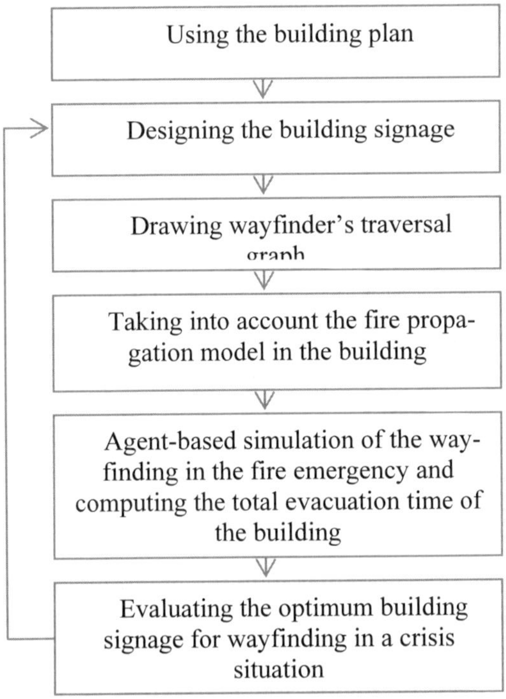

Fig. 2 The agent-based simulation of the wayfinding task

The agent would gain information from the building's signage which may be either directional or textual. By due attention to the cues and dynamic propagation of fire and smoke on the agent's interface, it would gain

Raubal, M. (2001). Agent-Based Simulation of Human Wayfinding: A Perceptual Model for Unfamiliar Buildings. PhD Thesis, Vienna University of Technology, Austria.

Richter, K. F., and Klippel, A. (2004). A model for context-specific route directions. In: Freksa et al. (Eds.), Spatial Cognition IV. Reasoning, Action, and Interaction: International Conference Spatial Cognition, pp. 58-78. Springer, Berlin.

Russell, S. and Norvig, P. (1995). Artificial Intelligence- A Modern Approach. Prentice- Hall International, London.

Siegel, A. and White, S. (1975). The development of spatial representations of large-scale environments. In Advances in Child Development and Behavior, Vol. 10, Reese, H., Ed. New York: Academic Press, pp. 9-55.

Timpf, S. (2005). Cognitive Wayfinding Agents in Public Transportation Networks, Geographic Information Science Center, University of Zurich, Switzerland.

Weisman, J. (1981). Evaluating architectural legibility: Way-finding in the built environment. Environment and Behavior, Vol. 13, pp. 189-20

Exploratory Spatial Data Analysis to Support Maritime Search and Rescue Planning

Cindy A. Marven, Rosaline R. Canessa and Peter Keller

Spatial Sciences Laboratories, Department of Geography, University of Victoria
PO BOX 3050 STN CSC, Victoria, BC V8W 3P5, Canada,
cmarven@shaw.ca

Abstract

Managers are often expected to analyze, report, plan, and make decisions using data that are aggregated to administrative areas historically delineated for other purposes. This enforced aggregation may misinterpret true patterns or complexities underlying the data, hindering recognition and communication of potentially important insights. The result may well provide misleading information on which to base decisions. Spatial data analysis tools are available that could allow managers to analyze and aggregate data more meaningfully and effectively for decision-making and planning, while still allowing them to report to the standard administrative units. These spatial analytical tools would be of importance to managers who are using data to prevent, plan for, or mitigate risk-related events.

The Canadian Coast Guard is offered as an example whereby managers are responsible for planning for the provision of maritime search and rescue emergency response using historical maritime incident data collected site-specific but aggregated to historical reporting units. We explore how spatial data analysis techniques, in combination with GIS, can provide a way to analyze incident data spatially regardless of existing reporting units, providing a better way to 'package' the data for use in emergency response planning and decision-making. We show how marine incident patterns over the region can be monitored to help planners anticipate emerging incident hot-spots or gauge the persistence of existing hot-spots. Finally, we show how a better understanding of incident patterns within existing administrative units can inform the development of new reporting boundaries that better reflect incident patterns.

1 Introduction

The availability of historical marine incident data with locational coordinates (e.g., latitude, longitude) poses both challenges and opportunities for managers involved in assessing the need for maritime search and rescue services. Traditionally, information about marine incidents is collected and reported by administrative areas. This information can be difficult to use in medium- to long-term search and rescue (SAR) planning because much of the spatial and temporal information is lost through data generalization unless the incidents are distributed uniformly over an area, which is usually not the case.

If the marine incident data coordinates are available, how can managers analyze and use them to support SAR planning process? How can managers distinguish between random patterns of SAR incidents and patterns which indicate hotspots to better provide direction in planning and resource allocation.

Using the Canadian Coast Guard SAR operations on the Pacific coast, this paper examines the challenges of analyzing SAR incident data for resource planning to minimise risk. The paper also examines the potential of using several exploratory spatial data analysis (ESDA) methods for analyzing marine incidents for supporting medium- to long-term maritime search and rescue planning. The use of these methods in epidemiology and criminology are discussed briefly followed by a discussion of how these methods might be suitable for maritime SAR planning.

2 Risk Analysis and Search and Rescue Planning

Maritime SAR planners have slightly different requirements than other maritime groups (for example, maritime insurance companies or marine vessel designers) interested in assessing risks associated with marine activities. Medium- to long-term SAR planning is necessarily oriented towards reducing risk to human life by anticipating the level of demand for SAR services and responding to demand in a timely manner. Assessing the demand for SAR services involves anticipating the quantity of incidents and where and when they are likely to occur. Estimating risks associated with various types of marine activities would be of great value to planners but has been hindered by lack of data and research effort.

In the maritime environment, there is often an absence of contextual data that would help managers estimate the risks associated with certain kinds of marine activities. Baseline data such as the marine activity levels over an area for different time periods are often unavailable or in formats that are difficult to integrate with the marine incident data. Unlike terrestrial transportation planning, maritime traffic (with the exception of shipping lanes for large commercial vessels) is not constrained to paths and is therefore not easily counted to estimate temporal peaks of traffic volumes.

Much of the existing research in the field of maritime transportation risk assessment has been focused on commercial shipping activities involved in transporting dangerous goods (Iakovou 2000, Merrick et al. 2000, Douligeris et al. 1997). These types of activities usually represent a very small proportion of marine incidents, albeit, when incidents occur, the impacts and implications for loss of life, property, or environmental damage are high. Still most SAR efforts are oriented towards other types of marine incidents such as those involving commercial fishing vessels or recreational boats. There is little research concerning risk in these areas, particularly with respect to SAR services.

3 Canadian Coast Guard Marine Incident Data

The Canadian Coast Guard is responsible for providing maritime search and rescue services for Canada. The agency collects data about marine incidents to facilitate SAR resource allocation planning and minimize risks. The data include information on:

- *Location*: incident coordinates (latitude and longitude); the SAR Area in which the incident occurs; and descriptive information about the incident location;
- *Time*: the date and time when the incident began, was responded to, or when response activities ended; and
- *Attributes*: incident seriousness; type and description of unit assisted; unit assisted activity (e.g., fishing, recreational); number of people on board; number of people saved and lost; incident cause; resources deployed; and limited environmental characteristics (e.g., wind speed, wave height).

The Pacific Region is divided into eight SAR Areas for reporting and planning purposes (Fig. 1). There are seven bounded areas (SAR Areas 301-307) and one area (SAR 308) that is unbounded but reflects the functional extension of SAR services. The seven bounded SAR Areas range in approximate area from 1,178 km^2 (SAR Area 301) to 67,058 km^2 (SAR Area 306). Incident data are primarily summarized and reported by SAR Area.

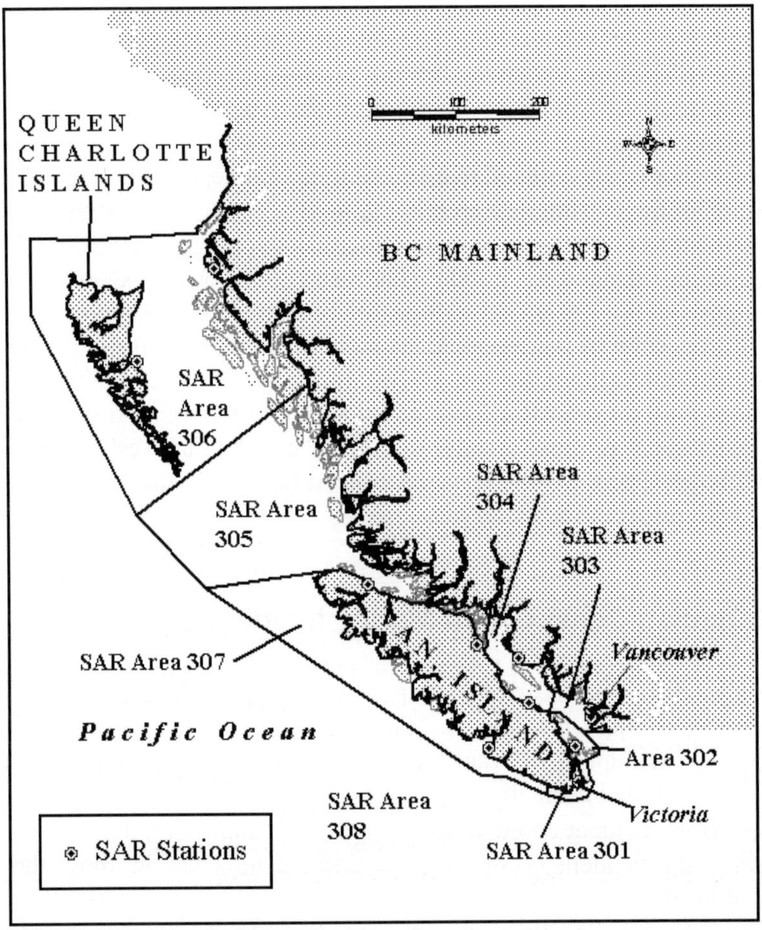

Fig. 1 Search and rescue areas, Pacific Region

Search and rescue incident data from 1993–1999 were acquired from the Canadian Coast Guard. Subsequent to validation for data consistency and completeness 11,457 incidents were examined[1]. Most of the incidents in the Pacific Region (6,402 or 56%) occurred along the south-eastern coast

[1] These data summaries will differ slightly from official counts of the Canadian Coast Guard due to data cleaning and variations in administrative boundaries.

of Vancouver Island in SAR Areas 303, 304 and 305 (Table 1). SAR Areas 301 and 303 report the highest density of incidents.

Table 1. Search and rescue incidents by Search and Rescue Areas, Pacific Region, 1993-1999.

SAR Area	Approx. Area [km^2]	Marine Incident Severity				Total Incidents	Incident density [Incidents/ km^2]
		Distress	Potential Distress	No Danger	False Alarm/Hoax		
301	1,178	85	160	581	398	1,224	1.04
302	2,032	83	157	646	315	1,201	0.59
303	2,584	220	341	1322	677	2,560	0.99
304	6,071	142	265	1214	445	2,066	0.34
305	53,150	162	286	1003	325	1,776	0.03
306	67,058	138	257	736	220	1,351	0.02
307	24,795	125	185	621	226	1,157	0.05
308	[a]	18	21	52	31	122	

[a] SAR Area 308 does not have offshore boundaries, therefore area cannot be calculated.

Further analysis of incidents involving specific vessels was undertaken. These include (1) motor craft (any vessel, usually private, with a motor attached, not including sailing vessels with auxiliary motor), (2) sail craft (vessels used predominantly for sailing with or without an auxiliary motor), (3) commercial fishing vessels (with or without housing over the wheel) and (4) small recreational craft (comprising kayaks, canoes, open boats, personal watercraft and sailboards). Motor craft are the most commonly assisted vessels in all SAR Areas except 306 and 308 (Table 2). In SAR Areas 305-307 incidents involving fishing vessels are comparatively higher than other SAR Areas. In Areas 306 and 308, the number of motor craft incidents is eclipsed by fishing vessel incidents.

Table 2. Search and rescue incidents by search and rescue area and vessel type, Pacific Region, 1993-1999

SAR Area	Motor Craft	Sail Craft	Commercial Fishing Vessel	Small Recreational Vessel
301	546	223	84	202
302	559	247	84	137
303	1112	477	199	312
304	1043	328	262	186
305	735	85	586	76

ticularly if the spatial units contain highly heterogeneous values – changing the boundaries, changes the values, which often changes analytical outcomes.

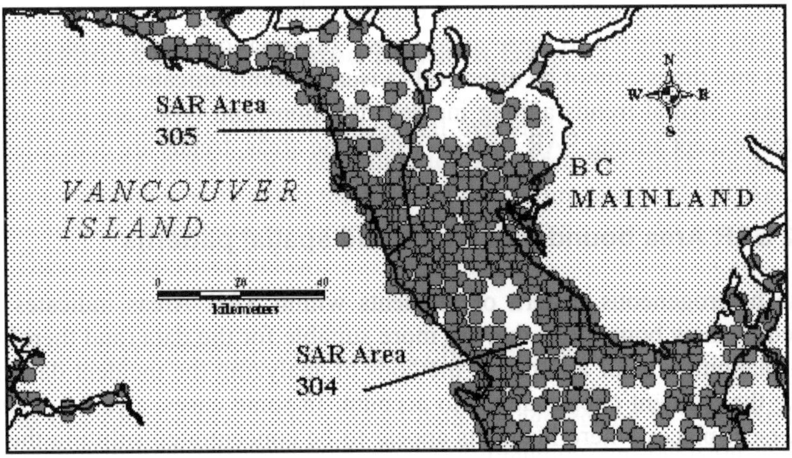

Fig. 3 An example of a SAR boundary bisecting a concentration of marine incidents between Search and Rescue Areas 304 and 305

4.3 Integrating other marine data

Using the incident data with other types of marine data is difficult because of the high degree of generalization and aggregation. If one were to use the aggregated data to model marine risk using marine incidents and exposure measures (e.g., levels of marine activity or others), the degree of heterogeneity within the areas may affect the results.

5 ESDA Methods in Criminology and Epidemiology

The existing approach of aggregating incidents to SAR Areas to support resource allocation planning could be improved upon using a combination of Geographic Information System (GIS) functions and ESDA techniques particularly those for point pattern analysis. Point pattern analysis identifies and tests for statistical consistency, trends and coherent groups among a set of point data which may otherwise appear to be random. These methods have been used to some extent in criminological research and epidemiological research, fields that often deal with incident data (crime incidents and disease incidents), but have not been explored for use in maritime SAR resource allocation planning.

The array of methods and approaches to spatial data analysis range along a continuum from those that are used to explore, describe and visualize patterns and relationships (exploratory) to those that confirm hypotheses of model outcomes (confirmatory). Tukey (1977) first described the idea of ESDA but most of the research pertaining to ESDA in the field has been accomplished since the late 1980s. ESDA involves the description of spatial data, the identification of statistical properties and the preliminary identification of data structure with the objective of encouraging hypothesis formulation from the data (Haining 1990). Exploratory methods can be used to determine what models might be appropriate for analyzing a point pattern and often are useful for examining the results of confirmatory spatial analyses. The types of methods reviewed in this study fall within the realm of ESDA.

Spatial clustering routines are a class of ESDA methods for point pattern analysis that have been used in criminological studies and in epidemiological studies. Cluster analysis has its base in taxonomic science where observations are grouped together based on the similarity of their attributes (Rapkin and Luke 1993, Griffith and Amrheim 1997) creating homogenous groups from an apparent heterogeneous set of values. Traditional forms of cluster analyses do not require entities to be geo-referenced and their location in space is not necessary for classifying the observations into groups. However, by geo-referencing observations and incorporating their location into the analysis, cluster analysis can be used for analyzing point patterns. Entities are assigned to clusters based on some proximity function or algorithm based on distance or adjacency of which there are many (Griffith and Amrheim 1997). There is no 'best' algorithm and the outcomes can be quite different. Griffith and Amrheim (1997) outline 'rules of thumb' for evaluating and choosing an appropriate algorithm for a particular study. Cluster analysis is an exploratory approach and explaining the theorized relationship between the resulting clusters and variables under study must be undertaken by a researcher familiar with the methods and data (Dillon and Goldstein 1984).

Cluster analysis methods have been used to identify 'hot spots' or places where crime incidents are concentrated. Anselin et al. (2000) credits the work of Sherman et al. (1989) for the regeneration of interest in the role of place in criminological theory with their work describing hot spots of predatory crime. Sherman (1995) defines crime hot spots as "small places in which the occurrence of crime is so frequent that it is highly predictable at least over a one-year period". Anselin et al. (2000) provide a similar definition suggesting that "a hot spot is a location or small area within an identifiable boundary with a concentration of criminal incidents that is chronic and concentrated over extended periods of time". The identification of crime hot spots has two main uses: one, to help law enforcement agencies determine where to focus intervention resources, and two, to build place-based theories of crime. Identifying places where the number of incidents is persistently high is important for the development of law enforcement intervention strategies. An underlying population can be used to identify areas of elevated risk. This helps researchers understand and identify factors that may explain reasons for higher crime concentrations.

In the field of epidemiology, spatial cluster methods have played a small and contentious role in the identification of disease incident clusters (Rothman 1990, Elliott and Wakefield 2001). In epidemiological studies, the focus is on finding areas or clusters where the risk of disease is elevated from a background population. Therefore, determining whether a disease cluster is significant is of greater importance that for many criminological applications. There are problems associated with attempting to determine whether a disease cluster represents an area where disease risk is higher than some baseline level such as the small numbers of incidents, pre-existing biases concerning the potential cause and site of a disease cluster, and difficulties in isolating potential causes (Rothman 1990).

6 ESDA Methods and Search and Rescue Planning

A number of ESDA methods targeting point pattern analysis could be useful for SAR planning. These include spatial clustering methods such as spatial and temporal analysis of crime (STAC) and nearest neighbour hierarchical clustering (NNH). Points can be generalized using kernel density estimation to estimate the intensity of a point pattern over a study region. There are also descriptive measures and statistics such as centrographic statistics and nearest neighbour analysis. Each of these methods are included in CrimeStat, a point pattern analysis software package (Levine 2004). The methods are exploratory; each method and choice of parameters provide a slightly different view of the data. This presents challenges for analysts and planners, in that a thorough understanding of each routine is needed to adequately inform the selection of methods, choice of parameters (e.g. minimum distance between incidents, minimum number of incidents to form a cluster) and the interpretation of results. This section of the paper introduces these techniques suggesting applications for SAR planning.

6.1 Spatial Clustering Methods

Cluster analysis models can be grouped broadly into two categories: hierarchical models and optimization models, with many variations existing within each of these basic models. A clustering method suitable for the analysis of point data called Spatial and Temporal Analysis of Crime (STAC) was developed by Block and Block (1995). This routine falls within the category of optimization methods, as it recombines clusters to avoid overlapped clusters. Hierarchical methods based on nearest neighbour distances and a *K*-means partitioning method generate overlapping clusters in which higher order clusters encompass two or more overlapping lower order clusters.

Analytical outcomes can be expressed graphically (standard deviational ellipse and mean center describing the cluster) and in tabular format (number of events, incident density, area of standard deviational ellipse, center, angle of rotation) and displayed using a GIS. Two or more point patterns can be compared if methods are applied and interpreted with caution. For example, if one is comparing point patterns, it is important to consider the number of events comprising each point pattern as well as to ensure that the software specifications are consistent. The size (area), shape and direction of clusters can be compared as can the number of points in each cluster.

Spatial and Temporal Analysis of Crime (STAC)

STAC is a scan-type spatial clustering method developed for use in crime pattern analysis (Block and Block 1995) that is generic enough to use for other point patterns. The clustering algorithm identifies areas in the study region that are the most densely populated with incidents. The STAC cluster routine requires that a grid be placed over the region (grid cell size and boundaries specified by the analyst). Using the grid intersections as nodes, the algorithm scans the region with a circle 1.14 (creating overlapping circles) times the distance specified by the analyst centred over the nodes; counting the events in the circles; ranking the top 25 according to the number of summed points; aggregating the circles that share points, and repeating this process until there are no overlapping circles or shared points among the clusters. The analyst specifies the minimum number of points per cluster.

Nearest Neighbour Hierarchical Clustering

Nearest neighbour hierarchical (NNH) clustering is a spatial clustering method used to identify and visualize groups of incidents that are closer together than could be expected if the distribution were completely spatially random. The analyst specifies a threshold distance between points and minimum number of points per cluster. If pairs of events are closer together than the threshold distance specified by the analyst, a cluster is started; other points are added if they are less than the threshold distance to the event-pair. If the minimum number of points per cluster specification is met, a cluster is formed and shown by a standard deviational ellipse and mean center point. The standard deviational ellipse does not show all the events in the cluster, just those within the number of standard deviations from the mean center of the cluster, specified by the analyst. If the centroids of two or more 1st order clusters are within the threshold distance specified, a 2nd order cluster will be formed, encompassing them. The threshold distance and minimum number of points per cluster need to be chosen in consideration of the process under analysis and the characteristics of the study region as well as other factors.

Application to SAR Planning

The NNH clustering and STAC clustering methods may be used to identify and visualize incident concentrations both within and across SAR Areas. Incidents could be analyzed by incident type, incident severity, vessel type, vessel activity, temporal period or by other partitioning critieria specified by the analyst. Integrated with additional information (e.g. environmental, activity level, weather, etc.), clustering methods could help inform decisions concerning SAR resource locations and the type of resource required; identify potential factors contributing to boating risk; generate questions and hypotheses about the relationships between incident locations and other potential contributing factors (e.g. "why are high concentrations of incidents happening here"?). The outcomes of cluster analysis also may be used to confirm expert or local knowledge.

6.2 Density Estimation

Kernel density estimation (KDE) is an exploratory approach that provides a way of estimating the intensity of a point pattern over a study region. It smoothes or generalizes the data from discrete points to continuous raster data. Kernel density estimation (or kernel smoothing) was described by

Diggle (1983). The kernel (k) is a bivariate density function, symmetric about the origin. A radius of a suitable length is chosen (the bandwidth, τ) and the distance from the kernel centre to points with a distance less than the bandwidth is used to calculate the amount they contribute to the intensity at the point at the center of the kernel. The height of the kernel at the center represents the point count estimate and the amount contributed towards the intensity. The study area is finely gridded and the kernel is moved from observation to observation to calculate the intensity within the distance prescribed by the bandwidth. The points within the 'region of influence' do not contribute equally to the intensity calculation– those closer to the center of the kernel contribute more than those near the perimeter (Bailey and Gatrell 1995, Fotheringham et al. 2000).

The kernel density estimation method may be useful to model incident density within the region, to show spatial variation of incident density and to compare density estimations for different groups of data (e.g., incident types, incident causes, vessel types, time periods) if the specifications are the same (e.g. type of function used, interval type (fixed or adaptive), etc.). The analytical outcomes provided by KDE include a GIS grid cell polygon layer with estimated values (absolute or relative density; probability) for each grid cell. These results can be visualized using routines in GIS or mapping software as choropleth maps, elevational models using the estimated value as the 'z' value; or isolines. Single KDE involves using just one incident distribution while dual KDE involves using two. For example, with a dual KDE analysis, commercial fishing vessel incidents for one time period can be compared with that of another time period. The change in value between the two layers can be visualized.

6.3 Centrographic Statistics and Nearest Neighbour Analysis

Centrographic Statistics

Some of the most basic descriptive measures for describing arrangements of points are centrographic statistics (Kellerman 1981). Centrographic statistics include the mean centre, the median centre, the centre of minimum distance, the standard deviation of x and y coordinates, the standard distance deviation, and the standard deviational ellipse. Centrographic statistics are based on the distances between events and do not rely on calculations that require a boundary. Centrographic statistics provide a visual and quantitative description of the spatial distribution of a set of incidents including its centre, dispersion, concentration, and the direction of the dispersion. Table 3 contains descriptions of several centrographic statistics summarized from Ebdon (1977), Kellerman (1981), and Levine (2004).

Table 3. Centrographic statistics

Statistic	Description
Mean centre	the mean of the X coordinates and the mean of the Y coordinates; shows the centre of the distribution of incidents
Median centre	the intersection between the median of the X coordinate and the median of the Y coordinate
Centre of minimum distance	shows the centre of the distribution where the sum of distances to all other points is minimized
Standard deviation of x and y coordinates	describes the dispersion and shape of a distribution (actually two separate statistics; units same as coordinate system)
Standard distance deviation	standard deviation of the distance of each point to the mean centre (a vector expressed in units such as meters, feet, nautical miles, etc.)
Standard deviational ellipse	gives an indication of the skewedness or (anisotropy) of the distribution; the two standard deviations in the X and Y directions are used to create an ellipse; X and Y axes are orthogonal to one another; the direction of the axes are rotated to minimize the sum of squares of distance of the points to the axes.

Centrographic statistics can show and quantify changes in the spatial arrangement of incidents over time and thus may illustrate large-scale spatial shifts in a distribution of incidents in any direction. They could also show and quantify spatial differences or similarities between two types of incident distributions and could be used at a variety of scales (e.g. sub-regional areas or the entire planning region). If these statistics were calculated for an operational area (or rescue vessel's operational area) it may help provide an indication of where search and rescue resources could be sited. These statistics could provide a broad indication of the distribution and concentration of events for comparison among groups.

Nearest Neighbour Analysis

Nearest neighbour analysis involves finding the mean distance between neighbouring incidents of a distribution. The nearest neighbour approach was first used by ecologists concerned with measuring the spatial distribution of plants (Clark and Evans 1954). A nearest neighbour index can be calculated by dividing the average nearest neighbour distance by the nearest neighbour distance that could be expected for the same area if the point pattern were randomly distributed.

A nearest neighbour analysis of marine incidents could indicate whether the average distance between marine incidents is different than that expected with a spatially random distribution. Performed on two or more categories or time periods (e.g. commercial fishing and recreational boating), a comparison of results could illustrate differences among incident patterns.

7 Integrating Marine Data

Point pattern analysis of marine incident data fulfils the role of exploratory methods in that the outcomes may suggest hypotheses about why incidents concentrate in certain areas at certain times. These methods may also provide a way to visualize incident distributions within and across administrative zone boundaries. However, to more fully explore why concentrations occur in particular places and times, it would be beneficial to integrate the ESDA results with other marine datasets.

Unfortunately, there are often few marine datasets to use with the ESDA analysis. This is a considerable hindrance in that while the incident data themselves can be analyzed for patterns, the critical steps of hypothesizing or questioning why the patterns exist are harder to investigate because supporting data are limited. For example, on the Pacific coast of Canada, there are few marine datasets available, spatial or non-spatial, concerning levels of recreational boating activity or boating activity patterns. A similar situation exists for commercial fishing activity although there are some data concerning the location of fishing grounds and fishing openings aggregated to large administrative areas. Unfortunately, however, there is often no information about where and when commercial fishing vessels started or ended their fishing trips (depending on the fishery). There are challenges integrating the data formats; for example, with respect to commercial fishing data, the number of fishing events that occurred within a large administrative area over a given time period may be known, but the actual locations or time the events occurred within the area are unknown. This uncertainty is difficult to model and integrate with the marine incident location layer. The dearth of marine datasets in marine risk analysis presents different challenges to that of land-based studies such as those undertaken for criminological research whereby demographic data for census areas, transportation corridor layers, land use information, and socio-economic data may be available in addition to the locations and times of crime events. In the latter case, the problems are often related to appropriately geo-coding events and choosing appropriate spatial units suitable for aggregating the data.

Information about navigational hazards, tides, navigational aids, commercial shipping lane locations, and other marine-related base data are represented on Canadian Hydrographic Charts. These layers could be overlain with the marine incident data to provide additional clues as to why incidents happen where and when they do. Weather and climate data exist but must undergo considerable manipulation and processing before they could be integrated with other data for analysis. The relationship between land and sea datasets and a means for modeling these relationships would need to be developed. For example, on the Pacific coast of Canada, most of the population is concentrated in the south-east portion of the province and islands. This is also where most of the recreational boating incidents occur. The relationships between the ports (commercial shipping, cruise vessels, commercial fishing) and incident concentrations would be interesting to explore as would that between terrestrial transportation links and incident concentrations (land access to the coast).

8 Conclusions

The ESDA techniques described in this paper are all suitable for point pattern analysis and may provide SAR planners with information about the distribution of marine incidents within their region. Spatial data management and visualization needs could be handled by a GIS with the spatial analysis performed with a software program such as CrimeStat (Levine 2004). Planners could then incorporate their analysis into plans and resource allocation decisions more effectively than they could incident statistics reported to administrative areas. The benefits of using these methods may include the opportunity to visualize and quantify spatial patterns of incidents within and across the administrative areas, something that would not be possible using conventional reporting or mapping methods. SAR resources could be sited more effectively, and ultimately, help reduce risks to human life.

Acknowledgements

The authors wish to thank the Canadian Coast Guard for providing SAR incident data. Funding was provided by the GEOIDE National Centre of Excellence.

References

Anselin L, Cohen J, Cook D, Gorr W, Tita G (2000) Spatial analyses of crime. In: Duffee D (ed) Measurement and Analysis of Crime and Justice. National Institute of Justice, Washington DC, 4:213-262

Bailey TC, Gatrell AC (1995) Interactive Spatial Data Analysis. Longman Group, Harlow, England

Block RR, Block CR (1995) Space, place and crime: Hot spot areas and hot places of liquor-related crime. In: Eck JE, Weisburd D (eds) Crime and Place, Vol 4. Criminal Justice Press, Monsey, New York

Clark PJ, Evans FC (1954) Distance to nearest neighbor as a measure of spatial relationships in populations. Ecology, 35:445-453

Diggle PJ (1983) Statistical Analysis of Spatial Point Patterns. Academic Press Inc., Toronto, Ontario

Dillon WR, Goldstein M (1984) Multivariate Analysis: Methods and Applications. Wiley, New York

Douligeris C, Iakovou E, Englehardt JD, Li H, Ip CM (1997) Development of a national marine oil transportation system model. Spill Science & Technology Bulletin, 4(2):113-121

Ebdon D (1977) Statistics in Geography: A Practical Approach (1 edn) Basil Blackwell, Oxford, England

Elliott P, Wakefield J (2001) Disease clusters: Should they be investigated, and if so, when and how? Journal of Royal Statistical Society, 164(1): 3-12

Fotheringham AS, Brunsdon C, Charlton M (2000) Quantitative Geography: Perspectives on Spatial Data Analysis. Sage Publications, Thousand Oaks, CA

Griffith D, Amrhein CG (1997) Multivariate Statistical Analysis for Geographers. Prentice Hall Inc, Upper Saddle River, NJ

Haining R (1990) Spatial Data Analysis in the Social and Environmental Sciences. Cambridge University Press, Cambridge

Iakovou ET (2000) An interactive multiobjective model for the strategic maritime transportation of petroleum products: Risk analysis and routing. Safety Science, 39:19-29

Kellerman A (1981) Centrography measures in geography. (Concepts and Techniques in Modern Geography) CATMOG 32. Geo Abstracts, University of East Anglia, Norwich.

Levine N (2004) CrimeStat: A spatial statistics program for the analysis of crime incident locations (v 3.0). Ned Levine & Associates, Houston, TX, and the

National Institute of Justice, Washington, DC. May (http://www.icpsr.umich.edu/NACJD/crimestat.html)

Merrick JRW, van Dorp JR, Harrald J, Mazzuchi T, Spahn JE, Grabowski M (2000) A systems approach to managing oil transportation risk in Prince William Sound. Systems Engineering, 3(3):128-141

Rapkin BD, Luke DA (1993) Cluster analysis in community research: Epistemology and practice. American Journal of Community Psychology, 21(2):246-276

Rothman KJ (1990) A sobering start for the cluster busters' conference. American Journal of Epidemiology, 132: 6-13

Sherman LW (1995) Hot spots of crime and criminal careers of places. In: Eck JE, Weisburd D (eds) Crime and Place, Vol. 4. Criminal Justice Press, Monsey, New York

Sherman LW, Gartin PR, Buerger ME (1989) Hot spots of predatory crime: Routine activities and the criminology of place. Criminology 27(1):27-55

Tukey JW (1977) Exploratory Data Analysis. Addison-Wesley, Reading, Massachussetts

A Model of Spatial Data Integration Interoperability on Oracle Spatial

Qiansheng Zhao[1], Quanyi Huang[2], Jiming Guo[1], Renqiang Wen[1], and Shaobo Zhong[2]

[1]School of Geodesy and Geomatics, Wuhan University, Wuhan 430079, China
[2]Center of Public Safety Research, Tsinghua University, Beijing 100084, China
zh_qsh@hotmail.com

Abstract

An Open GIS Consortium (OGC) white paper said that the future vision for sharing data might look like this: Each of the smaller counties or towns hosts its own online GIS. Each uses software and a data model selected to best meet its local needs, which is the objective of this paper studies to close with. This paper gives a model based on Oracle Spatial, within a local government or enterprise the spatial data is centralized storage, and with metadata interoperability, which enables organizations to use the right tool for the job while eliminating complicated data transfers and multiple copies of the same data throughout the enterprise or department. This paper has realized MapInfo and ArcGIS work together under the same oracle spatial database use trigger and storage process. And at another hand, with the situation of between the departments or enterprises this paper gives a three-tier structure solution: spatial data server, application server and application client. The application server is a mediation system, this model uses oracle application server as the mediation system, and through the application server the application client sends WMS or WFS request and get the map server for background application. The three-tier structure model exposes a GIS portal which is an online GIS for outside of the department. Any client can request the server if it accords with WMS or WFS specification.

1. Introduction

In the last 20 years, Geographic Information Systems (GIS) had been widely developed and applied to various fields, which include resources management, environment management, prevention of disaster, planning area, emergency response, scientific research, education and national defense etc in many sections and subjects. However traditional GIS application systems that can not interact with each other was considered as isolated island of information, both within and out of enterprises are hindered with the interoperability problem. This is because that different GIS application adopted different data modeling methods, and based on GIS software from different vendors.

Within an enterprise, Consider this scenario: your planning department uses MapInfo® software, your engineering group uses Autodesk® Map software, your parcel fabric is maintained with GeoMedia®, and the GIS group wants to use ArcGIS® to provide persistent topology. Actually, this situation is ubiquitous, especially in Land Planning and Management Department. For an example, MapInfo®, ArcGIS®, MapGIS® are all used in Wuhan Urban Planning and Land Administration Information Center. Many organizations have implemented a mixture of tools and technology – and with good reason. Different vendors have different strengths and capabilities that are driven by the history of the company, the markets they serve, and the demands of their users. In addition, your organizational history is important because you have significant investments in the systems you are using today. This situation also exits between different departments. For example, emergency response services need simultaneous access to many distributed GIS databases for a particular spatial feature (such as building information at the assessor's office in Oracle database, transportation data at the department of transportation's office in the GeoMedia format, environmental data at the environment agency in Shapefile format) (Peng 2003). Current GIS programs cannot satisfy the time requirements for these kinds of online GIS services. How to make use of different tools from different vendors while guarantee the interaction between the data formats? How to make the data be shared between enterprises?

The spatial data interoperability concept brought forward provide definite means for a huge amount the geospatial information for efficacious management, sharing and buildup in value usage, it make the GIS data share and interoperate between each other come true. During the past 20 years, the concepts, standards, and technology for implementing GIS interoperability have evolved through eight models: (1) Data format interoperability; (2) Metadata interoperability; (3) Data content interoperability;

(4) Database or query interoperability; (5) Component or application interoperability;(6) Services interoperability; (7) Semantic interoperability; (8) Integration interoperability; All eight of these approaches and related technologies are important and continue to play a significant role in GIS interoperability today.

Consider that only separate interoperability model can not satisfy the enterprise users ,they maybe need to integrate several interoperability technologies .This paper discusses the Integration Interoperability on oracle spatial, Oracle® offers a spatial data model that provides basic enterprise access to geospatial information. This spatial data model provides a standard structure for point, line, and area features. Text and oriented point feature representations have not yet been standardized, and as a result, are treated differently by major GIS product lines. To achieve interoperability among GIS applications, several major GIS software vendors are providing software kits for use in the Oracle environment. Recognizing this is a short-term solution, these companies, along with Oracle, are working together to advance the standard geospatial data model in Oracle.

This paper gives tow spatial data interoperability models respectively for intranet user within enterprise and for internet user between departments. Within the enterprise using two-tier structure model, those are the database server and application clients. This model depends on metadata, each GIS software system has their own metadata in oracle spatial, in order to make different GIS software work together under the same database, it must modify the metadata in the oracle spatial. This paper has realized MapInfo and ArcGIS work together under the same oracle spatial database use trigger and storage process.

Between the departments or enterprises this paper gives a three-tier structure model: database server, application server and application client. The application server is a mediation system, this model uses oracle application server as the mediation system, through the application server the application client sends WMS request and get the map server for background application. The three-tier structure model exposes a GIS portal for outside of the department. Any client can request the server if he or she accords with WMS specification.

2. Oracle Spatial Data Model

Oracle Spatial consists of a set of object data types, type methods, and operators, functions, and procedures that use these types. A geometry is stored as an object, in a single row, in a column of type

SDO_GEOMETRY. Spatial index creation and maintenance is done using basic DDL (CREATE, ALTER, DROP) and DML (INSERT, UPDATE, DELETE) statements. With Spatial, the geometric description of a spatial object is stored in a single row, in a single column of object type SDO_GEOMETRY in a user-defined table. Any table that has a column of type SDO_GEOMETRY must have another column, or set of columns, that defines a unique primary key for that table. Tables of this sort are sometimes referred to as spatial tables or spatial geometry tables. Oracle Spatial defines the object type SDO_GEOMETRY as: CREATE TYPE sdo_geometry AS OBJECT (SDO_GTYPE NUMBER, SDO_SRID NUMBER, SDO_POINT SDO_POINT_TYPE, SDO_ELEM_INFO SDO_ELEM_INFO_ARRAY, SDO_ORDINATES SDO-_ORDINATE_ARRAY);

Oracle Spatial also defines the SDO_POINT_TYPE, SDO_ELEM_INFO_ARRAY, and SDO_ORDINATE_ARRAY types, which are used in the SDO_GEOMETRY type definition, as follows: CREATE TYPE sdo_point_type AS OBJECT (X NUMBER, Y NUMBER, Z NUMBER); CREATE TYPE sdo_elem_info_array AS VARRAY (1048576) of NUMBER; CREATE TYPE sdo_ordinate_array AS VARRAY (1048576) of NUMBER.

The geometry metadata describing the dimensions, lower and upper bounds, and tolerance in each dimension is stored in a global table owned by MDSYS (which users should never directly update). Each Spatial user has the following views available in the schema associated with that user: USER_SDO_GEOM_METADATA contains metadata information for all spatial tables owned by the user (schema). This is the only view that you can update, and it is the one in which Spatial users must insert metadata related to spatial tables. ALL_SDO_GEOM_METADATA contains metadata information for all spatial tables on which the user has SELECT permission.Spatial users are responsible for populating these views. For each spatial column, you must insert an appropriate row into the USER_SDO_GEOM_METADATA view.

Oracle Spatial ensures that the ALL_SDO_GEOM_METADATA view is also updated to reflect the rows that you insert into USER_SDO_GEOM_METADATA. Each metadata view has the following definition: (TABLE_NAME VARCHAR2 (32), COLUMN_NAME VARCHAR2 (32), DIMINFO SDO_DIM_ARRAY, SRID NUMBER); in addition, the ALL_SDO_GEOM_METADATA view has an OWNER column identifying the schema that owns the table specified in TABLE_NAME.

3. Model based on Oracle Spatial

After discuss the Oracle Spatial data model above, it shows that Oracle Spatial support the uniform data storage, and especially some GIS vendors have support the spatial data model, for example ArcGIS, MapInfo and GeoMedia have developed respective tools for uploading spatial data as Oracle Spatial, but ArcGIS and MapInfo can not access each other within Oracle Spatial,i.e.. ArcGIS can not read or write the spatial data that managed by MapInfo, and MapInfo can not read or write the spatial data that managed by ArcGIS,They can not make interoperability within oracle spatial, this is because not only the table USER_SDO_GEOM_METADATA was updated when the data was uploaded into oracle spatial, but also ArcSDE or MapInfo updated their own metadata table respectively. The metadata table of MapInfo in oracle Spatial is MAPINFO.MAPINFO_MAPCATALOG,the field GEOLOC in which is the Spatial data type. And when upload oracle spatial data with ArcSDE, the related information will update the table SDE.LAYERS and SDE.TABLE_REGISTERY, the spatial field is SHAPE. While the client (ArcGIS Desktop or MapInfo) wants to access the spatial data, the first step is access the metadata table, after get related metadata information and then start reading and writing the real spatial data. In order to make they are interoperable inside the oracle spatial, we need to modify the metadata tables of ArcSDE and MapInfo. This method will be discussed particularly late in Section 3.2.

The other issue concerns 'internet GIS', there were also many initiatives and studies into the use of standard-based technologies for the interoperability of Internet GIS applications. The Open Geospatial Consortium(OGC)has promoted research projects and also proposed many implementation specifications of web interface such as Web Map Service (WMS),Web Feature Service (WFS),Web Coverage service (WCS), and so on, together with a GML data standard. There were several studies to experiment with those standard-based technologies in their applications, and one example was that by Peng and Zhang (2004), in which they implemented Internet GIS applications using standards such as GML, WFS, and SVG, etc.. This model based on oracle spatial will also include the part of internet GIS, this will be dealt in Section 3.2.

3.1 Main objectives of the model

This model was designed by considering three main objectives: (1) Implemented centralization enterprise spatial data, if data was centralized storage, it will be more convenient for data manage, updating, maintaining and data distributing. (2) enable organizations to use the right tool for the job while eliminating complicated data transfers and multiple copies of the same data throughout the enterprise. (3) Implemented data sharing between enterprises, especially as data sources for local emergency department. Because emergency department can not own and maintain the all data they need.

3.2 Model design and implementation

This model has tow parts, Fig. 1 shows the main top view of the model structure, one for intranet users which means this solution suits for local user or inside an enterprise. By far the paper writed, this part supports two types of softwear ArcGIS and Mapinfo, this will be discussed throughout in Section 3.2. And another part of the model for internet users which means this solution suits for the data sharing between government departments or enterprises; this will be dealt in Section 3.2.

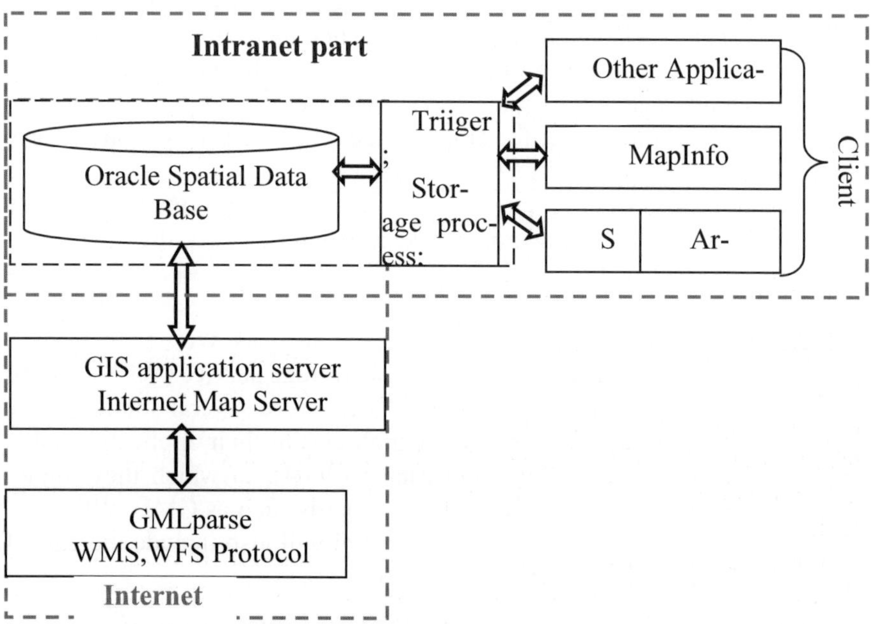

Fig. 1. Two parts of the structure model based on oracle spatial

Intranet Model

The data interoperability of intranet model based on spatial metadata which with in oracle spatial. So it is very important to analyze the respective metadata table inside the oracle spatial. The metadata table of oracle spatial named SDO_GEOM_METADATA_TABLE has described earlier.Now,the metadata tables of Mapinfo and ArcSDE will be discussed below.

The metadata table of Mapinfo inside oracle spatial named MAPINFO.MAPINFO_MAPCATALOG,which has 15 fields.In the interest of brevity,only 6 main fields discussed here: (1) SPATIALTYPE(Float): This column describes the Spatial Object Format of how the data is stored and indexed and the Spatial Object type(s) supported and not supported in the column. The digits to the left of the decimal point are the Spatial Object Format. The digits to the right represent the type of Spatial Object Type that can be stored in the column. (2)TABLENAME (char): The name of the table; (3) OWNERNAME (char): The owner name of the table; (4) SPATIALCOLUMN (char): The name of the column, if any, containing spatial features: SW_GEOMETRY (mappable using SpatialWare Type/IDS/UDO); NO_COLUMN (mappable using X–Y); MI_SQL_MICODE (mappable using MI Code); Or the name of the IDS/UDO, DB2, or Oracle column that is ST_SPATIAL datatype. Name of the Oracle 8i SDO_GEOMETRY column. SW_GEOMETRY;(5)MI_PRINX:this field does not exit in MapInfo table,it was created automated when uploading data; (6) DB_X_LL: The X coordinate of the lower left corner of the layer's bounding rectangle, DB_Y_LL: The lower left bounding Y value;DB_X_UR: The upper right bounding X value; DB_Y_UR: The upper right bounding Y value.

After uploaded Mapinfo table will create a unique index field named MI_PRINX,when spatial data was uploaded into oracle,the USER_SDO_GEOM_METADATA view was used to the information of mapinfo table.Using EasyLoader for uploading mapinfo data,the table MAPINFO.MAPINFO_MAPCATALOG will be created automated if it is not exit, and then write the metadata for the uploaded data in the created table.

When ArcGIS users access Oracle Spatial through ArcSDE, the metadata table SDE.TABLE_REGISTRY was used, the main fields of which are below: (1) REGISTRATION_ID:registry number,every spatial table has a unique number in the SDE.TABLE_REGISTRY; (2)TABLE_NAME:the name of the registried spatial table; (3) OWNER:the owner of the spatial table; (4) ROW ID_COLUMN:unique row ID number; (5) CONFIG_KEYWORD: data storage key word, is used

ger will be triggered to register the spatial data in the ArcSDE metadata table.

Because of ArcSDE supports LONG RAW, BLOG and Oracle Spatial for storage spatial data, and the type were determined by config_keyword.We choose SDO_GEOMETRY for config_keyword. We use ArcCatalog to upload a shape file named bld_poly.shp (the building),and at the same time, the trigger was fired to modify MapInfo metadata information. Figs. 5 and 6 are the two layers showed with both ArcMap and MapInfo.

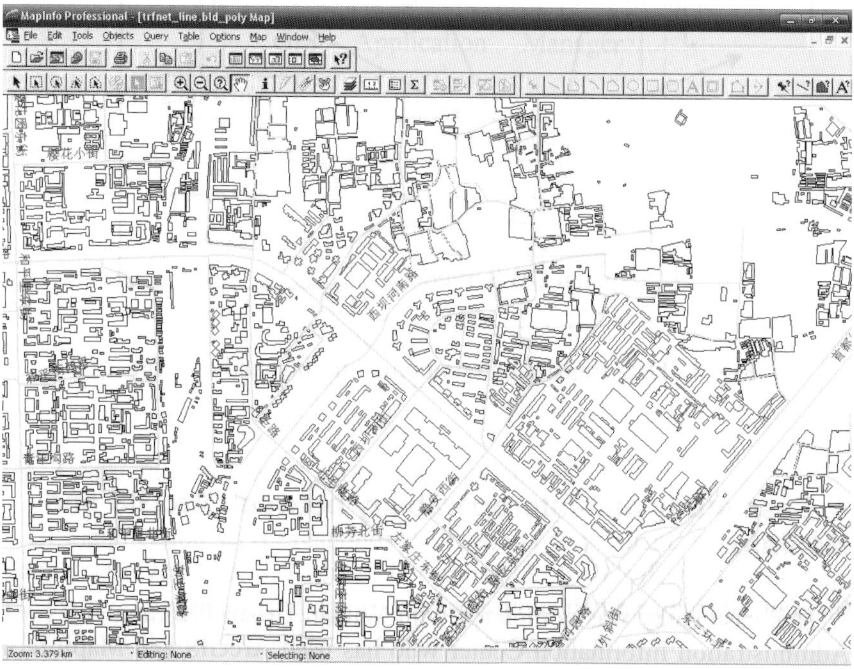

Fig. 5 Illustration of the data uploaded by ArcSDE and MapInfo in oracle spatial with MapInfo

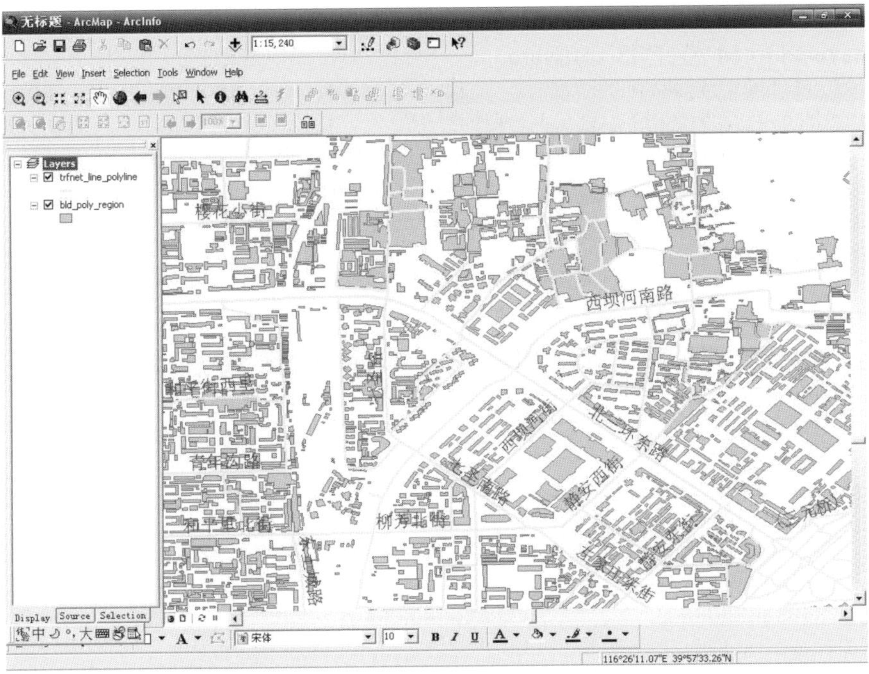

Fig. 6 Illustration of the data uploaded by ArcSDE and MapInfo in oracle spatial with ArcMap

5. Case study two-Interoperability WMS portals

This case study comes from two hypotheses, one supposes that a WMS portal from fire fighting department with fire fighting related data (distribution of 119, fire hydrant and so on), and the other portal from traffic department was supposed that supply transportation related data (street data, highway data).The emergency response department of BeiJing use the two portals' data to do some decision-making (Fig.7). The first step is to select and install a Web Feature Server. Currently there are several commercial WFS software programs available, such as MapServer, developed by the University of Minnesota which has WFS support, the Ionic Web Feature Server, GeoServer, GeoServerLite, and ArcIMS with WMS and WFS connector, developed by ESRI, which is not open-source software. Map

Server maybe a good choice, but in this case study, we choose ArcIMS with WMS and WFS connector because the related implemented projects have adopted ArcIMS software.

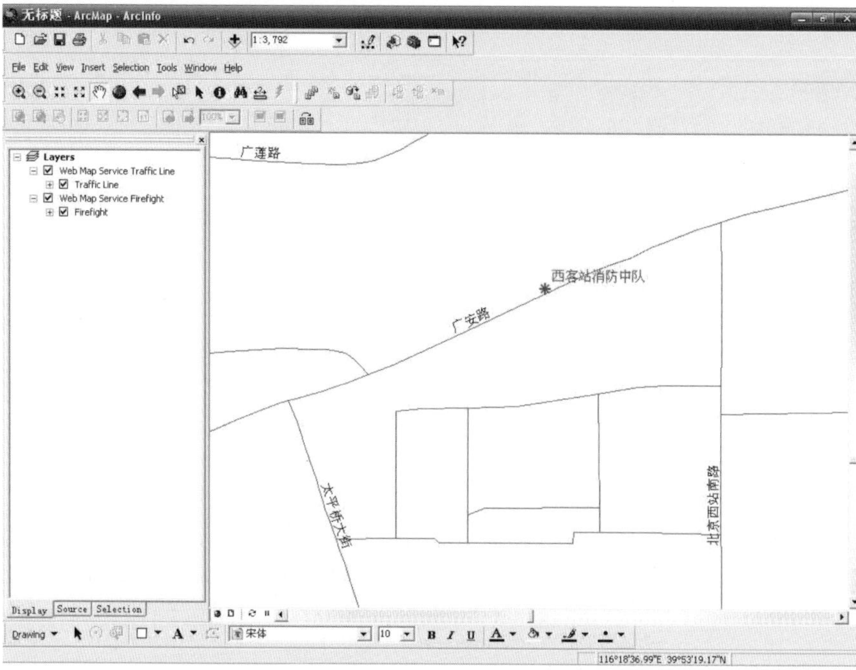

Fig.7 Integrate two WMS portals in a application

6. Conclusions

In this study, an oracle spatial based model has been suggested for spatial data integration interoperability inside and out side of an enterprise or local government departments. The design of the model was founded on the following main objectives: (1) Implemented centralization enterprise spatial data; (2) enable organizations to use the right tool for the job while eliminating complicated data transfers and multiple copies of the same data throughout the enterprise. (3) Implemented data sharing between enterprises, especially as data sources for local emergency department. The model has two main parts, one for intranet users which usually means inside an enterprise or local government department, and one for internet users means that between enterprises or government departments. The intranet part allows to select software to best meet its local needs and the internet part exposes a GIS portal which is an online GIS for outside of the enterprise or department .During the model design, two cases study were suggested, and they have been thoroughly tested for model validation and implementation.

Acknowledgments

The authors appreciate the financial supports from GLOBAL-SAFETY, and from Center of Public Safety Research, Tsinghua University.

References

Chang YS and Park HD (2006) XML Web Service-based development model for Internet GIS applications. International Journal of Geographical Information Science, 20, 371-399.

Cutter SL (2003) GI Science, Disasters, and Emergency Management. Transactions in GIS, 7, 439-445.

ESRI (2003) Spatial Data Standards and GIS Interoperability, available at http://www.esri.com/standars.last accessed on August 30,2005.

Harvey F and Tulloch D (2006) Local-government data sharing: Evaluating the foundations of spatial data infrastructures. International Journal of Geographical Information Science, 20, 743-768.

Intergraph (2005) Interoperability Framework, available at http://solutions.intergraph.com, last accessed on August 30, 2005.

MapInfo (2005) MapInfo User Guide, available at http://www.mapinfo.com.

OGC (2003) Data Models and Interoperability, available at http://www.opengis.org, last accessed on August 30, 2005.

Oracle (2003) Oracle Spatial User' guide and Reference.

Peng ZR (2003) A Framework of Feature-Level Transportation Geospatial Data Sharing Systems. Annual Meeting on Transportation Research Board, January 2003. Washington, DC

Peng ZR and Zhang C (2004) The roles of geography markup language (GML),scalable vector graphics (SVG), and Web feature service(WFS) specifications in the development of Internet geographic information systems (GIS), Journal of Geographical Systems, 6, 95-11

Collaboration enabled GIS Tools for Emergency Operation Centre

Zheng Chang and Songnian Li

Geomatics Program, Department of Civil Engineering, Ryerson University
350 Victoria Street, Toronto, Ontario M5B 2K3, Canada
{czheng, snli}@ryerson.ca

Abstract

Geographical information systems (GIS) have been widely used in emergency operation centre (EOC) for modeling, spatial decision-making and map distribution. However, the collapse of EOC itself, for example the destroyed EOC in the World Trade Centre on September 11, 2001 and the Hurricane Katrina in Louisiana had inevitably delayed the emergency response. Although virtual EOC (VEOC) provides partial solutions to this problem, most of the VEOC software products are limited in providing real-time collaboration and coordination using GIS tools. This paper presents the results of a research project, aiming at providing such collaborative GIS software tools over the Internet for the VEOC. The prototype software adopts a semi-replicated and distributed architecture and is deployed over Internet. Several collaborative tools such as telepointer, radar view, participant list and action status, chat and video are designed to support participant's awareness and collaboration. Using these tools, emergency personals, experts and government agencies at all levels and dispersed all over the country can share not only data sources but also the views, GIS modelling or process, even work environments hosted in a EOC or several EOCs, so that everyone with access privilege can evaluate the critical situations at the same time without physically present in the operations centre.

1 Introduction

The Emergency Operations Centre (EOC) is a facility designated for managing the disaster emergency. The main stages include: 1) giving policy direction and support to the incident commanders; 2) collecting, evaluating and displaying information about the incoming emergency situations; 3) preparing and creating Incident Action Plan; and 4) finalizing, approving, and implementing Incident Action Plan (KPB, 2004). Some critical problems were arising when the EOC in 7 World Trade Center collapsed on September 11, 2001 and so did the EOC in Louisiana, New Orleans during the Hurricane Katrina. The problems are also called "weakest link problems" since these events could collapse the entire emergency management system. These problems include: 1) the EOC and the hosted data and information system may be destroyed; 2) the emergency staff may not be able to reach the EOC from their normal locations in a major metropolitan; 3) the first response staff in EOC may not have enough knowledge to evaluate the situation, but the related experts can not access the systems to help them in EOC in time; and 4) a large scale disaster covering large areas could involve coordination and collaboration among multiple jurisdictions and disciplines beyond the stage of any EOC.

The Virtual EOC (VEOC) therefore is used to enable EOC staffs to participate in critical decision-making processes regardless of their physical location and to automate direct and control resources, assign and track tasks. In a word, an EOC is "virtual" when participants can share information, make decisions and deploy resources without being physically present in the EOC. Typically using web-enabled software, a VEOC allows participants to work from their normal workstation, from home, from the field (David, 2002), or ideally from any computers connected to the Internet.

Lots of VEOC software such as OpsCenter, WebEOC, E-Team, ESS Crisis and SoftRisk (DavisLogic 2006) appeared especially after September 11. A detailed review of these systems is not feasible given the limit of paper length. In short, all of these systems support the daily operations such as tracking, assigning, commanding and logging, etc. Many of them such as E-Team, ESS Crisis and WebEOC have GIS-enabled functions. However, most of these systems are restrict in simple map view and attribute query. More deeply logic integration between VEOC and GIS is very limited. Real-time collaboration and interoperation among users are not seen in these systems even in E-Team system which announces its real time collaborative ability.

This paper is not to design a fully functional VEOC system, rather to design a collaborative GIS software tools over the Internet for the VEOC. The technical background and related system review about collaborative GIS are presented in Section 2. The requirements of collaboration used in EOC are analyzed in Section 3 and followed by architecture and framework design in Section 4. The prototype of proposed system is introduced in Section 5, followed by some discussions in the Section 6.

2 Background and Related Work

This collaborative computing technique is also termed as Computer Supported Cooperative Work (CSCW) application, or groupware application. The following section will present a review of the background and the related work.

Computer supported cooperative work is a computer-based system that supports group of people engaged in a common task (or goal) and that provides an interface to a shared environment (Ellis 1991). Both in academia and in industry, various groupware prototypes and products have emerged and provided particular functionality to users. Each groupware system is designed to support a particular range of cooperative work situations. Cooperative work settings are very diverse in terms of task, duration, group, organizational context and culture (Hinssen 1998). Groupware systems are therefore often classified according to the type of collaboration that they support. In the current classification scheme, collaboration has a temporal and spatial dimension. The groupware systems can be either synchronous or asynchronous based on the time dimension and can be co-located or distributed based on the place dimension (Johansen 1998; Dix 1996).

Several synchronous distributed based collaborative frameworks, for example Habanero, Groove, GroupArc, NetMeeting, and so on, have been developed to support collaborative GIS functions. These early efforts focused on combining GIS with CSCW hardware systems and software (groupware), or at least applying CSCW concepts in developing collaborative GIS systems (e.g. Churcher 1996; Faber 1997; Jones 1997; Li and Coleman 2005; Nyerges 1997; MacEachren 2001). These developments used either an existing commercial GIS system or in-house developed GIS viewer, and integrated it with a groupware framework such as electronic meeting systems. For example, Habanero and Groove are complex groupware frameworks which support collaboration GIS functions. But the design and development of collaborative GIS have to be based on there groupware frameworks. These kinds of groupware frameworks are either too complex or too expensive to modify and/or extend a current single user GIS to a collaborative GIS.

The proposed collaborative GIS tools try to adopt a lightweight method, collaboration-transparent method, to integrate synchronous collaborative functionality and GIS into VEOC systems. This method makes it possible to change a single-user interface GIS application to a collaborative GIS application just through adding some collaborative component or plug-ins. Since GIS modules and Collaborative modules are highly independent, the collaborative modules can, therefore, be applied to other GIS applications with minor changes.

3 Collaboration in VEOC

In addition to EVOC logic and rules, more functions such as collaboration functions, GIS functions and other decision-making analysis functions are required to integrate into EVOC system. Analyzing the requirements of the entire EVOC system is beyond the scope of the research. In this section, we focus on the requirements of collaboration in an Internet-enabled GIS environment which supports EVOC.

3.1 Collaboration requirements analysis

The detailed collaboration requirements can be analyzed through a case study. When the gas pipeline, for example, is exploded in a city area, usually it is the first response expertise in the EOC to evaluate the situation and carry out related response such as command, operations, planning, logistics, and finance (KPB 2004). When the situation in the exploding site is too complex and beyond the evaluation abilities of the first response expertise in EOC, the second layer or a network of experts will be required to provide additional support. Unfortunately, this second layer's expertise who may be from provincial regional EOC, municipality and fire department of a city are usually remote and separate from the wealth of information and situational awareness provided by the emergency operations system. These supporting personnel frequently find themselves making decisions without the benefit of some critical information that is available locally, but no one knows to provide it (Intergraph 2004).

Moreover, when a major disaster happens, it may be impossible to get all the personnel from their normal locations to the emergency operation center in major metropolitan areas. Major disasters produce nearly instantaneous gridlock and congestion. The key personal cannot afford to be in traffic during the transit time which could be the most critical minutes and hours of the emergency.

The worst may happen if the EOC is destroyed like it happened in the EOC in 7 World Trade Center on September 11, 2001. The Following situation was described by ArcNews Online (2001).

"On September 11, 2001, New York City's Emergency Operations Center (EOC) was in all respects a state-of-the-art facility, ready for any crisis. Only three and one-half years old, the EOC was equipped with generators, backup generators, a water supply, and a ventilation system capable of filtering out 99 percent of airborne impurities as well as computer hardware, telephones, and radios with uninterruptible power supplies. All of the equipment was housed within steel-framed, reinforced exterior walls designed to withstand 200-mile-an-hour winds. But the EOC was not indestructible. When two jetliners slammed into 1 World Trade Center and 2 World Trade Center and the two skyscrapers collapsed including one at the EOC's home, 7 World Trade Center. By the end of the day, all of that state-of-the-art equipment and data was no more."

The new EOC was forced to move twice and finally rebuilt in a large ship terminal on Hudson River with heavy security on the third day (See Fig. 1). The first several days running was still facing problems such as short of data and tools to evaluate the situations.

Fig.1. The new EOC in Pier 92, a large ship terminal on the Hudson River after three days of September 11, 2001 (ArcNews Online 2001).

3.2 Detailed functionality

The collaboration requirements may be summarized into following three aspects according to the analysis of various EOC cases:

- Multi-user awareness and communication. The integrated multi-user text chat, shared video and audio, selective groupings of users, etc may work to share the same situation.
- Remote controls on the application. All the parties such as all level of EOC operators, field investigators, government agencies and so on, who are involved in the emergency, can remotely access, operate and manage the shared application.

- Shared view, shared data streams and data source. Users who join the same session can share the same view, data streams and data sources.

The above collaborative features can be implemented through so called collaborative interfaces or wedges which are used to help end users carry on collaborative work. These interfaces include telepointer, radar view, participant list and action status, chat and video, etc.

- Participants mean the users who take part in the same session. Participants usually can be aware of others' exist through a participant list. Participant list also shows participant's work status such as access privilege, floor control status, etc.
- "Floor control" refers to the management of interaction among participants in meetings. This comes from expressions such as "who has the floor" or "yielding in the floor" in formal meetings. For example in a shared white board, only one participant has the floor and draw at one time.
- Telepointer shows a participant's cursor when he/she controls the shared view and operate the user interface. All other participants in the same session can see the cursor. Usually, the cursor also shows user's information such as user's name or ID. Telepointer helps other participant be aware of the operator's action.
- The radar view displays an overview map of other participant's area of interest, especially in collaborative GIS. Viewports on the radar are represented as coloured transparent rectangles, and are used to convey awareness information about where the operator is currently working in the map area.

These tools together with the shared audio and video help multiple users improve awareness and presence, and thus enhance group spatial decision-making process in a VEOC system.

4. Framework Design

4.1 Transparent collaboration method

Most of the GIS applications are single user applications. It would be beneficial to just modify few codes to transfer the single user GIS applications into the collaborative (multiple user) ones.

Collaboration-transparent method is the potential solution. Li (1999) classified three main techniques to solve this problem (see Fig. 2). Approach (1) slightly modifies the application code to relate to the collaboration framework class. Approach (2) substitutes the underlying window system or graphics toolkit. Approach (3) interposes an agent to control the system event queue between the application and the window system. Each approach has its benefits and shortages. The proposed system uses a hybrid approach that combines (1) and (3). Some components use approach (1) and GIS related functions use approach (3).

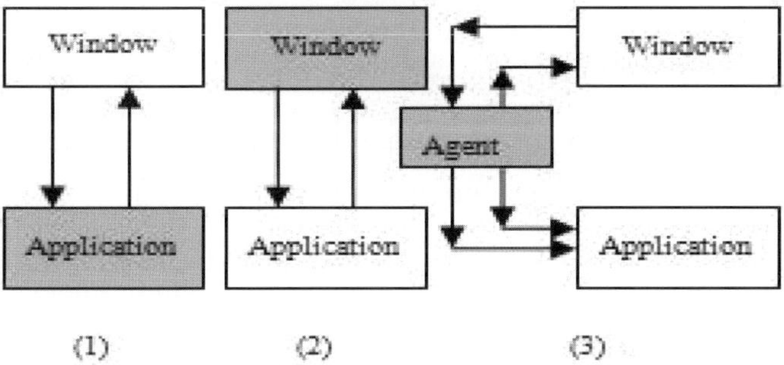

Fig. 2 Common technique for collaboration-transparent application (Li 1999).

The approach (3) is also called collaboration-transparent method. The agent could be a pseudo-layer in which every mouse event or keyboard event will be captured and sent to other clients. Through consistency mechanism, all the clients can work together and at same time ensure the consistency. The pseudo-layer is a Glass pane (Sun 2005) in a frame in the proposed client (see Fig. 3).

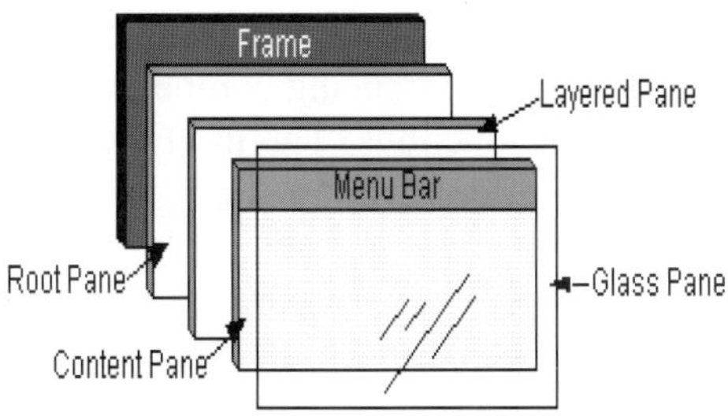

Fig. 3 Glass Pane in a Java system (Sun 2005)

The GIS component intercepts all the user events (e.g., mouse, keyboard, input focus events) using a transparent GUI component called Glass Pane, which is available in the Swing toolkit. It is at the topmost Zorder to cover the bean's GUI area and intercepts all the user's events, without occluding the underlying Java component. The Glass Pane can also be dynamically shrunk or expanded, thus allowing easier management of public and private areas of the same shared workspace.

A Collaborative component handles replica collaborative function including floor control, session management, message sending, parsing, and reconstructing and so on. This component uses approach (1) in Fig. 2. only command event, like button click, can be sent to other clients. Other events like mouse moving will not be responded.

A Single-user GIS application therefore can be modified to collaborative GIS application through the following three steps: (1) add the collaborative component which handles consistency into the single-user GIS application, (2) add the class pane over the main user GUIs in single-user GIS application, and (3) integrate the first two steps.

The client includes mainly two components: GIS component, and collaborative component. GIS component possesses mainly interactive GUIs and is the main workplace for the session participants to work together. Most collaborative operations, for example, view and zoom to the hotspot, annotation, buffering and so on, happen in this component. It is very important for the participants to view others actions even just moving mouse.

4.2 Architecture design

The proposed system is a light-weight java application composed of three tiers: replicated client tier, shared server tier and data tier (see Fig. 4).

The replicated client is downloaded from a known web server. Every user who wants to launch the system will get the same client. This client is composed of several components including GIS component, collaborative component, and multimedia conferencing component, etc. the GIS component includes all basic GIS functions. The multimedia conferencing component helps to communications among users by audio, video, and text chat. The collaborative component is responsible for the synchronous collaboration between the users. The transparency method, a Class panel, is used in this component. All consistency mechanism and multicast network communication are also included in this component.

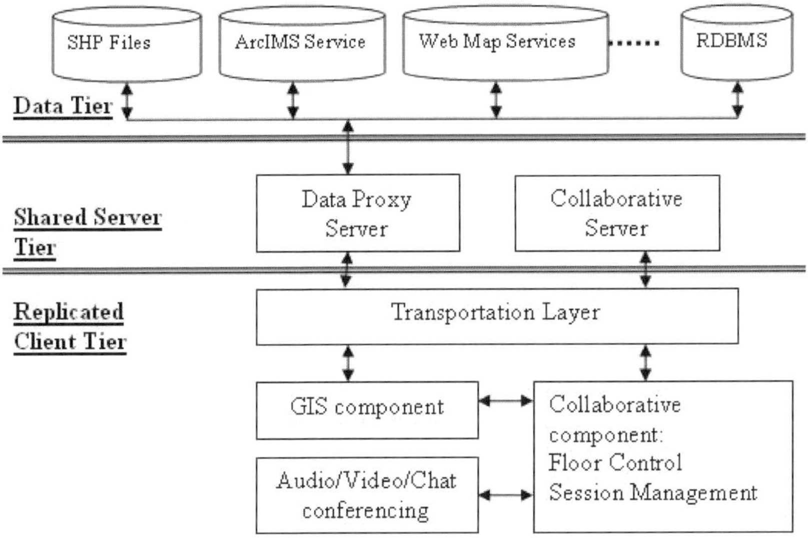

Fig. 4 Framework of proposed system

The shared server tier includes two servers: Data Proxy Server and Collaborative Server. These servers run on a known server, and are shared by all clients. The Data Proxy Server is mainly responsible for database source related issues including database source connection, data access and retrieve, etc. The Collaborative Server receive incoming messages and multicast them to others clients.

The data tier is the data sources that the system can handle through the Data proxy Server. There data sources include SHP files, rational DBMS, Web Map Services (WMS), Web Feature Service (WFS), etc.

The original single GIS application also can be extended into collaborative GIS application through adding two components separately in the Shared Server tier and the Replicated Client tier: Collaborative component and collaborative Server.

Following procedure presents how the system works: first of all, the Collaborative Server is launched. Second, a replicated client is launched and sends a request message to the Collaborative Server to query the system state. The system state includes how many clients in the system, what data sources are using, map extensions, and menu states, etc. After obtaining the state data, the client is ready to join the session and also is ready to share its operation with others. Then when the client operates its interface, the same operation is broadcasted to other clients with consistency mechanism. The consistency mechanism will be resented in the next section.

The sessions even can be distinguished in one machine through Session name.

GeoLink Server can create a new session through inputting these five parameters. GeoLink Server also deletes a launched server, disconnects and connects the server as well.

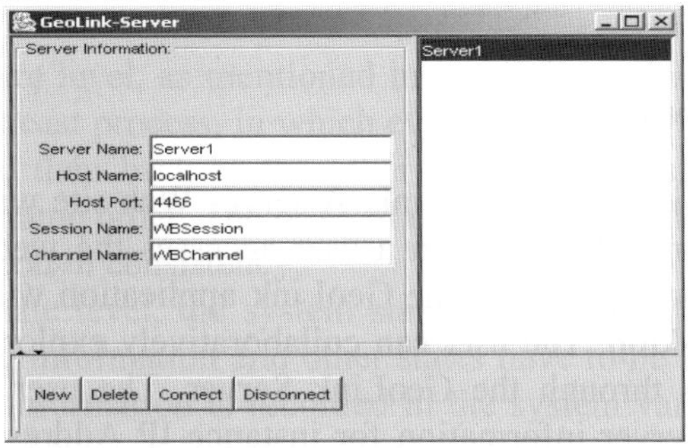

Fig. 6 Snapshot of GeoLink Server

5.3 Replicated client

GeoLink Client includes two important components, collaboration work place and GIS tools. Every client has the same collaboration work place in which other participants, shared files, shared data sources and text chat can be visualized (see Fig. 7). Shared files are those files that users want to share and upload them to a virtual directory where every user can download them. Shared data sources are GIS or other related data sources, like Shapefiles, or Web Map Service, which are used by GIS tools in the session. Text chat provides text communication among users.

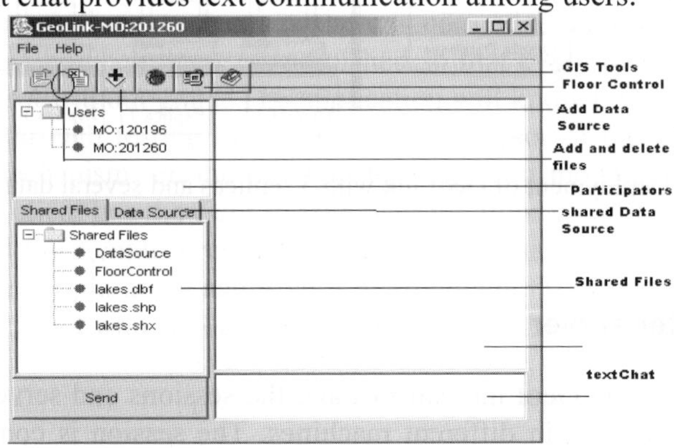

Fig. 7 The main interface of GeoLink Client

Add data source function (see Fig. 8 left) adds shared data sources into GeoLink system. Two kinds of data sources, Shapefile format and Web Map Service, are provided as the GIS data source. More data source types will be added later.

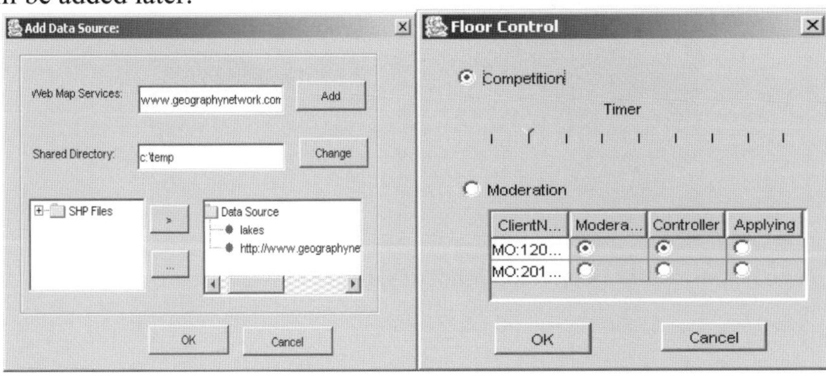

Fig. 8 The interfaces of shared data sources (left) and Floor Control (right)

Floor Control function (see Fig. 8 right) manage who control the session as Moderator, who control the operations of GIS tools as Controller and some state of the session.

There are two mechanisms to manage floor control: detection floor mechanism and grant floor mechanism, which refers to the interfaces of Competition and Moderation respectively. In the Competition option, there is a 10 second limit timer to control the gap of competition. For example, if the timer is set to 3 seconds, the controller, the user who is controlling the floor, will lost the control right after he does not have any operation on the floor after 3 seconds. The user has to re-catch the control right as other users. The first operation on the floor will catch the control right.

In the Moderation option, the controller can be assigned by moderator who may be the creator of the session. The moderator can see who has applied for the controller and make a decision. The moderator also can give his moderate right to others so that if he leaves the session can exist.

The shared GIS tools are basic GIS functions with collaborative features, such as shared view, telepointers, etc. Any user in the same session has the same map view and annotation. The new operations, such as annotation, zoom in, zoom out, and pan, etc. on the map will be updated to all the views synchronously. The controller shows his cursor and user name, which is termed as telepointers, to other users. Fig. 9 shows two users/clients (MO120196 and MO201260) launched in one machine sharing the same map view and observing other operations of the application.

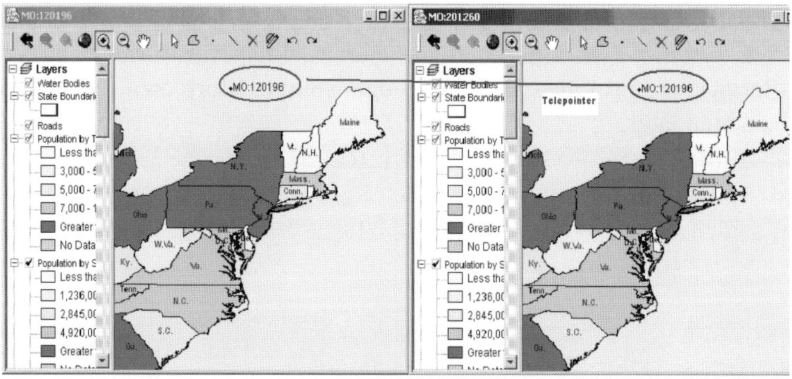

Fig. 9 Snapshot of shared GIS view

6 Discussion

This paper introduces a transparency method to create synchronous collaborative GIS tools supporting EOC. The system requirements are conducted through a case study in EOC. Through the case study, some core functional requirements in collaboration with GIS environments are obtained. A prototype for collaborative GIS tool is designed, developed, tested and deployed. The modules used in the prototype are extendable and scalable, which make it possible to change a single-user interface GIS application to a collaborative GIS application just through adding some collaborative component or plug-ins.

Because GeoLink is a concept-proof prototype, the research reported in this paper is subject to several possible constraints and limitations listed as follows:

- The system requirement analysis in this case study is approximate and not so well detailed that every piece of work is involved. The main requirements are focusing on GIS and collaborations. Others like workflows, audio/video conferencing which are also very important in designing EOC system are ignored at this stage.
- Based on the system design, any GIS tools should be available to develop the prototype, but just MapObject was tested for the development of GIS components. In the future, some open source GIS tools should be tested in proposed framework.
- The usability and functionality of prototype was not tested and evaluated in real VEOC system.

Acknowledgements

This research has been partially supported by a National Sciences and Engineering Research Council (NSERC) of Canada Discovery Grant awarded to S. Li and a GEOIDE Networks of Centres of Excellence project (SLM-DFM#15). The authors thank the reviewers for their constructive comments, which not only improved the quality of the paper but also help in further research effort of the project

References

ArcNews (2001) New York City-creating a Disaster Management GIS on the Fly. In: ArcNews, 001/2002

Churcher N, Churcher C (1996) GroupARC – A Collaborative Approach to GIS. In: the 8th Colloquium of the Spatial Information Research Center, University of Otago, New Zealand, July 9-11

David S (2002) Virtual Command Centers. In: RIMS, Risk Management Magazine, July 2002.

DavisLogic (2006) http://www.davislogic.com/EOC.htm#EOC%20Software (accessed December, 2006)

Dix A (1996) Challenges and Perspectives for Cooperative Work on the Web. In: Proceedings of the ERCIM workshop on CSCW and the Web, Sankt Augustin, Germany, February 7-9, 1996

Ellis C, Gibbs S, Rein G (1991) Groupware: Some issues and experiences In: Communications of ACM, 34(1), 38-58

ESS (2006) http://www.ess-home.com/products/crisis-management (accessed December, 2006)

ETEAM (2006) http://www.eteam.com/index.html (accessed December, 2006)

Faber B. G, Wallace W, Croteau K, Thomas V, Small L (1997) Active Response GIS: An Architecture for Interactive Resource Modeling. In: Proceedings of the GIS'97 Annual Symposium on Geographic Information Systems, Vancouver, B.C, March 1997, 296-301

Hinssen P.J.H (1998) What Difference Does it Make? The Use of Groupware in Small Groups. Telematica Instituut, Enschede, The Netherlands, 258 pp

Intergraph (2004) Technology Insights for Emergency Operations Centers, http://www.intergraph.com/resource_files/literature/DDFS013A0/DDFS013A0_print.pdf (accessed on July 2004)

Johansen, R (1988) Current User Approaches to Groupware, in: Johansen, R. (ed): Groupware, Free Press, New York, pp. 12-44.

Jones RM, Copas C, Edmonds EA (1997) GIS support for distributed group-work in regional planning. International Journal of Geographical Information Systems. 11(1), 53 – 71.

The first step in DPA is to compare the largest perpendicular distance between any point and the subtended baseline P_1P_{26} (d), to a pre-specified distance tolerance (Fig. 2). In this example, if T1 is the tolerance, no point on the trajectory is further than the tolerance distance T1 from the subtended baseline P_1P_{26}, and the baseline is deemed sufficient to represent the original trajectory (dashed line). If T2 is used instead for the tolerance, then furthest point (P_{15}) from the baseline exceeds it, and becomes the vertex of new line segments more closely following the original trajectory. For subsequent steps, the same procedure is iteratively carried out with the new segments, and so on until all points are within T2 of the closest segment of the final path.

It can be seen from Fig. 2 that this algorithm concentrates on the selection of points rather than on their deletion, and this algorithm is fast since choosing the furthest point as the line-splitting position improves calculation efficiency.

2.2 MARIN Douglas-Peucker Algorithm (MDPA)

While the DPA process of retaining the points that deviate too much from a line and eliminating the remainder yields good results, the algorithm may also eliminate information that is important for characterizing the line. Fig. 3 gives an example of a situation in which the final output from DPA yielded the straight line AB, which omits too much representative information about the vessel's movements.

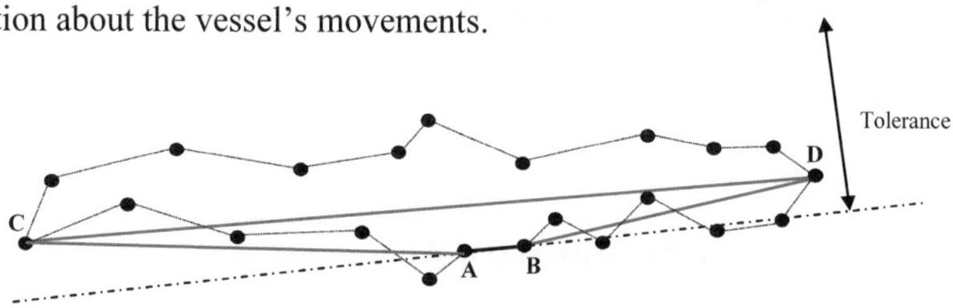

Fig.3. Over-Dedensification

The Maritime Activity and Risk Investigation Network (MARIN) remedied this DPA shortcoming by initially detecting points where the trip turns back. This can be determined by calculating the straight-line distance of every point from the launch point. If the distance from a specific point to the launch point starts to decrease relative to the previous one, that preceding point is defined as a "turning back" point and the trajectory is split there. The new launch point is that specific one, and then the process is applied recursively to the rest of the trajectory. As a result, the original trajectory will be separated into several segments according to the number of turning back points. For instance, point C and point D are the turning back points in Fig. 3, so the original trajectory is separated into three segments. Thereafter, the traditional DPA method is applied to every new segment. As shown in this example, the turns made by the trip are guaranteed to be preserved, which is essential information for the later classification of trip movements.

2.3 Two-Criteria algorithm (TCA)

The dedensified path yields two important measures; one is the segment distance between two successive points before the boat turns, and the other is the turning angle between the two adjacent segments.

In many cases, the turning angle is more important than the straight segment distance, because the distance may be influenced by more factors such as the geography; however, the turning angle patterns are largely associated with boat type. Moreover, the deviation of segment lengths is so large that it is not a suitable attribute for analysis stability (Wu 2006). Hence the turning angle is a more promising attribute for examination. DPA and MDPA only consider one criterion, which based on distance deviation tolerances. However, this Two Criteria Algorithm considers both angle and distance tolerances. As shown in Fig. 4, assuming a certain recreational boat traveled from point P_1 to point P_3 via point P_2, the two criteria of whether a boat makes a significant turn along its trajectory are:

- Turning angle α is greater than a deviation tolerance of angle, OR
- Perpendicular distance h is great than a deviation tolerance of distance.

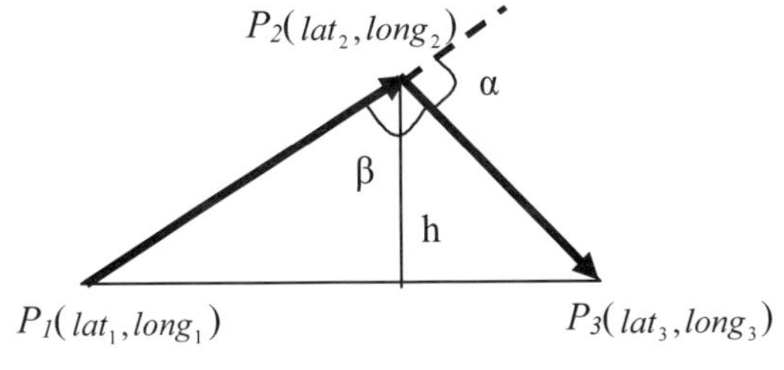

Fig.4. Parameters of the Two Criteria Algorithm

Based the above criteria, if the point P_2 is retained, the boat is considered to have turned. Otherwise this point is deleted, as would be warranted

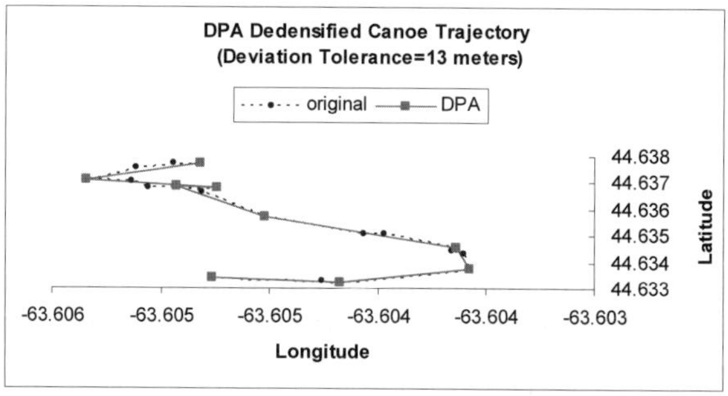

Fig.7. DPA Dedensified Boating Trajectory (DT=13)

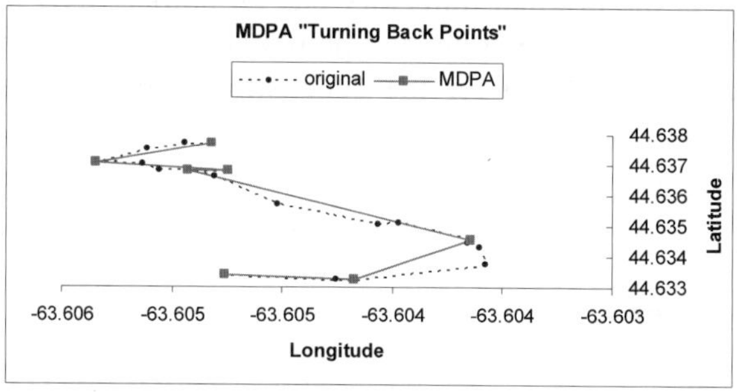

Fig.8. General Tendency from MDPA

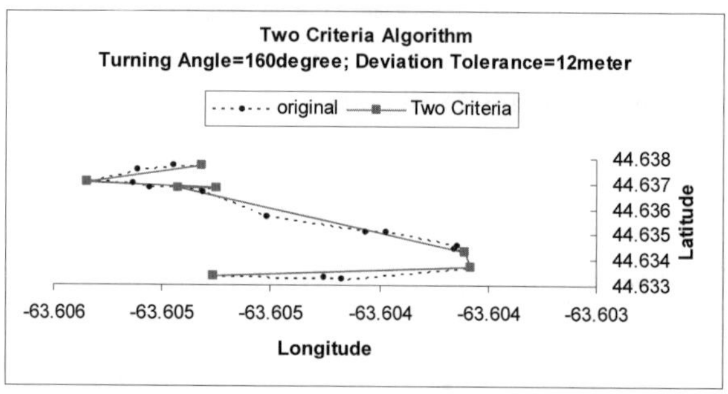

Fig.9. Two Criteria Algorithm

Table 1. Summary of Dedensified Paths for Different Algorithms

	DPA	MDPA	TCA
Deviation Tolerance	13 meters	31 meters	12 meters 160°
Number of Points	9	7	7
SAD (degree)	0.000324	0.001192	0.001255

3.2 Kayaks' Boating Trajectory

The features of kayaks are superficially similar to those of canoes; for example they are slower and are human-powered compared with other boat categories such as sailboats and motorboats. But this study found that kayaks typically travel longer distance than canoes. Fig. 10 shows an original kayak trajectory with 190 points. Because of the larger resolution, some parts of the trajectory are not clearly visible in this figure. Hence, Fig. 11 provides zoomed views of areas S1 and S2. The movements in these zones are relatively complicated, with several sharp turns in S1 (Fig. 11a) and a loop in S2 (Fig. 11b). Unlike the treatment applied to the canoe trajectory presented above with the aim of retaining the loop, it could be advantageous to treat these two kayak boating trajectory segments as wavers instead of principal movement features, since these two segments form tiny parts of the entire trajectory.

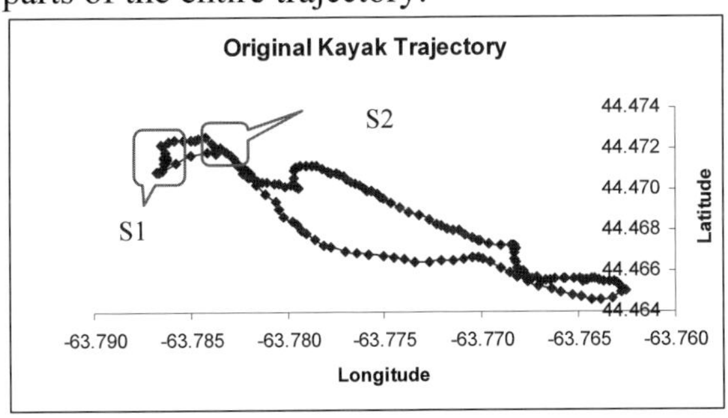

Fig.10. Original Kayak Trajectory.

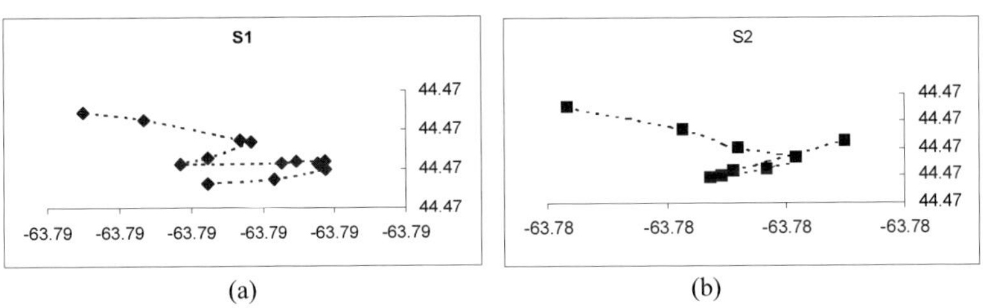

Fig.11. Trajectory Segments

ance to 50° and deviation tolerance to 14 meters (Fig. 17). However, TCA resulted in a bigger SAD. Actually this is a disadvantage of the TCA. It is myopic because TCA checks every three successive points to make decisions about keeping or deleting the middle point, while the Douglas-Peucker based algorithms examine all the points every step. When dealing with a relatively longer distance and smoother trajectory, TCA may result in a bigger value of SAD. Since the trajectory in Fig. 1 is relatively long and smooth, the angle subtended at the middle point by joining each set of three successive points was too small to be treated as a turning angle so that the application of two criteria algorithm deleted all of the points except the starting and ending points, resulting in the dedensified path shown as a dashed line in Fig. 1. Conversely, by choosing a suitable deviation tolerance, Douglas-Peucker based algorithms at least keep the point located at the position furthest away from the straight line (solid line in Fig. 1). The advantage of the TCA is its ability to capture the characteristics of the angles. The analyst can control the outcome by selecting a turning angle deviation tolerance to preserve turns of a specified sharpness, which furthermore can be tailored to different boat types.

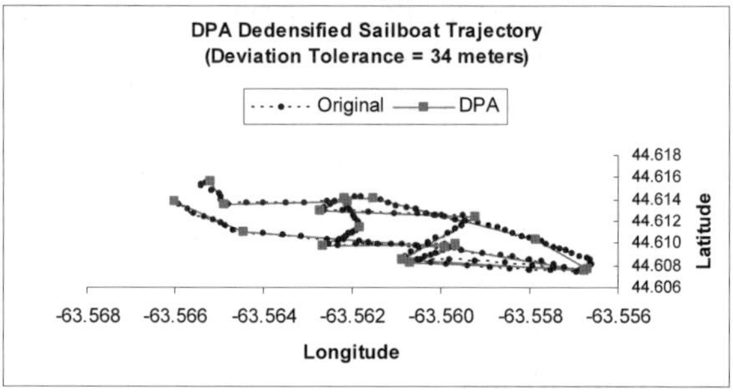

Fig.15. DPA Dedensified Sailboat Trajectory

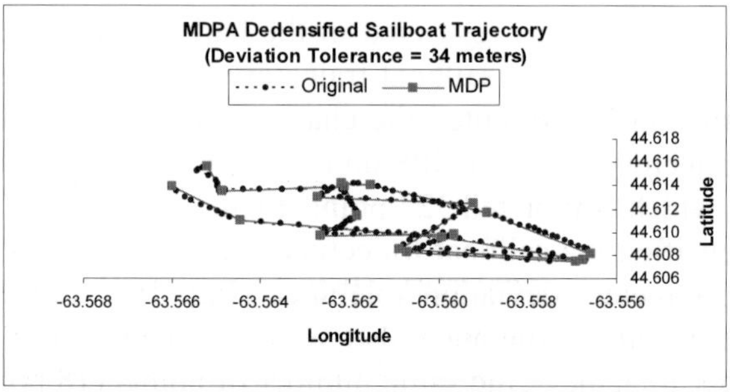

Fig.16. MDPA Dedensified Sailboat Trajectory

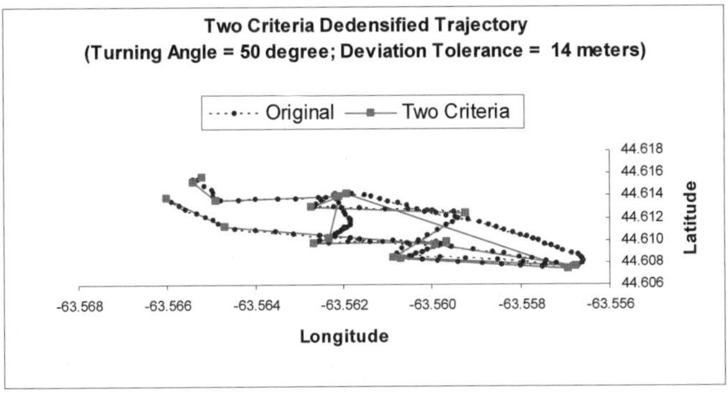

Fig.17. Two Criteria Dedensified Trajectory

4. Conclusion

Based on these outcomes from applying various dedensification procedures, one should apply these algorithms with caution. Their behaviour depends to some extent on the type of boat, which partially determines the characteristics of the simplified trajectory. Moreover the research aim is important. The type of attributes one wants to emphasize has a bearing on the choice of algorithms. For example, if one wants to pay more attention to turning angle, the two criteria algorithm should be selected. If it is deemed important to retain only the overall tendency of the movements relative to the scale of the entire trip, then the MDPA should be applied. Moreover, considering the number of points in the cleaned path, the most efficient algorithm is the DP algorithm, while the two criteria algorithm is the least.

Acknowledgements

This project is supported by the Canadian Coast Guard (CCG), the National Search and Rescue Secretariat (NSS), and the GEOIDE National Centre of Excellence (NCE). This research relies on excellent research assistance by members of the MARIN (Maritime Activity & Risk Investigation Network) research group in the Department of Industrial Engineering, Dalhousie University, in particular Christine Delbridge, Andrea Cameron, Lauren Plummer and R.C. Hilliard. Finally, we are grateful to the anonymous boaters who volunteered to employ our GPS units to monitor their vessel's movements.

References

Douglas DH, Peucker TK (1973) Algorithms for the reduction of the number of points required to represent a digitized line or its caricature. The Canadian Cartographer 10 (2): 112 –122.

Wu Y (2006) Characterizing Recreational Boating Patterns based on GPS Trajectory Points. Ph.D. thesis, Dalhousie University

Hierarchical Risk- Based Spatial Analysis of Maritime Fishing Traffic and Incidents in Canadian Atlantic Waters

Jamal Shahrabi[1] and Ronald Pelot[2]

[1] Faculty of Industrial Engineering, Amirkabir University of Technology, Tehran, Iran, * Dept. of Industrial Engineering, Dalhousie University, Halifax, NS, Canada, Jamalshahrabi@aut.ac.ir, Shahrabi@dal.ca,
[2] Department of Industrial Engineering, Dalhousie University, Halifax, NS, Canada, Ronald.Pelot@dal.ca

Abstract

Maritime traffic analysis is growing in importance for many reasons: risk management, accident prevention and response planning. Since most decisions are location-sensitive, one important consideration in maritime traffic analysis involves maritime risk analysis, including spatial analysis to identify hot spots. Hot spots areas are concentrations of incidents within a limited geographical area that appear over time.

In recent years, information - particularly spatially referenced information - and the tools for information analysis have become increasingly recognized as an essential part of the policy-making and decision- making processes to make the descriptive, explanatory, predictive and risk prevention models.

The increasing power of computing hardware and software and the increasing sophistication of the analytical geomatics techniques mean that new opportunities are available for spatial analysis to improve the quality of risk prevention plans.

Recent studies have shown that geospatial information is of fundamental importance to maritime risk analysis providing efficient risk management and GISs represent a powerful new technology that can address many information needs of risk managers and decision makers working with geographically referenced data.

This study used the increased capabilities offered by Geomatics techniques and geographic information systems to identify hazardous locations for maritime traffic in Canadian Atlantic waters. This research uses spatial analysis to examine risks associated with maritime commercial fishing vessels activities and incidents.

1 Introduction

Canadian coastal zone management comprises an extensive network of waterways, terminals and ports that play a critical role in both domestic and international commerce. Concern for preservation of the marine environment and improved vessel safety standards within Canadian waters has led to initiatives to prevent environmental disasters and reduce maritime casualties. The identification of relatively hazardous location can assist the Canadian Coast Guard (CCG) in risk prevention planning and strategic decision-making issues concerning the allocation of its search and rescue (SAR) resources, as well as better evaluation of Canadian fishing regulations and coastal zone management programs.

1.1 Maritime Risk Analysis

As fishing has long been a high-risk occupation, the high incidence of injury and loss in the US commercial fishing industry has been a focus of national concern for more than 20 years (Van Noy, 1995). For centuries, commercial fishing has been faced with the loss of life and vessels. The death rate for US commercial fishers in 2002 was seven times the national average for all industries, and fishing remains perhaps the most hazardous occupation in the country (Jin, 2001, 2002). Based on U.S. Coast Guard reports, the U.S. commercial fishing fatality rate in 1996 was 16 fisher deaths per 10,000 workers. This rate is 16 times higher than the fire and police protective service occupations rate. A U.S. Coast Guard report in 1999 says that between 1984 and 1998, 2074 U.S. commercial fishing vessel accidents resulted in total vessel losses. The annual property, injury, and other costs of U.S. commercial fishing vessel accidents are estimated to be over US$ 240 million, a figure which is more than three times the comparable cost of tanker accidents (ICF Kaiser Consulting Group, 1997).

The assessment and management of risk in maritime transportation is an important problem domain, specifically identified by the National Research Council (NRC 1986, 1991, 1994, 2000). Unfortunately there has not been much published on maritime risk analysis generally and on commercial fishing vessel safety specifically.

The few quantitative maritime risk analyses that were done all suffered from lack of fishing activity data. It is obvious that the accident records alone cannot determine whether the accident rate has increased or decreased, since the number of accidents is also affected by the number of fishing vessels and the number of trips per vessel. Data limitations have prevented researchers from investigating the probability of vessel accidents themselves, as for this kind of research data from both vessel accidents and safe fishing trips are required. While the former are often available, comparable data on safe fishing trips are not readily obtainable (Jin, 2001, 2002). One of the most important advantages of this study is the ability to use fishing activity data in addition to fishing incident data. The other advantage of this study is doing spatial analysis. Although few researches include some location variables they cannot be considered spatial analyses as they do not include any spatial factors - specifically neighbourhood variables.

1.2 Spatial Analysis and Risk Management

In recent years, quantified information – including spatially referenced information – and the tools for information systems and analyses have become increasingly recognized as essential for disaster and risk management. Because most real-world objects can be associated with a location, spatial aspects are potentially important to any problem. Generally speaking, spatial considerations concern geo-referenced data.

Consequently, quantitative data analysis is either spatial or aspatial. The key difference between these two types of data analysis is the inclusion or exclusion of spatial factors. All spatial analysis involves processing information about location. Spatial data derive from a variety of sources. The most commonly used spatial data are obtained from sources such as census, land use surveys, satellite imagery and aerial photographs. Recent technological advances in remote sensing and computer science provide access to tremendous amounts of spatial information on a daily basis.

Classical spatial analysis has been treated conceptually for about a century, and algebraically for several decades. It has played a central role in the quantitative scientific tradition in geography; until the recent development of Geomatics techniques and Geographic Information Systems (GIS) that provide efficient tools for analyzing large geographic databases, empirical studies of spatial autocorrelation were limited to small numbers of relatively simple geographical units.

Spatial analysis is defined broadly as a body of analytical techniques that require access to locational and attribute information of spatial objects which are the focus of analysis, and which are not invariant to changes in location. Spatial analytical techniques are dedicated to the analysis of the spatial order and associations of a phenomenon or variable. Spatial order delineates how geographic entities related to the phenomenon in question are organized in space, while spatial association describes the geographical relationships among phenomena.

Data that are tied to position on the Earth's surface, that is geo-referenced data, often serve as the empirical backbone of much of the research that is presently done in certain domains. Worldwide, nearly 70-80% of business data has a geographic component, so one can assume that a significant proportion of all business technical analyses do, or could, incorporate some spatial elements. Adding a spatial perspective has both advantages and disadvantages. The spatial element is beneficial, and often essential, for many types of evaluations, such as the identification of geographic clusters, service area delineation for facilities, as well as modeling which incorporates the impact of changes in populations. The most significant disadvantages of adopting a spatial perspective involve the data acquisition and maintenance, and the difficulty of processing the data because of dependencies and complex relationships. The set of methods for analyzing such data forms the subject matter of "spatial statistics". Compared with classical statistics, the observations analyzed within a framework of spatial statistics are rarely independent. Independence is a very convenient assumption that makes much of mathematical-statistical theory tractable. However, models that involve statistical dependence are often more realistic. Moreover, observations are correlated primarily due to their relative locational positions (referred to as spatial autocorrelation), resulting in spillover of information from one location to another (locational information).

Spatial statistics as one of the most rapidly growing areas of statistics; rife with fascinating research opportunities characterized by the National Research Council in the United States. Its board of mathematical sciences has targeted spatial statistics as one of twenty-seven topics of national concern in mathematics.

1.3 Geomatics and Risk Management

The adage "better information leads to better decisions and better management" is as true for GIS as it is for other information systems. Recent developments linking spatial statistical models with Geomatics techniques and GIS visualization tools provide a powerful medium for exploring and communicating solutions to the problems. The effective use of large spatial data volumes is dependent upon the existence of an efficient geographic handling and processing system to transform this data into usable information.

Work on GIS began in late 1950s, but the first commercially available GIS software came only in late 1970s from the ESRI labs in the U.S.A. Canada was a pioneer in the development of GIS as a result of innovations dating back to early 1960s.

One view of GIS is as an organized collection of computer hardware, software, geographic data, and personnel designed to efficiently capture, store, update, manipulate, analyze, and display all forms of geographically referenced information. Within GIS software, different types of data are easily related, thus making the data more easily accessible and providing a friendlier and more flexible user interface. However, various scholars have defined GIS in somewhat different terms; for instance Arlinghaus (1996) stated that: "GIS is a unique combination of computer hardware and software including high-resolution graphic analysis, large-capacity electronic storage devices, efficient strategies for data organization, high-volume communication channels, specialized algorithms for data integration and reliability analysis, specialized query computer languages. These components, together with massive amounts of highly complex geo-referenced data, are organized efficiently (through a sequence of electronic interfaces) to store, inventory, manage, search, manipulate, display (instantaneously), and analyze information contained in a geo-referenced database. The goal is to combine tabular attribute data with computerized maps in an enlightening way, achieving this goal by having a large storage capacity, a rapid response time, and a wide repertoire of analytical functions". Regardless of the scope, GIS technology integrates common database operations such as querying and statistical analysis with the unique visualization and geographic analysis benefits offered by maps. These abilities distinguish GIS from other information systems and make it valuable to a wide range of public and private enterprises for explaining events, predicting outcomes, and planning strategies including disaster and risk prevention planning.

Many organizations that have implemented a GIS have found that one of its main benefits is improved management of their own organization and resources. Because GIS has the ability to link data sets together by geography, it facilitates interdepartmental information sharing and communication. By creating a shared database, one department can benefit from the work of another -- data can be collected once and used many times. As communication increases among individuals and departments, redundancy is reduced, productivity is enhanced, and overall organizational efficiency is improved.

1.4 Geomatics and Maritime Traffic Risk Analysis

One of the primary applications of Geomatics and geo-information systems is in the field of transportation. This ranges from vehicle routing and scheduling issues, to risk management such as the transport of hazardous materials, to logistics and real-time tracking of vehicles. Although most of these applications have been developed for land transportation models, increasing attention is turning towards maritime traffic analysis.

The growing importance of this area relates to two simultaneous developments: the improved ability to quantify and analyze maritime traffic patterns, and the increasing need for this information. Quantifying the traffic flows is much easier than before due to the advent of powerful database packages for storing and organizing the information, as well as improved data acquisition tools and processes. For example, Geographic Positioning Systems (GPS) can be used for real-time tracking of ships and large fishing vessels. Even the availability of small, inexpensive hand-held GPS units permit easy sampling of small-boat traffic, such as those associated with recreational activities. Additional advances in this area of spatial data acquisition continue with the introduction of the Automated Identification System (AIS) which will further enhance maritime traffic management.

Ships report their position periodically as they approach their destination. Using these points, and a land-avoidance algorithm, the traffic pattern may be displayed and evaluated using a GIS.

For other types of maritime traffic such as fishing or recreational boating, reported or collected data can give a reasonable approximation of the durations, destinations, and directions of trips, which can be used as input information to generate representative simulated traffic patterns. The end result of this approach is a set of tracks, as shown in Fig. 1, with varying degrees of precision (Pelot 2002).

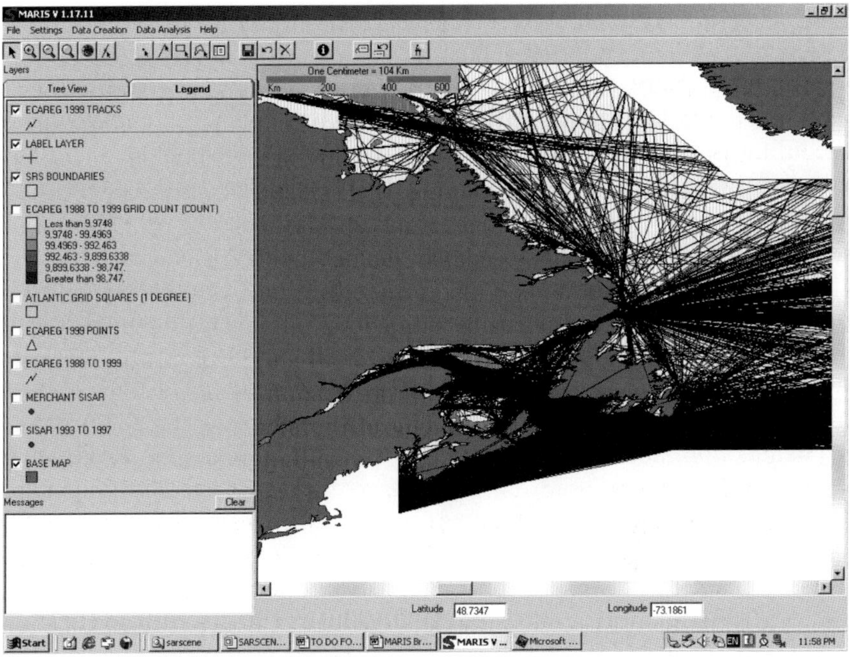

Fig. 1 Set of Vessel Paths Over Prescribed Time Frame

The paths are necessary for calculations and risk estimates, but this mass of black lines is not useful for conveying the density to decision-makers and risk managers, which can however be rectified by aggregating the frequency by grid squares as shown in Fig. 2.

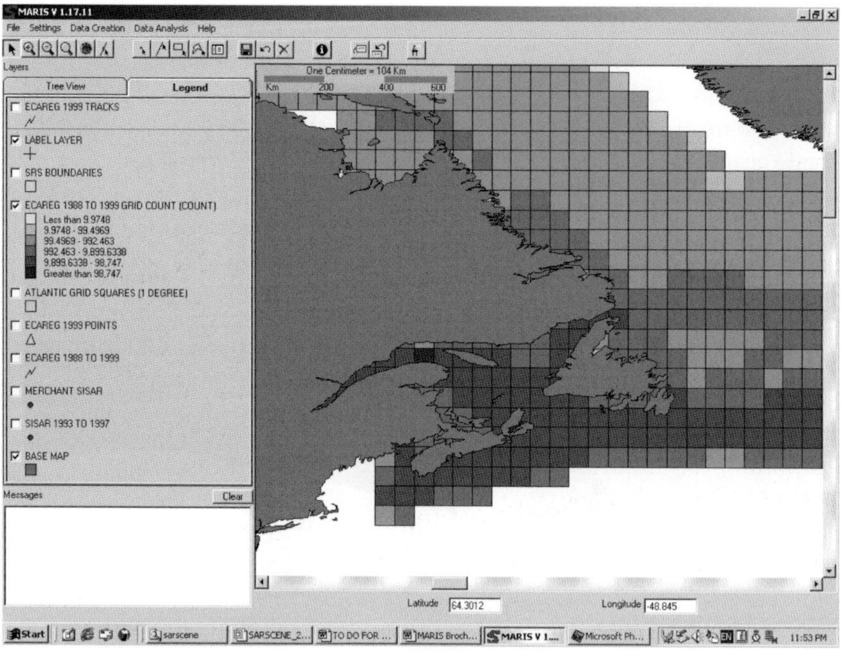

Fig. 2 Traffic density portrayal over prescribed time frame

1.5 Applications of Maritime Traffic Risk Analyses

As mentioned above, the demand for analytical models of maritime traffic risk is growing in tandem with the ability to generate them. Such models can be generally classified into macro- and micro-level analysis, depending on the scope and application. The micro models concentrate on individual vessel movements, or their operations in restricted areas. Typical topics include vessel stability analyses, human factors studies of ship operations, vessel traffic control systems, and detailed studies of ports and harbours.

The macro models, like the one presented here, take a wider geographical viewpoint, and may incorporate a more diverse set of maritime activities. This type of model can prove to be useful for several spheres of policy and decision-making for an effective risk management. From the perspective of accident risks, knowing the level and types of traffic, annual trends and geographic factors are important to help understand and mitigate these risks. Risks can be reduced through prevention programs and regulations, or at least better emergency response capability can be developed by the Coast Guard or similar organizations. For the coast guard to best target measures for prevention, regulation and search and rescue planning, it is essential to characterize activities to determine the highest risk factors. Studies can be conducted based on many different factors. As location is one of the most important issues in risk analysis, a study analyzing the risk based on location to identify "Hot Spots" has been carried out. Hot spots areas are concentrations of incidents within a limited geographical area that appear over time. There are several different statistical techniques designed to identify hot spots. These techniques are aimed at grouping cases together into relatively consistent clusters. All of these methods depend on optimizing various statistical criteria, but the techniques differ from one another in their methodology as well as in the criteria used for identification. To identify "hot spots" there are several types of cluster analysis techniques such as those based on point locations, hierarchical techniques, partitioning techniques, density techniques, risk-based techniques as well as zone-based techniques. In addition to the different types of cluster analysis, several different criteria can be used to customize these techniques. Definition of a cluster, choice of variables, measurement of similarity and distance, fixed versus variable number of clusters, the initial selection of cluster locations and the optimization routines used to adjust the initial seeds into final locations are some of those criteria. There are a large number of different cluster techniques for hot spot identification which can be applied to maritime search and rescue (SAR) incident and activity datasets, two of which are highlighted below.

2 Methodology

The objective of this study is to investigate incident clusters, and ultimately to identify hazardous regions by using two different identification techniques. This study examines activities and incidents associated with fishing vessel traffic in the waters of the four Canadian Coast Guard (CCG) SAR statistical areas in the Atlantic region including Cape Breton, Bay of Fundy, SouthWestern Shore, and Eastern Shore. This study is restricted to the 3-year period from 1997 to 1999, based on data availability. This yields 2,073,781 fishing activities, with 2002 reported incidents. Fishing vessel activities and incidents that occurred within this study area are mapped and examined using different statistical techniques in this paper.

2.1 Data Sources

1) System of Information for Search and Rescue (SISAR)

The Canadian Coast Guard maintains records of search and rescue maritime casualties in the System of Information for Search and Rescue (SISAR) database. This database includes casualty information such as casualty location, casualty type classification (e.g. collisions, groundings), and severity of the incident as well as vessel characteristics.

2) Zonal Interchange Fisheries (ZIF)

For the purpose of fisheries management, a standard set of information about fishing catch-efforts is collected. Each vessel is required to register catch information when landing fish at the end of each trip that it makes. The information in the ZIF database includes vessel and gear used, effort location, and time for each effort. Vessel information includes vessel ID, length overall, gross registered tonnage and homeport/port landed.

3 Analyses and Results

From a large number of different statistical techniques for hot spot identification, some have seemed more suitable for maritime incident and activity analyses. Each of these techniques may assist in different ways for short-term or long-term strategy, or for applying decisions in different ways. Two of the most promising techniques examined are the nearest neighbour hierarchical cluster analysis and risk-based nearest neighbour hierarchical cluster analysis.

3.1 Hierarchical Cluster Analysis

Hierarchical cluster analysis has been defined and developed by several people. This analysis is among the oldest cluster routines (Sokal and Sneath, 1963, Systat, 2000). Hierarchical cluster analysis is a statistical method for finding relatively homogeneous clusters of cases based on measured characteristics. Hierarchical models can be agglomerative (starting with many groups and reducing the number of groups by amalgamating them) or divisive (starting with one group and splitting into many based on dissimilarities). The agglomerative option starts with each case in a separate cluster and then combines the clusters sequentially, reducing the number of clusters at each step until only one cluster is left. Cluster analysis methods use common features of objects in a data set for dividing it into clusters in such a way that objects belonging to any one of the clusters should be similar, while objects associated with two different clusters should be as dissimilar as possible. In hierarchical hot spot analysis the spatial characteristics are measured to determine clusters. In hierarchical spatial cluster analysis two or more incidents are first grouped on the basis of closeness. Then the pairs are grouped into second-order clusters after which the second-order clusters are grouped into third-order clusters. This process is continued until either only one cluster remains, or the grouping criteria fails.

In comparison to other clustering analyses, the hierarchical clustering method has two general and specific advantages with respect to fishing incidents and fishing activity spatial risk analysis. The first advantage is that hierarchical clustering method is able to identify small hot spot areas where fishing incidents or fishing activities are concentrated. The second advantage of a hierarchical clustering method is that each order of clusters would be appropriate for different purposes of risk management such as incident prevention, search and rescue operation centers allocation, strategic management and long term planning. Fishing incident and fishing activity hot spot areas could be formed into very small areas, especially where potential sources of fish exist. In order to have micro and macro views of these incidents and activities, it is beneficial to generate all levels of cluster orders. For instance, in search and rescue allocation analysis based on the Coast Guard's budget and plans, the best location of search and rescue centers can be determined in a few alternative scenarios. It is very interesting when the relationship between lower order and higher order clusters is illustrated by this method.

In this study two hierarchical clustering techniques (nearest neighbour hierarchical clustering and risk-based nearest neighbour hierarchical clustering) are employed (Shahrabi, Pelot, 2003).

3.2 Nearest Neighbour Hierarchical Clustering

The nearest neighbour hierarchical spatial clustering routine is a one of the most commonly used hierarchical techniques of hot spot analysis. This method is a clustering routine that identifies groups of incidents that are spatially close. The minimum number of points that are required for each cluster should be specified by an analyst. The nearest neighbour clustering method defines a threshold distance and compares the threshold to the distance for all pairs of points. Only points that are closer to one or more other points than the threshold distance are selected for clustering. The threshold distance is a one-tailed confidence interval around the random expected nearest neighbour distance. The confidence interval defines a probability for the distance between any pair of points. The first-order clusters will be formed by grouping selected points that are closer together than the threshold distance. This procedure will be continued if the centres of first-order clusters are closer than the threshold distance and, last until either all points are covered by one unique cluster or the clustering criteria fail.

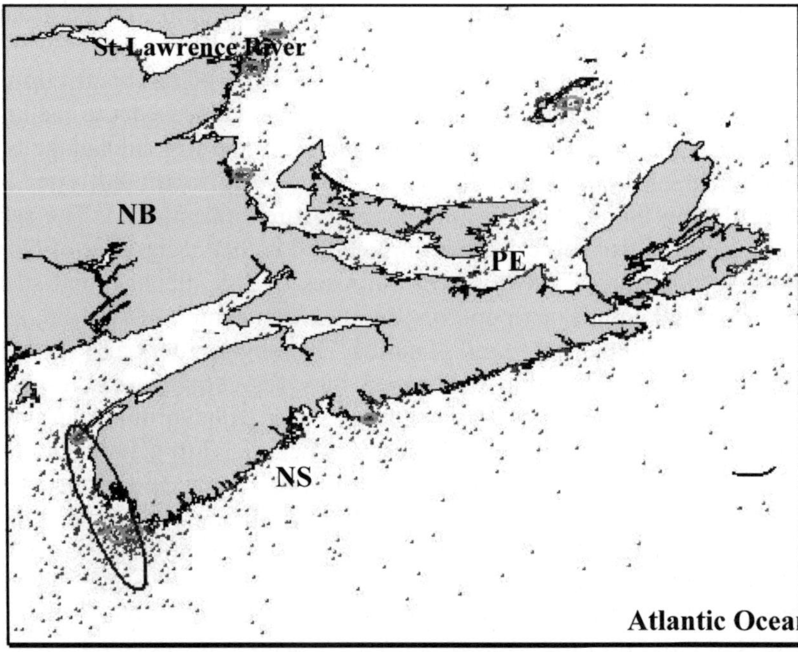

Fig. 3 1st Order and 2nd Order Nearest Neighbour Hierarchical Clustering of Fishing Incidents.

The nearest neighbour hierarchical clustering method identifies 10 clusters, nine 1st order and one 2nd order. Hot spot that covers most of the incidents compared to the others is located in the South-West region of Nova Scotia. The cluster that has noticeably the highest density of all clusters is located near the St-Lawrence River. The biggest cluster by area also located in the South-West portion of Nova Scotia. Five of the nine 1st order clusters are located in the waters North of Nova Scotia, while four are in the South-West of Nova Scotia. These four 1st order clusters are covered by the unique 2nd order cluster of this analysis that represents the largest hot spot area. Based on these results the areas with most concentration of fishing incident are identified.

3.3 Risk-Adjusted Nearest Neighbour Hierarchical Clustering

Risk-adjusted nearest neighbour hierarchical clustering method can be categorized as either hierarchical clustering or risk-based techniques. Generally speaking, risk-based clustering techniques identify clusters in relation to an underlying base 'at risk' variable, such as activity, population and so on.

The risk-adjusted method is a combination of the hierarchical clustering analysis and kernel density interpolation techniques. As previously discussed, the nearest neighbour hierarchical clustering method identifies clusters that group those incidents that are closer together than a threshold distance. It could be expected that some of these incident clusters are formed simply due to a high concentration of activity. As was previously examined and presented (Shahrabi, Jamal 2003), fishing activity is not arranged randomly over a study area, but is highly concentrated in some regions due to differing factors, such as potential fish stocks and near congested ports. For search and rescue purposes, the concentration of fishing incidents is an interesting issue in itself. Search and rescue centers have to be involved in hot spots areas with high fishing incidents regardless of whether greater fishing activity also takes place at those locations. For instance prevention activities like special search and rescue programs may be aimed at hot spot areas, whether there is low or high amount of activity there. For these purposes the preceding method was most useful.

However, for other purposes, the concentration of fishing incidents relative to the fishing activity is of interest. The demands for search and rescue services and response to emergency calls is sensitive to the amount of fishing activity since more calls are expected from those regions with more fishing activity. For this purpose, the probability of fishing incidents relative to fishing activity is needed, rather than the volume of fishing incidents alone. To identify clusters representing locations with high probability of fishing incidents relative to fishing activities, a risk-adjusted nearest neighbour hierarchical clustering is applied. This method defines clusters of fishing incident points that are closer than what would be expected on the basis of fishing activity. This technique dynamically adjusts the threshold distance in the hierarchical clustering method according to the distribution of a second, baseline variable – which in this study is fishing activity.

In this method to make fishing incident clusters, the threshold distance would be adjusted for varying fishing activity densities. This technique changes the threshold distance inversely proportional to the fishing activity density of the location in such a way that in those areas with high fishing activity, the threshold should be short, and conversely in the regions with low fishing activity, since there is less fishing activity per unit of area, the threshold distance should be considerably larger.

This method overlays a standard grid and uses an interpolation kernel density algorithm to estimate the expected number of fishing incidents per grid cell if the actual incident data were distributed according to the fishing activity. Then clusters with the minimum required number of points are made by grouping the fishing incident points that are closer than would be expected on the basis of the fishing activity. The clusters made by this method are more concentrated than would be expected from both chance and the distribution of the fishing activity. These clusters can be considered as clusters with a probability of fishing incidents relative to fishing activity.

Since this method is derived from the nearest neighbour hierarchical clustering method, the risk-adjusted nearest neighbour hierarchical clustering has the same advantages as previously discussed for the latter technique. Fig. 4 shows the risk-adjusted nearest neighbour hierarchical clusters of fishing incidents relative to fishing activity in the study area in Canadian Atlantic waters.

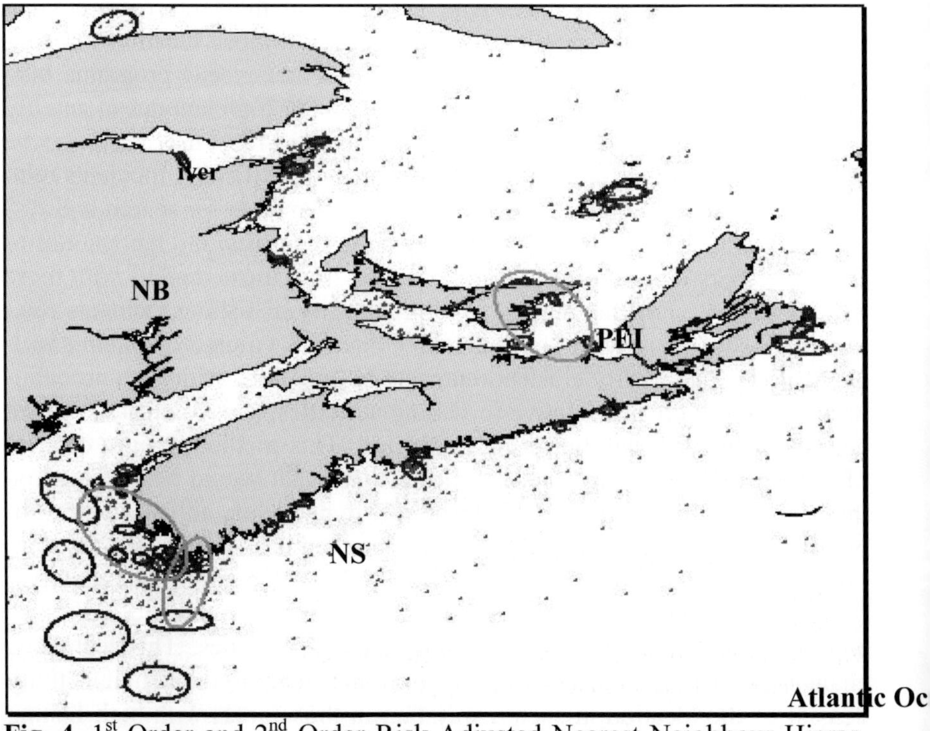

Fig. 4. 1st Order and 2nd Order Risk-Adjusted Nearest Neighbour Hierarchical Clustering of Fishing Incidents relative to fishing activity.

In Fig. 4 there are 47 clusters; 44 first-order and 3 second-order ones. Most clusters are formed in South Western Nova Scotia, of which the largest number is specifically located close to Clark's Harbour. Two of the three second-order clusters are formed in this area that represents a high concentration of fishing incidents relative to fishing activity. The other second-order cluster is located on the East Coast of Prince Edward Island. It is seen that the majority of hot spot areas are very close to shore, and are smaller and more condensed than others. In comparison to first order clusters it is seen that clusters further from shore are formed only in South Western Nova Scotia.

Acknowledgments

We would like to thank the Canadian Coast Guard and the GEOIDE National Centre of Excellence for financial support for this study. We appreciate the professional computing support provided by the MARIN programmers and researchers at Dalhousie University.

References

Arlinghaus SL (1996) Practical Handbook of Spatial Statistics. Canadian Coast Guard, Search & Rescue Branch, System of Information for Search and Rescue (SISAR). Ottawa, Canada

Department of Fisheries and Oceans, Scotia Fundy management region, Canada. Zonal Interchange Fisheries (ZIF).

ICF Kaiser Consulating Group, Soza and Company, and Marine Research Associates, (1997). The economic impacts of accidents on the marine industry. Washington, DC, Government Printing Office.

Jin Di, J, Kite-Powell H, Talley W (2001). The safety of commercial fishing: Determinations of vessel total losses and injuries, Journal of Safety Research, Vol. 32, 209-228.

Di J, Kite-Powell H, Talley W, (2002). A model of fishing vessel accident probability, Journal of safety Research, Vol. 33, 497-510

National Research Council (1991) Spatial statistics and digital image analysis panel on spatial statistics and image processing, Board on Mathematical Sciences,

Van NM (1995) Toward a systematic approach to safety in the commercial fishing industry, Journal of Safety Research, Vol. 26(1), 19-29.

Pelot R (2002) Maritime Activity and Risk Analysis System, Dalhousie University, Halifax, Nova Scotia, Canada.

Shahrabi J, and Ronald P (2003) Evaluation of hot spot analyses applied to maritime SAR incidents, GEOIDE Annual Conference, Victoria, BC, Canada

Shahrabi J (2003). Spatial and temporal analyses of maritime fishing and shipping traffic and incidents, Ph.D. Thesis, Industrial Engineering Department, Dalhousie University, Halifax, Nova Scotia, Canada.

Sokal RR and Sneath PHA (1963). Principles of numerical taxonomy. W. H. Freeman and Co.: San Francisco.

Systat, Inc., (2000). Systat 10: Statistics I. SPSS, Inc.: Chicago

A Fuzzy Relation Analysis Method Implemented in GIS for Modeling Infrastructure Interdependency

Qiuming Cheng

Department of Earth and Space Science and Engineering, Department of Geography, York University, Toronto, Ontario, Canada M3J 1P3
State Key Laboratory of Geological Processes and Mineral Resources, China University of Geosciences, Wuhan, China, qiuming@yorku.ca

Abstract

Infrastructure Interdependency (II) involved in emergency and disaster management processes is a complex phenomena. Understanding and modeling II is an emerging interdisciplinary field of study which has attracted substantial interests from modelers and geomatics professionals. This paper introduces a new method for modeling and analyzing interdependencies of critical infrastructures. The modeling processes include the following four components: (1) an asymmetrical fuzzy relation matrix representing direct relationships between nodes in infrastructure networks; (2) an asymmetrical fuzzy relation matrix representing direct and cascade relationships between infrastructure networks; (3) a mathematical transformation converting asymmetric relations in to transitive relations with properties of direct relations stronger than indirect relations; and (4) a method for ranking infrastructures in terms of relative importance of infrastructures.

1. Introduction

Understanding interdependencies between infrastructures is essential for effective response and management of resources for rescue, recovery, and restoration. Due to the complexity of the dynamics of infrastructures, understanding the dependencies of infrastructures especially their indirect dependencies and cascade effects is complex. Quantitative modeling of these types of interdependencies is a relatively new field and of a great deal of challenges. Although it has been substantial efforts and some excellent works developed for representing and analyzing infrastructure interdependencies, a lot more investigation still need to be done to provide models and techniques to meet the practical needs. There are a several types of work worthy of mention here that involve formulation of models for quantification of relationships between infrastructures and identification of key issues involved in the modeling independencies of infrastructures. Two main ways were developed for representing relationships between infrastructures: graphics (Rinaldi et al. 2001) and matrix (Dunn and Wigert 2004). Graphic representation of independencies is often intuitive and easy to understand whereas matrix representation provides a more organized form which can be further extended for quantitative analysis and modeling. As part of a joint project “Modeling infrastructure interdependency for emergency management using a network-centric spatial decision support system approach” supported by the Natural Science and Engineering Research Council of Canada (NSERC) and the Public Safety and Emergency Preparedness Canada (PSEPC), this paper introduces a new fuzzy relation analysis model for representation and modeling of multiple infrastructures interdependency. A dependency matrix can represent interdependencies and their strengths between infrastructures. The cascade influences of one infrastructure on another through direct and indirect dependencies have been recognized during several recent disasters and this has promoted the effort for modeling and simulating the cross infrastructure effects using various techniques including quantitative means and information technology (Pederson et al. 2006).

Characterizing infrastructure interdependencies needs a quantitative framework for formulating the problems. There have been several attempts along this line. For example, Dudenhoeffer, Permann and Manic (2006) introduced a graphic presentation of infrastructures in which nodes represent infrastructure components and edges for the relations between nodes. A dependency matrix provides a potential formulation for quantitative representation of interdependencies between infrastructures and analysis of their impact. For example, the Critical Infrastructure Protection Task Force of Canada used a dependency matrix to relate the interdependency among six sectors identified as crucial: Government, Energy and Utilities, Services, Transportation, Safety, and Communications (Dunn and Wigert 2004). The current paper will further explore an asymmetric fuzzy relation matrix to represent the interdependencies of infrastructures and a fuzzy relation analysis method to analyze their impact. The emphasis will be given on quantitative assessments of direct and indirect interdependencies and the cascade effects of one node on another in a set of interrelated infrastructure networks. The paper is organized first to introduce necessary mathematical principal and models, and then to use a simulated example to demonstrate the implementation of the methods and interpretation of the results.

2. Asymmetrical Fuzzy Relation Matrix

Cross-tabulation matrix has been commonly used in statistics for representing relations between two variables or multiple variables and it has been also applied in many applied fields for quantification of interdependencies among variables, for example, used as confusion matrix in remote sensing image processing for post-classification comparison and multi-date composite image change detection (Jensen 1996). The cross-tabulation matrix is a fundamental method in the analysis of land change detected using remotely sensing images (Jensen 1996). In mathematical geology cross-tabulation matrix has been commonly used for characterizing interdependencies of qualitative and quantitative variables such as interdependencies among controlling factors of structures, lithology, geofluid, and alteration on mineralization (Cheng 1990, 1991, 1996). This paper will introduce a new asymmetric relation matrix and a fuzzy relation analysis technique for investigating cascade effects of infrastructures from each other. The modeling processes include the following components: (1) an asymmetrical fuzzy relation matrix is constructed using knowledge driven or data driven approaches to represent the direct relationships between two nodes in infrastructure networks; (2) asymmetric incidence coefficients are constructed on the basis of the fuzzy relations between two nodes as defined in (1) and are used as measures of interdependency between infrastructure networks; (3) a mathematical transformation can be applied to the asymmetric matrix constructed in (2) so that the new matrix showing the relationships with a transitive property that the strength of the direct relationship of any given two infrastructures is stronger than or equal to those of indirect cascade relationships; and (4) a method to estimate the overall interdependencies among the infrastructures from the transformed direct relationships.

2.1 Fuzzy Relation Matrix

To represent interdependencies between infrastructures, we use a cross-tabulation matrix or transition matrix with rows displaying nodes of infrastructure network, $I_p = \{a_1, ..., a_m\}$ and columns displaying nodes of infrastructure network, $I_q = \{b_1, ..., b_n\}$. Denote $edge(a_i, b_j)$ *as* a measure representing a relation between the nodes a_i *and* $b_j, a_i \in I_p, b_j \in I_q$. The networks may be connected if edge(a_i, b_j) $\neq$ 0 or disjoint if edge(a_i, b_j) $\equiv$ 0. It may be symmetrical if edge(a_i, b_j) $\equiv$ edge(b_j, a_i) or asymmetrical edge(a_i, b_j) $\neq$ edge(b_j, a_i). The larger the value of edge(a_i, b_j) is, the stronger the relationship between the two nodes. Methods to define the measure edge(a_i, b_j) can be data-driven or knowledge-driven which will be discussed in later section of the paper. For representation convenience, we will denote the edge(a_i, b_j) as e_{ij} ($e_{ij} \geq 0$), and the interrelations between all nodes of two infrastructure networks (I_p and I_q) can be represented by the following relation matrix

$$\begin{pmatrix} e_{11} & \cdots & e_{1n} \\ \cdot\cdot & & \cdot\cdot \\ e_{m1} & \cdots & e_{mn} \end{pmatrix} \tag{1}$$

where m and n are the number of nodes in I_p and I_q, respectively.

To introduce the asymmetric relation analysis method we first introduce several notations and statistics. Let $e_i. = \sum_j e_{ij}$, $e._j = \sum_i e_{ij}$ and $e.. = \sum_i \sum_j e_{ij}$ represent the row sum, column sum and total sum of matrix (1), respectively. A normalized relation between nodes a and b can be represented as $P_{ij} = e_{ij}/e..$, $0 \leq P_{ij} \leq 1$, which can be regarded as probability or fuzzy membership function. This normalized measure has the following properties: $P_{ij} = 0$, iff $e_{ij} = 0$ implying a non-relationship or disconnection; $P_{ij} = 1$, iff $e_{ij} \neq 0$ and the relations for other pairs of nodes having non-relationship with $e_{kl} = 0$ ($k \neq i$ and $l \neq j$) implying only the two nodes are related with a strength e_{ij}.

2.2. Asymmetric Incidence coefficients to measure the asymmetric dependency

According to Cheng (1990), the following asymmetric incidence coefficients can be defined on the basis of the cross-tabulation (1) (Cheng, 1990, 1991, 1996).

$$Q_{pq}^2 = \frac{n}{(n-1)m}\sum_{i=1}^{m}\sum_{j=1}^{n}\left(\frac{e_{ij}}{e_{i\cdot}}\right)^2 - \frac{1}{n-1} \tag{2}$$

$$Q_{qp}^2 = \frac{m}{(m-1)n}\sum_{i=1}^{n}\sum_{j=1}^{m}\left(\frac{e_{ij}}{e_{\cdot j}}\right)^2 - \frac{1}{m-1} \tag{3}$$

where Q_{pq}^2 and Q_{qp}^2 are called the forward and backward incidence coefficients of the variables I_p and I_q, measuring the asymmetrical dependencies of the two infrastructure networks, respectively. They have the following properties: (1) $0 \le Q_{pq}^2 \le 1$, $0 \le Q_{qp}^2 \le 1$, (2) $Q_{pq}^2 = 1$, if and only if I_p is the function of I_q with probability 1, implying the states of nodes in I_p completely control the states of the nodes in I_q. Similarly, $Q_{qp}^2 = 1$, if and only if I_q is the function of I_p with probability 1, and (3) $Q_{pq}^2 = 0$, if and only if I_p and I_q are independent and I_q has a uniform distribution with $P_{\cdot j}$ =1/n,or $e_{\cdot j} = 1/ne_{\cdot\cdot}$ $j = 1, \ldots, n$. Similarly, $Q_{qp}^2 = 0$, if and only I_p and I_q are independent and I_p has a uniform distribution with $P_{i\cdot}$=1/m, or $e_{i\cdot} = \frac{e_{\cdot\cdot}}{m}$ $j = 1, \ldots, m$.

The measure Q_{pq}^2 characterizes the dependency relationship of infrastructure network I_q on I_p, and similarly, Q_{pq}^2 characterizes the dependency relationship of infrastructure network I_p on I_q. Usually, it has $Q_{pq}^2 \neq Q_{qp}^2$ implying asymmetrical relationships between infrastructure networks. Therefore, these indexes defined in Eqs. (2) and (3) can be used in the study of interdependencies of infrastructures. For a group of infrastructure networks I_p, p = 1, 2, …, k, the interdependencies can be represented as an asymmetric matrix of the asymmetrical index (Q_{pq}^2).

$$\begin{pmatrix} Q_{11} & \cdots & Q_{1k} \\ \cdot\cdot & & \cdot\cdot \\ Q_{k1} & \cdots & Q_{kk} \end{pmatrix}$$

(4)

This matrix will be further analyzed using fuzzy relation method to evaluate the overall dependencies of the networks.

3. Fuzzy Relation Analysis Method for Assessment of Overall Dependencies of Infrastructure Networks

The asymmetric matrix in Eq. (4) characterizes the interdependencies of all pairs of infrastructure networks including their direct dependencies and indirect dependencies. Usually dependencies between two nodes are well understood, but the dependency between two infrastructure networks becomes more complex and those among multiple infrastructure networks become even worse. In a region of interest, interdependencies of infrastructure networks should include not only the direct impacts of one on another but also indirect impacts or cascade impacts of one on another through a chain effect. Therefore, the key effects to analyze the interdependencies are the chains of influence that cross multiple infrastructures. These chains linking two infrastructures, for example a and b, can compose of intermediate infrastructures and these can be denoted as a array diagram as {(a,c), (c,d), ...(y,b)}. This particular path represents the cascading consequence of an infrastructure network b impacted by network a, further denoted (aDb). These paths may not be unique in terms of effect and they may change over time. The intertwining of networks in this fashion represents a complex system where emergent behaviors are rarely fully understood (Pederson et al. 2006). Rinaldi, Peerenboom and Kelly (2001) provide a visual representation of this intertwining and the potential cascading effects. However, effective methods for quantitative assessments of these types of cascading effects still need to be developed.

With the measures defined in asymmetrical matrixes in Eqs. (1) and (4) and the fuzzy relation analysis method to be introduced in this paper, the following questions can be answered:

1. What are the rankings of the infrastructure networks in terms of interdependencies?

2. What are the sets of nodes essential in controlling the cascade influence on the whole system?
3. Given a set of infrastructure networks and a critical function, what is the subset of critical nodes {x, y, z , …} across all networks that will significantly impact on the interdependencies of the set of infrastructure networks due to direct or indirect dependency?

In order to answer the above questions, we need to introduce some concepts and notations. Suppose that a set of k infrastructure networks denoted as E = {I1, I2, …, Ik}. A fuzzy relation on E can be defined as follows (Kaufmann 1975; Zimmermann 1991):

Definition 1, R(x, y) is called a fuzzy relation on E, if ∀ x and y ∈ E and there exists a value R(x, y) corresponding to x and y with

(1) $0 \leq R(x, y) \leq 1$,

(2) R(x, y) = 1, indicates x and y having the strongest possible degree of relation; and

(3) R(x, y) = 0, indicates x and y being not related at all.

Thus, the value of R(x, y) represents the intensity of the relationship between x and y. From this definition, Q_{pq}^2 and Q_{qp}^2 defined in Eqs. (2) and (3) are fuzzy relations on infrastructure networks.

Definition 2, Fuzzy relation R(x, y) is called a preorder relation if

(1) R(x, x) = 0, all x ∈ E and

(2) ∀ x, y and z ∈ E, it is true that $R(x, z) \geq \min\{R(x, y), R(y, z)\}$.

The first condition in the Definition 2 implies that the measure between any infrastructure network and its self is equal to zero, indicating either no dependency or that it is not meaningful to consider such dependency. The second condition in the Definition 2 indicates that the direct dependency between two infrastructure networks is stronger or equal to the weakest dependency in the chains connecting these two infrastructure networks. Therefore, in the case of preorder relation between two infrastructure networks, the direct dependency is the dominate relation between the two infrastructure networks which might be based to evaluate the overall dependencies of infrastructure networks.

Definition 3, R(x, y) is called a perfect antisymmetric relation if ∀x, y∈ E with x ≠ y, if R(x, y) > 0 then R(y, x) = 0.

This definition emphasizes directional representation of the relations between infrastructure networks. The antisymmetric relation represents the "absolute" measure of the relation between x and y.

Definition 4, R(x, y) is called a fuzzy partial order relation if R(x, y) is a perfect antisymmetric preorder relation.

In order to rank infrastructure networks according to their interdependencies, all types of relations including both direct and indirect relations have to be considered. However, if the relation R(x, y) is fuzzy partial order relation, then the infrastructure networks can be ranked by their direct relations only due to their dominating relations between the infrastructure networks.

The relations defined in Eqs. (2) and (3) do not always meet the requirement of a partial order relation. In other words, direct dependencies may not be always stronger than indirect dependencies. Therefore, we need to introduce the following mathematical transformations that can be used to convert ordinary fuzzy relations into partial order relations.

Theorem 1, If R(x, y) is a fuzzy preorder relation then

$$P(x,y)=\begin{cases} R(x,y)-R(y,x), \text{if } R(x,y)>R(y,x) \\ 0, \text{otherwise} \end{cases} \quad (5)$$

is a fuzzy partial order relation.

Theorem 2,

The transitive closure of any fuzzy relation

$$\hat{R}=R\cup R^2\cup\ldots\cup R^k \quad (6)$$

is a transitive relation satisfying the second condition of Definition 2. With these two theorems a partial order relation can be constructed from any fuzzy asymmetric fuzzy relation R(x, y). This can be achieved in two steps: (1) according to Theorem 2, any fuzzy relation can be converted into a transitive relation; and (2) according to Theorem 1, a new partial order relation can be formed on the basis of the new transitive preorder relation. Therefore, the fuzzy relation indexes defined in Eqs. (2) and (3) can be implemented and infrastructure networks can be ranked using the following ranking scheme (Cheng 1990, 1991, 1996).

Definition 5, Supposing R(x, y) is a partial order relation on E, the weight of any element with respect to others in E can be determined by

$$d(x)=\max_{z\in E}R(x,z)-\min_{y\in E}R(y,x) \quad (7)$$

which has the following properties:

$-1\le d(x)\le 1$,

d(x) = -1, if and only if $\exists$ y $\in$ E with R(y, x) = 1 and $\forall$ z $\in$ E, R(x, z) = 0; This property implies that the minimum weight is due to the property of element x that there is an infrastructure y which completely control x but x does not control any other infrastructures;

d(x) = 1, if and only if $\forall$ y $\in$ E with R(y, x) = 0 and $\exists$ z $\in$ E, R(x, z) = 1; This property implies that the maximum weight is due to the property of element x that there is an infrastructure z which is completely controlled by x but there is no any infrastructure networks that control x.

Therefore, the value of d(x) can be used as a measure to rank infrastructure networks in terms of overall asymmetric dependencies. The larger the value d(x) is, the more important of the infrastructure network implying that it has strong influences on other infrastructure networks and is weakly effected by any other infrastructure networks.

4. Assessment of Interdependencies of Infrastructure Networks

4.1 Interdependencies between infrastructure networks

In this section the method introduced in Section 3 will be used to assist in assessment of interdependencies of infrastructure networks. It will use an arbitrary example to demonstrate the application of the method. Suppose four infrastructure networks are considered in the area of interest and they are denoted as $I_1 = (a_1, a_2, a_3)$, $I_2 = (b_1, b_2)$, $I_3 = (c_1, c_2, c_3)$ and $I_4 = (d_1, d_2)$, respectively. These infrastructure networks are dependent from each other during a type of disaster. To establish fuzzy relations between the nodes of these infrastructures two different ways can be followed: data-driven and knowledge driven approaches. In data driven approach some historical data have to be collected to show the statistics of the relations between any two pairs of nodes which can be used as measure to represent the relation between two nodes. In a knowledge driven approach, the relations between any two pairs of nodes can be assigned according to experts' experiences and judgments. For example, a group of experts can be interviewed for evaluating the relations between any two nodes which can be used as a measure of relation between two nodes. In this example, we assume that several events have generated data showing the actual relations between the infrastructures. Suppose 8 events of similar types have been recorded and the numbers of coincidences of these infrastructure networks are shown in Table 1. Let us take infrastructure networks I_1 and I_2 as an example to illustrate the method for estimating the asymmetric relation between infrastructure networks. The frequency distributions of I_1 and I_2 values can be obtained as shown in Table 2, an asymmetric matrix showing the numbers of coincidences between all pairs of nodes of the two infrastructure networks during the 8 events. The higher coincidences between two nodes indicate stronger relationship between the nodes. Therefore, these values will be based to construct the fuzzy relations as defined in Eqs. (2) and (3).

Table 1 Data from Four Infrastructure Networks in 8 Events

	Infrastructure networks			
Events	I_1	I_2	I_3	I_4
1	a_1	b_1	c_1	d_1
2	a_1	b_1	c_2	d_1
3	a_3	b_2	c_3	d_2
4	a_3	b_2	c_3	d_2
5	a_2	b_1	c_1	d_1
6	a_2	b_2	c_3	d_2
7	a_3	b_2	c_3	d_2
8	a_2	b_1	c_3	d_2

Using Eqs. (2) ad (3) with m = 3 and n = 2, we obtain $Q^2{}_{pq} = 0.70$ and $Q^2{}_{qp} = 0.31$. These results show that the dependency of I_4 on I_1 is stronger than that of I_1 on I_4. From the data in Table 2, we can calculate the conditional probability of the states of nodes in I_4 given the states of nodes in I_1. For example, conditional probabilities $P\{b_1|a_1\} = 1$ and $P\{b_2|a_1\} = 0$, $P\{b_1|a_2\} = 1/3$ and $P\{b_2|a_2\} = 2/3$, $P\{b_1|a_3\} = 0$ and $P\{b_2|a_3\} = 1$, indicate that the states of the nodes in I_4 are highly dependent on the states of nodes in I_1. In other words, knowing the states of nodes in I_1 gives lot information about the states of the nodes in I_4. On the other hand, the conditional probabilities of states of the nodes in I_1 given the states of nodes in I_4: $P\ \{a_1|b_1\} = 2/3$, $P\ \{a_2|b_1\} = 1/3$, $P\ \{a_3|b_1\} = 0$ and $P\ \{a_1|b_2\} = 0$, $P\ \{a_2|b_2\} = 2/5$, $P\ \{a_3|b_2\} = 3/5$ imply that knowing the states of nodes in I_4 gives relatively less information about the states of nodes in I_1. This is in agreement with the results that the dependency calculated for I_4 on I_1 ($Q^2{}_{pq} = 0.70$) is much stronger than that of I_1 on I_4 ($Q^2{}_{qp} = 0.31$).

Table 2 Contingency Table Showing Relations between I_1 and I_4

		Nodes of I_4		
		b_1	b_2	Σ
Nodes of I_1	a_1	2	0	2
	a_2	1	2	3
	a_3	0	3	3
	Σ	3	5	8

Table 3 Asymmetric Relations (Q2) between Infrastructure Networks I1 to I4

	I_1	I_2	I_3	I_4
I_1	1.00	0.70	0.53	0.70
I_2	0.34	1.00	0.52	0.62
I_3	0.51	0.78	1.00	1.00
I_4	0.31	0.68	0.67	1.00

Table 4 Transitive Closure of Relations between Infrastructure Networks I1 to I4

	I_1	I_2	I_3	I_4
I_1	0.51	0.70	0.67	0.70
I_2	0.51	0.62	0.62	0.62
I_3	0.51	0.78	0.67	1.00
I_4	0.51	0.68	0.67	0.67

Table 5 Fuzzy Partial Order Relations between Infrastructure Networks I1 to I4

	I_1	I_2	I_3	I_4
I_1	0.00	0.19	0.16	0.19
I_2	0.00	0.00	0.00	0.00
I_3	0.00	0.15	0.00	0.33
I_4	0.00	0.06	0.00	0.00

Similarly, we can calculate the values of asymmetric relations between the other infrastructure networks and the results are shown in Table 3. The asymmetric relations shown in Table 3 are not only direct relations between all pairs of infrastructure networks but also the cascade relations among these infrastructures. Take I_1 and I_4 as an example; the two infrastructure networks are associated through the direct and indirect relations of various forms such as:

$$I_4 \overset{0.31}{\rightarrow} I_1$$

$$I_4 \overset{0.68}{\rightarrow} I_2 \overset{0.34}{\rightarrow} I_1$$

$$I_4 \overset{0.67}{\rightarrow} I_3 \overset{0.51}{\rightarrow} I_1$$

$$I_4 \overset{0.67}{\rightarrow} I_3 \overset{0.78}{\rightarrow} I_2 \overset{0.34}{\rightarrow} I_1$$

Among these paths, the direct dependency of I_1 on I_4 is 0.31 which is not greater than the minimum indirect dependencies such as 0.34 for $I_4 \rightarrow I_2 \rightarrow I_1$, 0.51 for $I_4 \rightarrow I_3 \rightarrow I_1$ and 0.34 for $I_4 \rightarrow I_3 \rightarrow I_2 \rightarrow I_1$. This example shows that the asymmetric relations in Table 3 do not have the transitive property implying that the relations between I_1 and I_4 can not be evaluated using direct relations along. It must consider the indirect cascade relations. However, according to Theorem 2, these relations can be converted into transitive relations by applying a closure transformation. The transformed results are shown in Table 4. The relations represented in Table 4 have transitive property, for example

$$I_4 \xrightarrow{0.51} I_1$$

$$I_4 \xrightarrow{0.68} I_2 \xrightarrow{0.51} I_1$$

$$I_4 \xrightarrow{0.67} I_3 \xrightarrow{0.51} I_1$$

$$I_4 \xrightarrow{0.67} I_3 \xrightarrow{0.78} I_2 \xrightarrow{0.51} I_1$$

$$I_4 \xrightarrow{0.68} I_2 \xrightarrow{0.62} I_3 \xrightarrow{0.67} I_3 \xrightarrow{0.78} I_2 \xrightarrow{0.51} I_1$$

This indicates that the direct relations between I_4 and I_1 are dominating in comparing with the minimum indirect relations. Therefore, the direct relations can be based to evaluate the overall dependencies of the infrastructures. According to Theorem 1, the transitive relations in Table 4 can be easily converted into partial order relations. The results are shown in Table 5. Since the relations in Table 5 are fuzzy partial order relations, the weights of each infrastructure network can be estimated according to relation in Eq. (7) as: $d(I_1) = 0.19$, $d(I_2) = -0.19$, $d(I_3) = 0.17$ and $d(I_4) = -0.27$ showing the descent orders of $I_1 \rightarrow I_3 \rightarrow I_2 \rightarrow I_4$. These results demonstrate that among the four infrastructure networks, the first infrastructure is the most important one since the states of its nodes impact strongly on nodes in other infrastructure networks but less influenced by other infrastructures, on the contrary, the forth infrastructure network is most less important infrastructure network in terms of interdependencies since it does not strongly control other infrastructures but it is strongly influenced by other infrastructures. These results are useful for infrastructures management, for example, a priority might be needed to give the first infrastructure, otherwise, its failure may cause strong cascade effects on other infrastructures, whereas the failure of the fourth infrastructure network may be local without strong cascade impacts.

4.2 Interdependencies between nodes and infrastructure networks

This section will explore the method to evaluate interdependencies between a set of nodes and infrastructure networks, which may be useful for instance to answer these two types of questions: what are the set of nodes essential in controlling the cascade influence on the whole system? Given a set of infrastructure networks and a critical function, what is the subset of critical nodes {x, y, z , …} across all networks that will adversely impact on the interdependencies of the set of infrastructure networks due to direct or indirect dependencies? We can define the following indexes to quantify the interdependencies between a node and an infrastructure network:

$$Q_{iq}^2 = \frac{n}{(n-1)} \sum_{j=1}^{n} \left(\frac{e_{ij}}{e_{i\cdot}} \right)^2 - \frac{1}{n-1} \tag{8}$$

$$Q_{pj}^2 = \frac{m}{(m-1)} \sum_{i=1}^{m} \left(\frac{e_{ij}}{e_{\cdot j}} \right)^2 - \frac{1}{m-1} \tag{9}$$

Similarly, we can use

$$Q_{ij}^2 = \left(\frac{e_{ij}}{e_{i\cdot}} \right)^2 \tag{10}$$

$$Q_{ji}^2 = \left(\frac{e_{ij}}{e_{\cdot j}} \right)^2 \tag{11}$$

to measure the dependencies between two nodes. These indexes in Eq. (8) to (11) provide asymmetric relations to quantify the interdependencies between a node and an infrastructure network or between a node and a node. Together with Eqs. (2) and (3), these indexes can quantify interdependencies between node and node, node and network, and network and network, respectively. To demonstrate the calculations and properties of indexes in Eqs. (8) to (11), we will once again use network I_1 and I_4 as examples. From Table 2, we can calculate the values of dependency indexes between nodes of I_1 and network I_4 as $Q_{a_1 I_4}^2 = 1, Q_{a_2 I_4}^2 = 0.111, Q_{a_3 I_4}^2 = 1$ and the values of dependency indexes between nodes of I_4 and network I_1 as $Q_{I_1 d_1}^2 = 0.333, Q_{I_1 d_2}^2 = 0.28$, respectively. These results show that the dependencies of I_4 on nodes a_1 to a_3 are generally stronger than the dependencies of I_1 on nodes d_1 and d_2. Moreover, it shows that the dependencies of I_4 on a_1 and a_3 are the strongest relations with $Q_{a_1 I_4}^2 = 1 \; and \; Q_{a_3 I_4}^2 = 1$.

Table 6 Asymmetric Relations between Nodes in I_1 and I_4

		Nodes of I_4	
		b_1	b_2
Nodes of I_1	a_1	1	0
	a_2	0.11	0.44
	a_3	0	1

Table 7 Asymmetric Relations between Nodes in I_4 and I_1

		Nodes of I_1		
		a_1	a_2	a_3
Nodes of I_4	b_1	0.44	0.11	0.00
	b_2	0.00	0.10	0.44

Similarly from Table 2, we can calculate the values of dependency indexes between the nodes of I_1 and the nodes of I_4. The results are shown in Tables 6 and 7, respectively. From these two tables, we can see that the relations between nodes in I_1 and I_4 are asymmetrical, for example, $Q^2_{a_1b_1}=1, Q^2_{b_1a_1}=0.44$, $Q^2_{a_2b_1}=0.11, Q^2_{b_1a_2}=0.44$, $Q^2_{a_2b_2}=0.44, Q^2_{b_2a_2}=0.10$, and $Q^2_{a_1b_2}=0.00, Q^2_{b_2a_1}=0.00$. The results indicate that the dependency of node b_1 on a_1 ($Q^2_{a_1b_1}=1$) is much stronger than that of node a_1 and b_1 ($Q^2_{b_1a_1}=0.44$), the dependency of node b_1 on a_2 ($Q^2_{a_2b_1}=0.11$) is much weaker than that of node a_2 on b_1 ($Q^2_{b_1a_2}=0.44$) and nodes a_1 and b_2 have no dependency. Similarly using the indexes defined in Eqs. (10) and (11) and the method introduced in section 3, we can rank the nodes according to their relative importance in terms of interdependencies with other nodes and a set of nodes with top rankings can be identified as the set of nodes providing crucial cascade influences on the whole system.

5. Conclusions and discussions

The interdependencies of infrastructure networks can be represented and characterized using the asymmetric relation matrixes introduced in the paper in three levels: between nodes and nodes, nodes and networks, and networks and networks. The relations defined in the asymmetric relation matrixes between nodes characterize the strength of edges linking nodes from a set of networks. The values can be calculated statistically by data driven approaches from actual distributions of nodes with historical data or can be assigned by knowledge driven approaches according to experts' opinions. From fuzzy asymmetric relations between all pairs of nodes, the asymmetric relations between nodes and networks can be calculated and the overall asymmetric relations between infrastructure networks can be estimated using the methods introduced in this paper. Further using the fuzzy relation analysis method, we can evaluate the overall dependencies of nodes on nodes, networks on nodes, and networks on networks. Networks and nodes can be ranked according to their relative importance in the overall cascade interdependencies. Understanding the relative overall interdependencies of nodes and networks is the central goal of this study and the results generated using the methods introduced are useful in decision making for resources allocations and emergency management. The fuzzy relation analysis method can be used both for evaluation and for simulation of interdependencies of infrastructures. It is straight forward to implement the methods in a GIS.

Acknowledgements

This research is finically supported by a project "modeling infrastructure interdependency for emergency management using a network-centric spatial decision support system approach" awarded jointly by the Natural Science and Engineering Research Council of Canada (NSERC) and the Public Safety and Emergency Preparedness Canada (PSEPC). Some academic activities related to this research were financially aided by a Distinguished Young Researcher Grant awarded by the China Foundation of Science (40525009).

References

Cheng Q (1990) Order analysis method and its application in geology, In Proceedings of International Workshop on Statistical Mineral Resource Prediction, Chinese University of Geosciences, Wuhan, China, 20–25 October, 2, 64–70.

Cheng Q (1991) Asymmetric association of qualitative variables and its applications in mineral resource appraisal, Journal of Statistics and Applied Probability, 6(2): 152–161 (in Chinese with English abstract).

Cheng Q (1996 Asymmetric fuzzy relation analysis method for ranking geoscience variables. Nonrenewable Resources, 5(3): 169–180.

Dudenhoeffer D, Permann M, and Manic M (2006) CIMS: A framework for infrastructure interdependency modeling and analysis, in Proceedings of the 2006 Winter Simulation Conference, Piscataway, NJ.

Dunn, M and Wigert I (2004) International CIIP Handbook 2004: An Inventory and Analysis of Protection Policies in Fourteen Countries. Zurich: Swiss Federal Institute of Technology: p. 243.

Jensen J (1996) Introductory digital image processing: a remote sensing perspective (2nd edn). UpperSaddle River, NJ: Prentice Hall.

Kaufmann A (1975) Introduction to the theory of fuzzy subsets: v. 1, NY, Academic Press, 416p.

Pederson P, Dudenhoeffer D, Hartley S, and Permann M (2006) Critical infrastructure interdependency modeling: a survey of U.S. and international research, a unpublished report prepared by the Critical Infrastructure Protection Division of Idaho National Laboratory for the Technical Support Working Group under Work for Others Agreement 05734, 23p.

Rinaldi S Peerenboom J and Kelly T (2001) Identifying, understanding, and analyzing critical infrastructure interdependencies, IEEE Control Systems Magazine, 12, 11-25.

Zimmermann H-J (1991) Fuzzy set theory and its applications, Kluwer Academic Publishers, Boston, 399p.

Modeling of Flood-Related Interdependencies among Critical Infrastructures

Sharmin Sultana and Zhi Chen

Department of Building, Civil, and Environmental Engineering
Concordia University, 1455 de Maisonneuve Blvd. W.
Montreal, Quebec, Canada H3G 1M8
zhichen@bcee.concordia.ca

Abstract

This study presents an integrated approach for modeling the flood-related interdependencies among critical infrastructures and their vulnerabilities. The developed method is based on the development of a Petri-Net model and fragility curves analysis. Specifically, infrastructure interdependency is simulated using Petri Nets, and consequences to infrastructures following a flood are quantified through developing fragility curves. Both empirical and analytical fragility curves are developed using a regional flood hazard database, hydraulic modeling of dam failure, and Monte Carlo simulation. The developed method has been applied to a case study. Cascading disruptions in the interconnected water infrastructures due to a dam failure are simulated in this study. Reasonable results have been obtained, which indicate that the proposed modeling system can address the interdependency among critical infrastructures and serve as a decision tool for flood-related emergency response and management.

1. Introduction

Flood occupies the highest rank among the natural disasters in respect of their adverse impacts (Isomina et al. 2005). Flooding from levee failures due to Hurricane Katrina set in motion an unanticipated failure of multiple infrastructure systems in the City of New Orleans USA that resulted from the complex interactions among interdependent infrastructures (Leavitt and Kiefer 2006). The major floods in Europe hampered the functioning of electricity, transportation systems, buildings directly, and indirectly hampered the gas pipes, telecom, emergency response systems which were the result of the interactions among these infrastructures (Rahman 2005). Consequently, study of interdependency among critical infrastructures has become more important to address the cascading effects when one infrastructure is failed and thus to help the related emergency management in decision making.

Interdependencies among multiple infrastructures connected as a 'system of systems' can significantly increase the overall complexity due to the change in the system(s). Previously, infrastructure interdependency and the related vulnerability assessment were addressed by few studies. Rinaldi et al. (2001) classified the types of infrastructure interdependency as physical, cyber, geographical and logical; failure mechanisms were classified as cascading, escalating and common cause failure. Several approaches were explored for the quantitative analysis of infrastructure interdependency. Nozick et al. (2005) developed a mathematical framework of the interconnected infrastructures network and described the algorithms using the Markov-Semi Markov model to estimate the performance. Ezell et al. (2000) developed a holistic and systematic probabilistic model to simulate expected and extreme risk for a water infrastructure system's interconnectedness and interdependencies using event tree-fault tree analysis techniques. Based on the Leontief economic model, Haimes et al. (2005) developed the 'Inoperability Input-Output Model (IIM)' to describe the degree of interconnectedness among critical infrastructures and the cascading failures through the interconnected systems. Very few studies dealt with the structural perspective whereas most of the research works addressed economic flow among the infrastructures or other criteria to determine their interdependencies.

Recently, network-based analysis was employed to study the behaviors of the interconnected engineering infrastructures. For example, Petri Net is a system analysis method proposed by Carl Adam Petri in early 1960s (Petri 1962), which is a useful tool to determine the interdependencies among the infrastructures in a network. Gursesli and Desrochers (2003) used the graph based Petri-Net for identifying vulnerabilities of the infrastructures in a network system; interdependencies among those infrastructures due to the failure of one infrastructure were simulated. Petri-Net based coordination mechanisms were performed for multiple workflows by Raposo et al. (2000). There was lack of studies on the examination of the flood-induced interdependencies among critical infrastructures.

Issues of infrastructure interdependency are directly related to environmental or economical risks or impacts. For the risk analysis of infrastructures, vulnerability assessment with fragility curves has been practiced widely. Hwang et al. (2000) presented a method for evaluating seismic damages to the bridges and highway systems in earthquake prone area like Memphis and Shelby County, Tennessee, by classifying the bridges and developing the fragility curves for each class of bridges. Chock (2005) developed fragility curves of Hawaii residential buildings using a GIS supported hurricane damage database, post hurricane aerial photography, property tax databases of building parameters and Monte Carlo simulation based wind-speed mapping to simulate the fragilities and associated risks of a wide variety of buildings. Developing fragility curves for infrastructures was studied as an important component along with the other steps for a national-scale flood risk assessment by Hall et al. (2003). Previous studies indicated that integration between a net-work method for infrastructure independency analysis and a fragility-curve method for the related impact quantification will help to examine complex flood-induced cascading failures of civil infrastructures.

In this research, an integrated modeling approach will be developed to study the vulnerability of an infrastructure system and the consequent cascading effects. It is intended to address flood hazard and its impacts on the related urban infrastructure system, consisting of a gravity dam, hydroelectric power plant, water storage facility and treatment plant, irrigation dam, water and power distribution pipeline system and highway. Structural analysis and flood hazard data will be included to develop fragility curves for assessing the vulnerability of the gravity dam. Interdependencies among the interconnected infrastructures following a dam failure will be simulated through a constructed network based Petri-Net model.

2. Development of a Net-Work Based Modeling Approach: Petri-Net model and Fragility Curves

2.1 Modeling framework

According to Moselhi et al. (2005), the critical infrastructure systems can be represented as nodes in a network where they are connected through a set of links representing the logical relationship among them; damage or malfunction in one node affects the functioning of the connected succeeding node. With such a concept, the cascading impacts throughout the network can be assessed. Therefore, assessment of infrastructures vulnerability can be combined with a network method to study the safety of a regional interconnected infrastructure system. This analysis leads to the framework of a new modeling system, which can be implemented with the following procedures:

Quantitative analysis: This step includes (1) a hydraulic analysis of flood frequency and probability of occurrence of a certain level of flood and (2) a structural analysis of the infrastructures to determine their failure modes for selected flood levels. Thus, the vulnerability of the infrastructure in relation to floodwater will be predicted with the development of fragility curves.

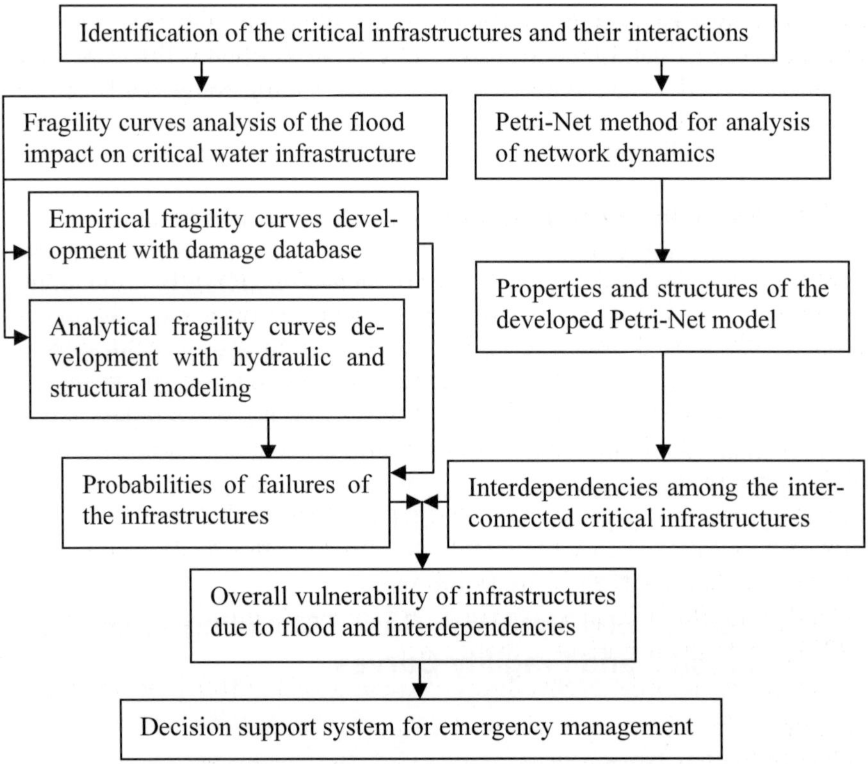

Fig. 1. The modeling framework of flood induced infrastructure interdependency

Qualitative analysis: It will be analyzed how one infrastructure is affected by another infrastructure(s) in a network. This step can be formulated as: (1) identifying the critical infrastructures in the area of study; (2) developing a network of the identified infrastructures by setting them as the network nodes and linking them logically; and (3) simulating the dy-

namics of the developed network using available traditional network models (e.g. Petri-Net model, agent based model, system dynamics).
Decision analysis for emergency management: When the quantitative measurement of vulnerability of each infrastructure at a certain flood level is known, then the overall vulnerability can be determined from the predicted dependency through a Petri-Net model.

Specifically, a set of fragility curves will be developed for the hydraulic gravity dam using a flood damage database and structural modeling. The fragility curves will convey the information about the vulnerability of dam infrastructure through the probability distribution for various levels of flood hazard. Considering the extreme vulnerable condition, i.e. the failure of dam, the cascading impacts on the interconnected infrastructures in the flood plain will be captured through the development and analysis of a network based model. In this research, a Petri-Net method is developed for modeling of its dynamic characteristics. The modeling components can be summarized in Fig.1.

2.2 Development of Fragility Curves

Definition of basic damage states, corresponding fragility curves and conditional probabilities for estimating damage matrices were discussed before (Filliben et al. 2002). Fragility curve is defined as a mathematical expression that represents the conditional probability of reaching or exceeding a certain damage state for an infrastructure at a given hazard level. In this research, both the empirical and analytical fragility curves analyses will be developed.

In this study, historical damage database is used to draw empirical fragility curves and calculate the conditional probabilities of exceeding certain damage states. The steps are as follows:

- Conducting flood frequency analysis by Gumbel's method (Subramanya 2001) with the historical hydraulic data
- Determination of the probabilities of water levels, $P[WL]$ for which the empirical fragility curves will be drawn
- Classifying the damage states, e.g. 10% damage state etc.
- Collecting the records of the number of events of the corresponding damage states for the selected water levels
- Determining the probabilities of occurrence of the damage levels for the water levels, $P[damage \bigcap WL]$

- Calculating the conditional probabilities of the damage levels,
$$P[damage|WL] = \frac{P[damage \cap WL]}{P[WL]} \quad (1)$$
- Plotting the empirical fragility curves with water levels in X-axis and corresponding probabilities of exceedence in Y-axis

For analytical fragility curves, usually the outputs of the structural failure model of an infrastructure for a hazard are used as the inputs for generating the curves. In this study, several failure modes of the hydraulic dam will be analyzed for flood water levels. The steps are: (i) modeling of infrastructure failure modes for flood water level with the Monte Carlo simulation; (ii) classifying the damage states; (iii) determining the probabilities of exceedence of the damage states; (vi) repeating the steps for different water levels; and (v) developing fragility curves with the probabilities of damage states.

2.3 A Brief of Petri-Net

Petri-Net has been applied to study the behavior of concurrent, asynchronous, distributed, parallel, non deterministic, and/or stochastic systems (Murata 1989). A basic Petri-Net structure *(C)* is a four-tuple, $C = (P, T, I, O)$, where, P stands for places, T for transitions, I for input functions and O for output functions. Another element of the Petri-Net is 'token' which are placed in the places; tokens flow throughout the network during the execution of the network. Placement of the tokens is called 'marking' of the network. Any Petri-Net has an initial distribution of the tokens which is called initial marking. For example, a network of systems is as follows:

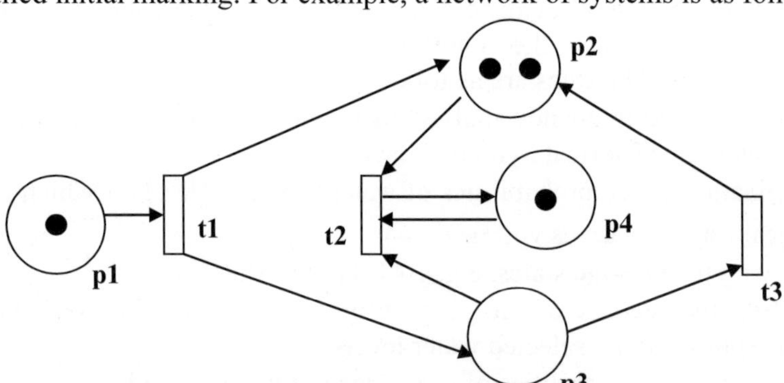

Fig. 2. Petri Net graph representing the relationships

$C = (P, T, I, O)$; $P = \{p_1, p_2, p_3, p_4\}$; $T = \{t_1, t_2, t_3\}$; $I(t_1) = \{p_1\}$; $O(t_1) = \{p_2, p_3\}$; $I(t_2) = \{p_2, p_3, p_4\}$, $O(t_2) = \{p_4\}$; $I(t_3) = \{p_3\}$; $O(t_3) = \{p_2\}$; the initial marking of the Petri-Net is, [1, 2, 0, 1].

Graphically, the above network can be represented in Fig.2. The marking shows the 'tokens' distribution which gives the existing condition of the net. A transition is enabled if each of its input places has a minimum number of tokens equal to the weighting of arcs from the place to the transition. The transition fires by removing tokens from its input places and creates new tokens which are distributed to its output places. In the above Petri Net, only transition t1 is enabled initially.

Flood hazard not only poses impacts on one single infrastructure, also the interconnected infrastructures become vulnerable due to their high degree of interconnectedness. The proposed infrastructure interdependency modeling approach is further delineated through a case study of flood and its impacts on the floodplain infrastructures system in the section below. The infrastructures will be graphically represented with the Petri Nets where the places of the net will represent the states or conditions of the infrastructures, transitions will represent the impacts of one infrastructure on another; the structural analysis will simulate their interdependencies. In summary, the fragility curves of a critical infrastructure will show the probabilities of its various vulnerable conditions, and the Petri-Net model will simulate the cascading impacts when the dam is failed.

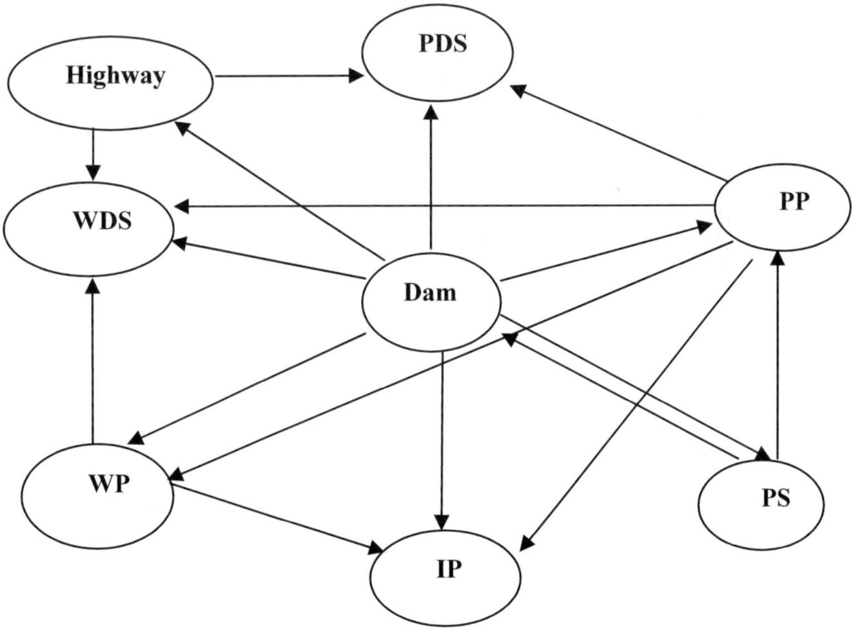

Fig. 3. The study system and interdependencies

3. Application to Modeling Flood-Induced Interdependencies and Impact Assessment

3.1 An Overview of the Study System

The developed modeling approach is applied to a case study in a flood plain area. The study system refers to a region with a gravity dam, water supply infrastructures, water and power distribution pipes, and highway. The water infrastructures include a hydraulic power plant, penstock for receiving water from the reservoir created by the dam and supplying water to the power plant, a water plant, and an irrigation plant. The infrastructures system of the case study area is depicted in Fig.3.

Here, PS stands for 'penstock', PP for 'power plant', WP for 'water storage and treatment plant', IP for 'irrigation plant', PDS for 'power distribution system', and WDS for 'water distribution system'. In the water plant, there is a water treatment facility, which provides drinking water supply through the water distribution pipes into the remote areas, and a water storage tank providing water for the irrigation plant. Information about these infrastructures are stated briefly in below, which will lead towards the development of the Petri-Net model for this study area.

Dam

When dam is damaged or collapsed due to a flood, it causes the invasion of floodwater into the adjacent flood plain area. High pressure from floodwater will lead to the rupture of penstock. Floodwater causes the malfunction of power plant, water plant, irrigation plant, water and power distribution pipes as these infrastructures become inundated. Highways are also inundated by floodwater.

Penstock

A hydroelectric dam is usually so designed that huge water is received through the penstock for power generation.

Power plant

The power plant provides supply of electricity to the water plant, irrigation plant; power distribution pipes convey electricity from the power plant to the remote regions. Functioning of these infrastructures also require electricity supply.

Water facility

The operation of water storage and treatment plant requires a constant supply of power; otherwise, when the reservation in the water distribution pipes is depleted; drinking water supply becomes unavailable. Stored water cannot be conveyed to the irrigation plant.

Highways

Power distribution pipes and water supply pipes are carried underneath the highways to remote places; inundated highways may also cause the ruptures or damages of these infrastructures.

3.2 Modeling Process and Results

Vulnerability Assessment of dam with Fragility Curves

(A) Empirical fragility curves of dam

The breaching conditions of the dam are considered as damage and flood levels are denoted as hazard. Following the steps described in section 2.2, the probability data of different damage states can be generated. Height of the dam in the study area (Fig.3) is 30 m, top width is 2 m, and base width is 18.5 m. The damage states and the resulting fragility curves are given in Fig.4. It shows that, probability of '10% damage' state initiates at 26.5 m water level and reaches to 1 after 28.2 m; for '25% damage' level, the range is (26.8 - 28.8 m); for '50% damage' level, the range is (27.4 - 29.2 m), and for '75% damage', the range is (28 - 29.8 m).

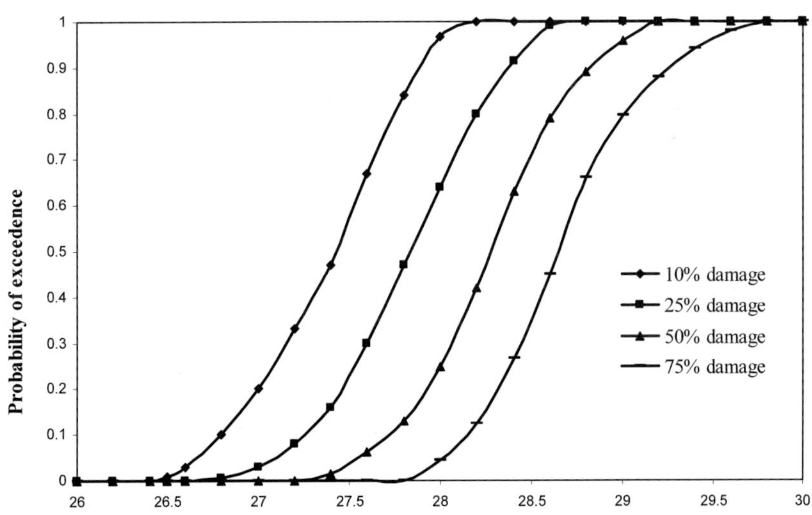

Fig. 4. Empirical fragility curves of dam for flood hazard

(B) Analytical fragility curves

For developing the analytical fragility curves of the dam, first the hydraulic modeling of dam failure is performed; probabilities data of different damage levels will be then generated from the model outputs.

Hydraulic modeling of dam failure

The above dam under study is made up of two blocks; the upper block is 10 m high and the lower block is 20 m high. The slope in the upstream side is H:V (1:20) for the blocks, slopes of the first block and second block in the downstream side are 3:10, and 6:10 respectively. Unit weight of the dam material is 23.6 KN/m^3. The coefficient of friction is μ = 0.65. The dam is analyzed to determine its failure modes for different water levels. In the structural failure modeling, we consider the following failure modes of the dam: [1] dam is failed due to its overturning with respect to the toe of the dam; and [ii] dam is failed from sliding due to the shear failure along the intersection of the blocks.

Hydraulic modeling of dam failure is conducted (Linsley & Franzini 1992):

$$M_o = H_h * \overline{x_h} + u * \overline{x_u} ; \tag{2}$$

$$M_R = W_c * \overline{x_c} + H_v * \overline{x_v} ; \tag{3}$$

$$FS_0 = \frac{M_R}{M_o} ; \tag{4}$$

For the shear failure, if it is assumed that there is no bond between the blocks, then it can be written,

$$F_f = \mu(W + H_v - u) \tag{5}$$

$$FS_s = \frac{F_f}{H_h} \tag{6}$$

$$IFS_o = \frac{1}{FS_o} \tag{7}$$

$$IFS_s = \frac{1}{FS_s} \tag{8}$$

where, M_o = overturning moment (KN-m/m); M_R = righting moment (KN-m/m); FS_o = factor of safety for overturning against the toe of dam; F_f = friction force along the contact plane (KN/m); FS_s = factor of safety against sliding along the contact plane of the dam blocks; H_h = horizontal hydrostatic pressure (KN/m); H_v = vertical hydrostatic pressure (KN/m); u = uplift pressure (KN/m); W_c = weight of gravity dam (KN/m); $\overline{x}$ = distance of the center of gravity from the toe of the dam (m); μ = coefficient of friction along contact planes; IFS_o = inverse of FS_o; IFS_s = inverse of FS_s.

For both types of failures, factor of safety greater than 1 is desirable. The inverse of the factor of safety is used in this study. If its probability exceeds 1, a deteriorated condition of the dam is implied.

Thus, the analytical fragility curves development for the study case follows the steps: i) classifying damage states or failure modes of the dam. Here, we have two failure states; ii) hydraulic modeling of dam failure modes to determine IFS_o and IFS_s for a set of flood water levels, iii) calculating the exceeding probabilities of the failure states, and iv) determining the analytical fragility curves with these values. In this study, the Monte Carlo simulation is performed for the unit weight of dam to generate the probabilities. The damage states and the developed fragility curves are given in Fig.5.

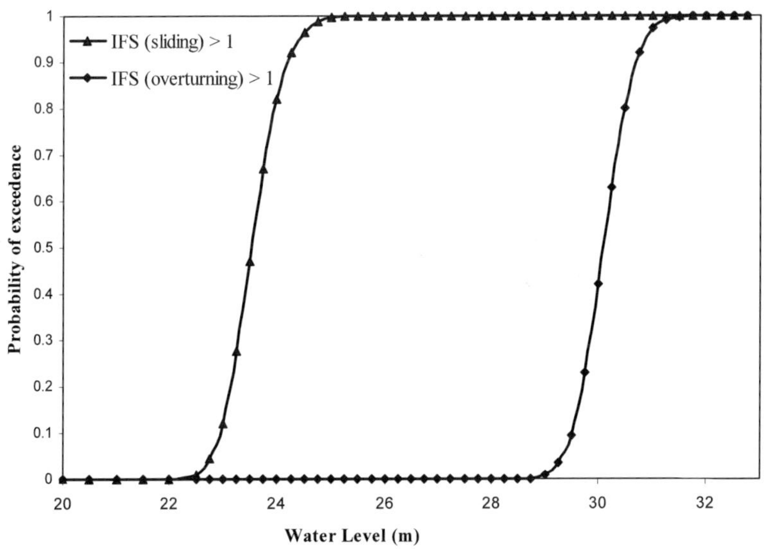

Fig. 5. Analytical fragility curves of dam for different flood levels

As shown in Fig. 5, probability of the damage state '$IFS_o > 1$' starts at the water level of 29 m and reaches to 1 at the 32 m water level; for '$IFS_s > 1$', probability starts from 22.5 m and reaches to 1 at 25.5 m water level. It is noted that the dam is in higher risk for sliding than overturning of the dam based on the results from the hydraulic and fragility curves analysis.

Overturning failure occurs with full probability at the flood level of 2 m above the dam, whereas, sliding failure occurs at the 4.5 m free board below the dam.

The failure modes of the dam and its vulnerability have been assessed with fragility curves development. The consequent interdependencies among the interconnected infrastructures due to the extreme vulnerability, i.e. dam failure, will be simulated in below using the Petri-Net tool.

3.2.2 Petri-Net modeling of flood affected infrastructure interdependencies

With the interdependencies and interrelationship information, the places and transitions are defined accordingly for the study system. The conditions or states of the infrastructure components are denoted as 'places', the events of impacts or disruptions as 'transitions', occurrence of an event as 'firing', and holding of a condition as the 'token'. After defining these terms, Petri-Net model is developed including the following steps: i) determining the basic model with the initial marking state where all the infrastructures are in service; ii) choosing the first infrastructure which is destructed by a hazard and determining the subsequent marking states; iii) choosing one of the remaining infrastructures (Fig. 3) and assigning 'tokens', 'transitions' around it considering its interactions with the other infrastructures; and iv) applying the same procedure for the remaining infrastructures.

For the study system (Fig.3), the places and transitions of the model are listed in Tables 1 and 2. A set of 'dummy places' are introduced in the Petri Net to resolve the problem of the determination of place invariants, which results from the self-loop of the net used for holding the real condition in the net (Murata 1989).

Table 1. List of places

Place	Description
p1	No flood and dam is functioning
p2	Collapsed dam
p3	Penstock rupture
p4	Inactive power plant
p5	Inactive water plant
p6	Inactive irrigation plant
p7	Disrupted power distribution pipes

p8	Disrupted water distribution pipes
p9	Inundated highways
p10	Collapsed dam (dummy place for penstock)
p11	Collapsed dam (dummy place for power plant)
p12	Collapsed dam (dummy place for water plant)
p13	Collapsed dam (dummy place for irrigation plant)
p14	Collapsed dam (dummy place for power distribution system)
p15	Collapsed dam (dummy place for water distribution system)
p16	Collapsed dam (dummy place for highways)
p17	Penstock rupture (dummy place for dam)
p18	Penstock rupture (dummy place for power plant)
p19	Inactive power plant (dummy place for water plant)
p20	Inactive power plant (dummy place for irrigation plant)
p21	Inactive power plant (dummy place for power distribution system)
p22	Inactive power plant (dummy place for water distribution system)
p23	Inactive water plant (dummy place for irrigation plant)
p24	Inactive water plant (dummy place for water distribution system)
p25	Flooded highways (dummy place for power distribution system)
p26	Flooded highways (dummy place for water distribution system)

Table 2. List of transitions

Transition	Description
t1	Flood occurred and dam is collapsed
t2	Collapsed dam affects penstock
t3	Collapsed dam affects power plant
t4	Collapsed dam affects water plant
t5	Collapsed dam affects irrigation plant
t6	Collapsed dam affects power distribution system
t7	Collapsed dam affects water distribution system
t8	Collapsed dam affects highways
t9	Penstock rupture affects dam

t10	Penstock rupture affects power plant
t11	Inactive power plant affects water plant
t12	Inactive power plant affects irrigation plant
t13	Inactive power plant affects power distribution system
t14	Inactive power plant affects water distribution system
t15	Inactive water plant affects irrigation plant
t16	Inactive water plant affects water distribution system
t17	Flooded highways affects power distribution system
t18	Flooded highways affects water distribution system

(A) Petri Net model development

The places, transitions and tokens are set in the model according to the information stated before for the study system. The notations are listed in the tables. The developed model is shown in Fig.6.

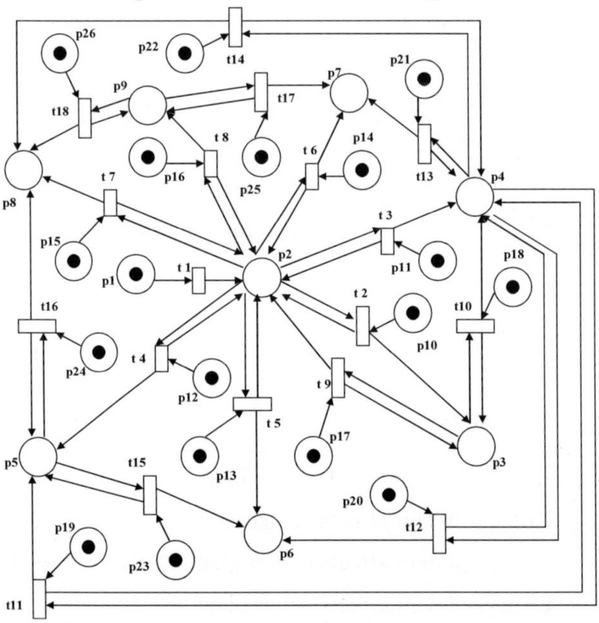

Fig. 6. Petri Net model of the infrastructures system in the study area

(B) Model execution and analysis

Initially, there is no flood and all the infrastructures are functioning properly in the developed model. The initial marking of the net is,

M_0=

1 0 0 0 0 0 0 0 0 1 1 1 1 1 1 1 1 1 1 1 1 1 1 1 1 1

We start the model execution with firing t1, i.e. dam is collapsed. Thus, the token in p1 flows to p2, which is available for firing the transitions t2 through t8.

Structural properties of the developed model: invariant analysis

Invariant analysis is used for determining the structural properties such as liveness, boundedness, consistency, conservativeness of Petri-Net, which are independent of the initial marking (Murata 1989), including place (P) invariants and transition (T) invariants analyses. A P-invariant indicates the set of places in which the weighted sum of the tokens in those places remain constant for all markings. On the other hand, a T-invariant indicates the sequence of transitions whose firings result the net to return to its initial condition. If the number of places is m and number of transitions is n, then the incidence matrix is, $C = n * m$. To refer the flow of tokens from and to a place in the network, -1 and 1 are assigned, respectively. The procedure of the assignment of tokens in the remaining rows of the matrix is the same. Thus, an incidence matrix [C] can be obtained for the developed Petri-Net model.

P-invariant analysis

The minimal P-invariants are capable of representing the interdependencies among the interconnected infrastructures (Gursesli and Desrochers 2003). If the P-invariant *'x'* is a $m*1$ column vector, then, solution of $C * x = 0$ equation gives the following minimal P-invariants.

IP_1	0	0	1	0	0	0	0	0	0	1	0	0	0	0	0	0	0	0	0	0	0	0	0	0	0	0
IP_2	0	0	0	0	0	0	0	0	1	0	0	0	0	0	0	1	0	0	0	0	0	0	0	0	0	0
IP_3	1	1	0	0	0	0	0	0	0	0	0	0	0	0	0	0	1	0	0	0	0	0	0	0	0	0
IP_4	0	0	0	1	0	0	0	0	0	0	1	0	0	0	0	0	0	1	0	0	0	0	0	0	0	0
IP_5	0	0	0	0	1	0	0	0	0	0	0	1	0	0	0	0	0	0	1	0	0	0	0	0	0	0
IP_6	0	0	0	0	0	1	0	0	0	0	0	0	1	0	0	0	0	0	0	1	0	0	1	0	0	0
IP_7	0	0	0	0	0	0	1	0	0	0	0	0	0	1	0	0	0	0	0	0	1	0	0	0	1	0
IP_8	0	0	0	0	0	0	0	1	0	0	0	0	0	0	1	0	0	0	0	0	0	1	0	1	0	1

T-invariant analysis

If the T-invariant *'y'* is a $n*1$ column vector, then the solution of $C^T * y = 0$ equation gives the T-invariants. The solution gives empty matrix in this analysis. Namely, it is not possible to return to the initial condition in this case study.

(C) Interpretation of the interdependencies from the model results

From the invariant analysis of the developed Petri-Net model, eight minimal P-invariants (IP_1 through IP_8) were found, which can simulate the infrastructure interdependency in the study area.

For example, IP_1 interprets that token flow between place 3 and place 10 is always 1 through the simulation of the model, i.e. M (p3) + M (p10) = 1; checking the model, we notice that, penstock is ruptured (place 3) due to the dam collapse; thus, collapsed dam induces the rupture of penstock.

Similarly, other invariants simulate the interdependencies among the infrastructures which are stated below:

- IP_2 gives, M (p9) + M (p16) = 1. This simulation implies that the highway gets inundated directly from the collapse of dam where the 'token' flows between the places representing the 'collapsed dam' and 'inundated highways'.
- IP_3 gives, M (p1) + M (p2) + M (p17) = 2. If the penstock doesn't function properly, the water height may increase to induce dam failure.
- IP_4 gives, M (p4) + M (p11) + M (p18) = 2 (constant). The hydraulic power plant is highly dependent on the penstock pipes which carries huge amount of water from the reservoir. When the penstock collapses from the dam collapse and high water pressure, the plant can no more sustain its service.
- IP_5 gives, M (p5) + M (p12) + M (p19) = 2 (constant). Here, the water plant might become inactive due to the flood water after the dam collapses as well as the power plant is unable to supply the power for its function.
- IP_6 gives, M (p6) + M (p13) + M (p20) + M (p23) = 3 (constant). It indicates that the function of the irrigation plant can be hampered by the floodwater from the collapsed dam; shutdown of the power supply from the power plant and shortage of water supply from the water plant are also responsible for its inactivity.
- IP_7 gives, M (p7) + M (p14) + M (p21) + M (p25) = 3 (constant). Power distribution system generally contains a large system of pipes to carry the electric power to the territory. Floodwater from collapsed dam inundates the highways underneath which the pipe systems run through. Again, if the power supply is depleted due to the inactivity of the power plant, the system has no commodity to supply.
- IP_8 gives, M (p8) + M (p15) + M (p22) +M (p24) +M (p26) = 4 (constant). Water distribution system also contains a large system of distribution pipes which receives the water supply from the water plant; the

water treatment activities in the water plant require enough supply of electricity which comes from the hydraulic power plant. The distribution pipes also run through under the highways; flooded highways from dam collapse might cause the rupture of the pipe systems. Disturbances in these infrastructure systems obviously have impacts on the serviceability of the water distribution system.

From the above analysis, it is found that most of the interconnected systems are vulnerable to the collapse of dam. Particularly, power plant affects significantly the functioning of the dependent infrastructures e.g. water plant, irrigation plant, and water and power distribution systems. These predications about flood-affected infrastructure interdependency are useful steps of disaster preparedness for emergency management.

4. Concluding Remarks

Significant amount of issues has been raised in North America to assess the interdependencies among the critical infrastructures for disaster preparedness. In this study, an integrated method has been developed and examined for determining the flood-affected infrastructure interdependencies. This development includes a development of both empirical and analytical fragility curves of a critical infrastructure to address its vulnerability characteristics for flood hazard assessment; and a Petri-Net model for predicting infrastructure interdependency induced by the failure of the infrastructure.

The developed modeling approach has been applied to a case study involving a hydraulic dam, a penstock system, highways, a power plant, and a water storage and treatment plant. Reasonable results have been obtained, which include a fragility assessment of critical infrastructure system, a quantification of hazard-induced infrastructure interdependencies, and a decision analysis for disaster preparedness. It indicates that a combined Petri-Net and Fragility analysis approach is useful for addressing flood-induced cascading impacts on critical infrastructures.

References

Bobbio A (1990) System modeling with Petri Nets. Colombo AG, de Bustamante AS. (eds) System Reliability Assessment. Kluwer p.c., pp 102-143

Chock GYK (2005) Modeling of hurricane damage for Hawaii Residential construction. J Wind Engg. and Industrial Aerodynamics 93: 603–622

Ezell BC, Farr JV, Wiese I (2000) Infrastructures risk analysis model. J Infrastructure Systems 6(3): 114-117

Filliben JJ, Gurley K, Pinelli J-P, Simiu E (2002) Fragility curves, damage matrices, and wind induced loss estimation. 3rd Int. Conf. on Comp. Sim. in Risk An. & Haz. Mit.. June 19-21, Sintra, Portugal, 119-126

Gursesli O, Desrochers AA (2003) Modeling infrastructure interdependencies using Petri Nets. IEEE:1506-1512

Haimes YY, Horowitz BM, Lambert JH., Santos JR, Lian C, Crowther KG (2005) Inoperability Input-Output Model for Interdependent Infrastructure Sectors. I: Theory and methodology. J Infrastructure Systems 11(2): 67-79

Haimes YY, Horowitz BM, Lambert JH, Santos JR, Crowther KG, Lian C (2005) Inoperability Input-Output Model for Interdependent Infrastructure Sectors. II: Case studies. J Infrastructure Systems 11(2): 80-92

Hall JW, Dawson RJ, Sayers PB, Rosu C, Chatterton JB, Deakin R (2003) A methodology for national-scale flood risk assessment. Proceedings of the Institution of Civil Engineers - Water Maritime and Engg 156(3): 235-248

Hwang H, Jernigan JB, Lin YW (2000) Evaluation of seismic damage to Memphis bridges and highway systems. J Bridge Engg 5(4): 322-330

Istomina MN, Kocharyan AG, Lebedeva IP (2005) Floods: genesis, socioeconomic and environmental impacts. J Water Resources 32(4): 349-358

Leavitt WM, Kiefer JJ (2006) Infrastructure interdependency and the creation of a normal disaster: the case of Hurricane Katrina and the City of New Orleans. J Public Works Management & Policy: 10(4): 306-314

Linsley RK, Franzini JB (1992) Water resources engineering. 4th edn McGraw-Hill, 1992, NY

Moselhi O, Hammad A, Alkass C, Debbabi M, Haider M (2005) Vulnerability assessment of civil infrastructure systems: A network approach. 1st CSCE Specialty Conference on Infrastructure. Tech, Management and Policy, Toronto, 2-4 June

Murata T (1989) Petri Nets: properties, analysis and applications. Proceedings of the IEEE 77(4): 541-580

Nozick LK, Turnquist MA, Jones DA, Davis JR, Lawton CR (2005) Assessing the performance of interdependent infrastructures and optimizing investments. Int. J. Critical Infrastructures 1(2/3): 144–154

Office of Critical Infrastructure Protection and Emergency Preparedness (2003) Threats to Canada's Critical Infrastructure. Government of Canada, Number: TA03-001

Peterson JL (1981) Petri net theory and the modeling of systems. Prentice-Hall, Inc., Englewood Cliffs, N. J. 07632

Petri CA (1962) Kommunikation mit Automaten. Ph.D. thesis, Germany

Public Safety and Emergency Preparedness Canada (PSEPC) Modernization of the Emergency Preparedness Act. Consultation paper, July 2005

Rahman S (2005) Impact of natural disasters on critical infrastructures. The 1st Bangladesh Earthquake Symposium, 14-15 December, 2005

Raposo AB, Magalhães LP, Ricarte LM (2000) Petri Nets Based Coordination Mechanisms for Multi-Workflow Environments. Int J Computer Systems Sci & Engg 15(5): 315-326

Rinaldi SM, Peerenboom JP, Kelly TK (2001) Critical infrastructure interdependencies. IEEE Control System Magazine, 11-25

Simpson DM, Rockaway TD, Weigel TA, Coomes PA, Holloman CO (2005) Framing a new approach to critical infrastructure modeling and extreme events. Int J Critical Infrastructures 1(2/3): 125–143

Subramanya K (2001) Engineering Hydrology. 2nd edn, McGraw Hall, New Delhi

Challenges for the Application of GIS Interoperability in Emergency Management

Rifaat Abdalla[1], C. Vincent Tao[1] and Jonathan Li[2]

[1]GeoICT Lab, York University, 4700 Keele Street, Toronto, Ontario M3J 1P3, Canada, rifaat.abdalla@gmail.com, tao@yorku.ca
[2]Department of Geography, Faculty of Environmental Studies, University of Waterloo, 200 University Avenue West, Waterloo, Ontario N2L 3G1 Canada, junli@uwaterloo.ca

Abstract

This paper highlights application challenges for GIS interoperability for emergency management with emphasis on critical infrastructure sectors. In the first part, this paper provides a comparative analysis of emergency management operations in the City of Vancouver; the City of Toronto, the Kitchener Waterloo Region, and the Dufferin County. A variety of qualitative research methods were employed for gathering information from key decision-makers involved with emergency management. The second part of this paper presents a scenario-based case study, which aims to provide a demonstration of the utility of GIS interoperability, for disaster management. This paper also discusses the strengths and weaknesses of leveraging GIS interoperability for disaster management.

1 Introduction

Geographic information system (GIS) interoperability aims to ensure a process that will allow for the use of data, information, and services across organizational boundaries, and to make information available for all of the levels involved in disaster management activities. One of the first studies that aimed to characterize GIS interoperability was by Bishr (1998) who has demonstrated the need for working on data that are scattered over several independent databases. This goal has been achieved and thc internet has played the key role in connecting systems over a common network protocol. According to Erharuyi and Fairbairn (2003), real-time ability to visualize map locations, determine the scale of emergency, identify and evacuate at-risk populations, and expedite and direct response efforts are key areas in which GIS interoperability is much needed. As a result, a great deal of interest towards disaster management applications using internet GIS is rapidly emerging. Advancements in internet technology, in the form of high speed broadband, have made GIS interoperability more effective and accessible.

This paper will address issues related to GIS interoperability, and will examine the status of GIS interoperability in disaster management, with emphasis on specific Ontario municipalities (i.e., Toronto, Kitchener Waterloo and Dufferin County). The second part of this paper will present an example (i.e., a demonstration case study) that highlights benefits that GIS interoperability could bring to the disaster management community.

1.2 GIS Interoperability

GIS interoperability is unique in that it allows one to ensure easy and compatible spatial data access over the internet (Visser et al. 2002). Farley (1999) indicates that interoperability is important for disaster management applications, because it helps to eliminate system and data heterogeneity. The importance of this has made interoperability a central research and development initiative. According to Goodchild (2003) interoperability aims to eliminate incompatibilities in data formats, software products, and spatial conceptions and data quality standards. The distinction between data transfer and interoperability is very important. Table 1 summarizes the differences between interoperability and transfer.

Table 1. Differences between the process of data transfer and data interoperability (after Glover 1995).

Aspect	Transfer	Interoperability
Scope	Data, no process	Data and Process
Data Unit	Dataset	Object (Dataset or Lower)
Communication	Bind (One Way)	Negotiated (Two Way)
Integration	In Target System	In Server or During Communication

The concept of GIS interoperability, as shown in (Table 1) is a process of using standard protocols to control inter-system access over the Internet. Disasters and their impacts may extend across large areas which, when dealing within a GIS context, invariably involves large volumes of data. Thus, simulating natural disasters may require advanced and high-capacity computing capabilities. Satellite imagery, digital elevation models and vector data available to municipalities and to the provincial authorities may exceed the processing capacity of a desktop computer. This is one of the areas where the strength of network-centric computing and systems interoperability is required. Interoperability can also allow efficient share of data / information in near real time when one is experiencing an emergency.

1.3 Interoperability Specifications

The Open Geospatial Consortium (OGC) is a US-based Industry consortium, with members from government and academia that has a mandate to develop and foster interoperability specifications (OGC 2002a). OGC has initiated interoperability specifications. Three of these specifications represent the backbone for GIS interoperability, which are:

Web Map Service (WMS) specification, which standardizes the process of requests and response between the server and the client for efficient data sharing.

- Web Feature Service (WFS) which controls the process of basic edits over the internet.
- Web Coverage Service (WCS) which standardizes the process of exploring coverage data.

2 Methodology

A comparative analysis of applications, resources and needs was conducted in order to examine the actual need for interoperable GIS applications for disaster management. A focus was placed on the City of Toronto, since it is one of the largest metropolitan centers in Canada. The study compared Toronto with similar selected medium and small-sized cities in Canada. The City of Vancouver was chosen on the basis of its comparable size, similar administrative structure and the similarity of operational activities related to emergency management.

A questionnaire was carefully designed in order to address three issues, namely: 1) usability of network-centric GIS; 2) GIS interoperability appli-

2.5 Comparison with Dufferin County

Dufferin County is located in south-central Ontario, Canada. Orangeville, the County seat, is located approximately 75 km north-west of Toronto. It consists of three towns: Mono, Orangeville and Shelburne; and five rural townships: Amaranth, East Garafraxa, East Luther Grand Valley, Melancthon, and Mulmur. Although the region is small, it has had an Emergency Management Coordinator since 1989.

Dufferin County is a typical small region with limited resources. Emergency management activities in this region can be directly related and compared to large regions. Dufferin County does not have a specialized GIS officer to deal with emergency mapping issues. All GIS work related to emergency management and other activities is managed by a public works drafts person. Table 2 shows a comparative operational analysis for the municipalities assessed in the study.

Table. 4 Comparison of available resources for the studied municipalities

Resources	Toronto	Vancouver	Waterloo	Dufferin
Emergency Mapping	Yes	Yes	Yes	No
GIS General Operations	Yes	Yes	Limited	No
Emergency Operations Center	Yes	Yes	No	No
Specialized GIS Personnel	Yes	Yes	Yes/IT Dept.	No
Need for External resources	No	No	Limited	Yes

3 Demonstration Case Study

This section demonstrates the optimal solution that GIS interoperability might provide for enhancing emergency management operations. The designed demonstration scenario stems from the various problems with data and systems heterogeneity found when analyzing the utility of GIS interoperability for emergency management in four local governments. It is important to mention that the main objective of this section is to demonstrate how GIS interoperability can provide a solution for disaster management. Hypothetically an emergency response can be simulated in order to show how GIS interoperability functions in such a situation and what this could provide to emergency management decision-makers. In the present case, when the various flood scenarios under different flood levels have been simulated and the damages can be assessed, the results obtained can be published and visualized in a 3D web environment using GeoServNet.

A hypothetical flood scenario in the Don Valley was used to demonstrate the solution that GIS interoperability can provide for emergency management. The Don River is a system that flows through the Greater Toronto Area (GTA). The origin of the Don River is in the Oak Ridges Moraine where the headwaters are fed by numerous aquifers. The river then flows for 38 km to Lake Ontario, providing drainage for 360 square km of land. The section used for this study is part of the Don Valley watershed system in the province of Ontario, as shown in Figure 2. DMTI Topographic digital maps in the form of shapefiles and DEM were obtained from the York University Map library (DMTI 2004). Topographic sheets of the area were used as a reference for conducting this study. Data describing flow statistics were simulated.

4 Emergency Management Actors and Roles

Based on the Ontario Emergency Management Doctrine (EMO 2004), different ministries are responsible for dealing with different natural hazards. This section provides a generic overview of what the exact benefits are that each ministry can gain by utilizing GIS interoperability. Due to information privacy and security issues, it was not possible to simulate this model as part of a real exercise. However, the adopted techniques have closely mimicked the simulation exercise environment.

The TRCA is responsible for flood simulation, floodplain mapping and water surface measurement. All hydraulics and hydrology data are under the custody of the TRCA. In such an event the TRCA will provide flood models and data through interoperable access to all decision-makers involved. Products like the flood plain map shown in Figure 2 can be simulated and made available to all actors involved. Real and semi-real time situational awareness models can be generated and accessed on demand based on stakeholder needs. Access to tabular and text information is also possible.

City of Toronto emergency services is responsible for providing services in a variety of situations, ranging from simple road maintenance closures to extreme disaster situations. The city produces and hosts its own data, and aims to make data standardized and interoperable with other parties. The City of Toronto can provide data about land use categories affected as a result of the flood and about what kind of infrastructure sectors can be impacted by the flood, as shown in Figure 4. The city can also provide real-time field information by utilizing interoperable mobile devices for data collection and transmission.

The Toronto Police Service is responsible for maintaining law and order in the city. In emergency situations the city police department can utilize GIS interoperability in a very efficient way. This includes accessing flood data provided by the city's emergency mapping department and that provided by the Toronto Region Conservation Authority (TRCA), pinpointing areas that are most vulnerable to traffic problems. Based on this information they can deploy officers to organize traffic in that region. Another example of how GIS interoperability can be utilized is that police can transmit real-time field data of traffic problems, or any related security problems, on the fly for all stakeholders involved in a particular emergency management scenario.

In emergency situations the city fire department can play role similar to that of the city police, however, in addition the fire department can assist with evacuation operations for particular locations in the city. Also the fire department can access data and information provided by the city and the TRCA to identify the most vulnerable areas and to concentrate its efforts there.

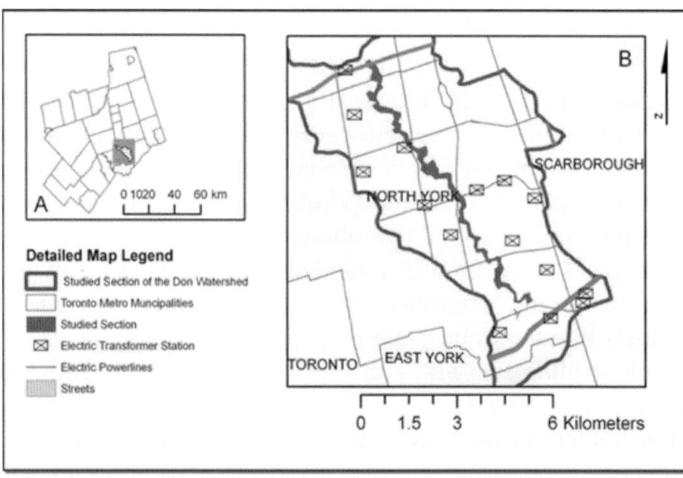

Figure -1. Infrastructure proximity; (A) Location of a focus section of the Don Valley within the GTA, Toronto, Canada; and (B) Detail of infrastructure proximity to flood extent

Toronto EMS can utilize GIS data and information for predicting areas at high risk of experiencing an emergency and for planning how they can dispatch their services to these areas. Another important utility of GIS interoperability for EMS is that, in emergency situations, it is not easy to prioritize your response to calls from different parts of the city. Knowing where a call is coming from relative to the emergency is invaluable to prioritizing response.

A key role for the EMO is to monitor emergency situations, and to not get involved unless it is required. Towards this end, GIS interoperability could provide EMO with improved situational awareness models by assembling data from all different departments and to make these available for basic analysis and visualization as shown in Figure 3. The legend on the clip was captured from the web-based program. It is interactive and can be shown by either activating the legend feature or by pointing to the particular item in the map.

5 Strengths and Weaknesses

GIS interoperability can provide enormous benefits to the disaster and emergency management community. It is clear from the analysis of the questionnaire that the level and capability of GIS interoperability for emergency management is not very well established or utilized at the municipal level, in particular in the small municipalities studied. Network-centric computing is necessary for processing large volumes of data and is achieved through the sharing of the computing power of other systems within the network.

A limitation of a non-interoperable environment is that users from different decision-making levels are not able to access the same information at the same time. In such a case it is not a simple task to secure simultaneous data access and processing. This makes the elimination of system and data heterogeneity potentially very useful for disaster management applications. Such applications make GIS interoperability a highly demanding mechanism for disaster management coordination.

Strengths

GIS interoperability is very useful in disaster and emergency management. Disaster and emergency management stakeholders can benefit from GIS interoperability by gaining access to standard, fast, accessible, shared GIS for achieving efficient decision-making processes. This section will highlight some key implementation benefits that the emergency management community can gain by leveraging GIS interoperability.

Establishing advanced interoperable applications and services is a key future aim of GIS and technology development plans at the City of Toronto. A second aim is that of providing the required geomatics services to support City of Toronto emergency management detection, prevention, planning, response, and recovery challenges.

The City of Vancouver is working on improving and extending the current level of interoperability in order to improve the efficacy of their Disaster Management coordination. It is also working to improve network-centric GIS computing through a three year evaluation and upgrading plan. Furthermore, the city is considering to extend their visualization and analysis applications to include 3D GIS. The City of Vancouver is also looking for university contribution in the area of data management and sharing.

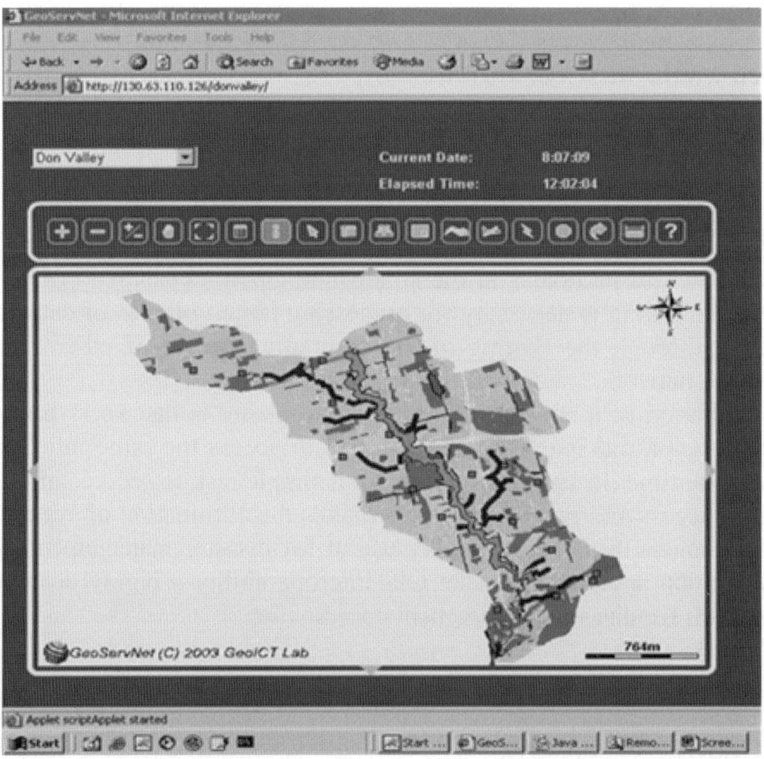

Figure -2. GeoServNet web-interface; (a) GeoServNet controls; and (b) Don Valley flood scenario visualization and analysis

Emergency service must aim to be timely and accurate, whereby geospatial information and geoprocessing services are easily accessible and capable of being shared across jurisdictions and multiple security levels. Also, the city plans go further to advance GIS interoperability and capabilities through the hosting of a web map service (WMS/WFS) site. The City of Toronto GIS Emergency Mapping Department is also aiming to get assistance from universities for GIS development for Disaster Management applications with regards to standards and specifications development. The City of Toronto is intending to use web-based GIS for disaster management and emergency response projects for operational plans and in the event of an emergency, this constitutes a Geomatics-centric Incidence Response. The specific benefits resulting from GIS interoperability are summarized in the following several sections.

GIS interoperability allows for open access to multiple data sources. This benefit helps decision-makers to access multiple servers, thereby obtaining data and services that do not necessarily occur in the same location or even within the same organizational boundary for that matter. Through GIS interoperability, the emergency management community can move towards one or several closely related standard data formats that can be used in data transfer and information sharing.

Standards-based GIS interoperability helps to provide a simple and accessible means of integration to decision-makers. This is crucial since emergency management operations stand to benefit considerably from a process that allows for timely gathering, modeling and analysis of information. If the system is not simple and accessible it will not be easily implemented during disaster management and emergency response. GIS interoperability, through its standardized protocols has allowed GIS users to utilize a simple and standard service.

Within the context of interoperability, transparency refers to the process of accessing and sharing data between systems without the use of complicated, cumbersome protocols. This differs from the concept of 'common format' in that a common format is not necessarily easy to use. Through a transparent system, emergency management stakeholders can readily access GIS data and systems that are external to their system domain.

GIS interoperability can be expanded or contracted by adding or eliminating nodes. This flexibility in scalability in systems and services is useful in emergency management operations, which, due to their dynamic, fast-paced nature, require that previously unexpected situations be accommodated.

Weaknesses

To date, considerable effort has been expended toward providing efficient, interoperable solutions for disaster and emergency management. However,

GIS interoperability remains a stimulating field and research is still underway into addressing many aspects of interoperability. The following subsections address specific problems and challenges that remain in this field.

From an operational perspective, there are particular issues related to the degree to which data conversion hinders efficient data interoperability. This may arise if, for example, a particular department is using engineering data that is stored in a Computer Aided Design (CAD) format and another department is using data stored in ArcGIS shapefile formats. These two formats can be made to be compatible by converting CAD data into the shapefile format. The time required for this process depends on data size and system capabilities.

Different processes and techniques are used by different departments. This can create an obstacle to achieving GIS interoperability. A streamlined process and system capabilities are required for effective GIS interoperability. Different departments have specialized staff expertise and technical capabilities, this makes it a difficult and delaying process to comply with GIS interoperability requirements. Another issue is related to bandwidth load. This is very important in securing efficient network performance without overloading a particular network.

Data maintenance and update is another issue that represents an obstacle to achieving GIS interoperability. Where there is no clear policy that identifies roles and responsibilities for each node in an interoperable system, data update, maintenance and management can be a challenge.

Access rights to sensitive information, such as infrastructure and emergency management, represent another issue. Much disaster management data and information are, for a variety of reasons, not meant for public access and as a result are categorized as 'classified.' What are the criteria that should be used to categorize data and information to a specific clearance level? This is a key question and challenge that can represent a serious obstacle to GIS interoperability. Another crucial issue is related to corporate technology procurement policies, which can contribute to delayed implementation.

It is evident from the study that the City of Toronto produces and owns its own data while at the same time using data from other neighboring regions. The level of data sharing in this form doesn't constitute 'interoperability'. However, this could be enhanced by using data exchange standards to make these data useable not only to Toronto and neighboring regions, but to any other institution involved in emergency management operations. The same benefit could be gained by the City of Vancouver, which utilizes its locally produced data along with other non-proprietary data.

Waterloo and Dufferin regions utilize external data sets, indicating a clear need for data standards and GIS interoperability, since both regions use the same web-service and have different problems that could be eliminated through proper GIS data sharing. The region of Waterloo has indicated that, when utilizing the Environment Canada web service, that they had experienced problems related to the level of detail and to layer matching. This indicates heterogeneity that could have been avoided through the use of a standard data model and format.

The Dufferin region emergency manager faced a problem when dealing with the online web GIS service, finding it to be too technical in nature. In contrast, large cities like Toronto and Vancouver have specialized technical departments that are capable of dealing with technical GIS issues. GIS interoperability can help in eliminating the detailed technical expertise required to utilize desk-top GIS packages and can provide a considerable additional benefit that it can be leveraged over the internet for many users simultaneously.

As the case with the City of Toronto indicated, large cities produce, utilize and share data with other users. However, this is performed through procedural data exchange not through a systematic, dynamic process. GIS interoperability achieves this capacity in that it can allow for decentralized data handling and access from different servers.

Waterloo and Dufferin counties are able to access and visualize data from an interoperable web service, which provides them with access to data that they do not own. This level of GIS interoperability helps to provide an efficient solution to the challenge of timely access to emergency management data. At the same time it reduces duplication of data sets in more than one site, thereby saving storage resources.

6 Conclusions

This paper has shown that GIS interoperability is effective at coordinating the emergency management decision-making process by allowing simultaneous data and information access and updates. Two scenarios have outlined how network-centric GIS can best be used in modeling and visualization of disaster management scenarios. Through its integration with other modeling systems, GIS functionality provides decision-makers with tools that can improve the decision-making process. In this sense, a virtual emergency management center can provide near real-time data access to decision-makers in different locations.

One of the limitations of this study is that it has only two integrated earthquake shakemap models and hydraulic simulation models as input into a GIS database. Other models that provide detailed information about cascading effects, escalating failures and systems dynamics could also be included as input into GIS databases and would provide very advanced visualization capabilities. Utilizing other spatially enabled modeling components would significantly help with expanding our understanding of the complex issue surrounding critical infrastructure interdependencies.

The temporal dimension is very important in quantifying and understanding damage magnitude, peak flood spatial extent and the impact on each of the interdependencies between various infrastructure sectors and systems. In this dissertation, the temporal aspects of disaster management scenarios have been discussed, and only the peak flood extent has been used in building the knowledgebase. Explicit study of the temporal aspects associated with disaster management would be of great benefit to the disaster management community.

References

Allen, E., Edwards, G. and Bedard, Y., 1995. Qualitative causal modeling in temporal GIS. Spatial Information Theory, 988: 397-412.

Bishr, Y., 1998. Overcoming the semantic and other barriers to GIS interoperability. International Journal of Geographical Information Science, 12(4): 299-314.

Briggs, D., 2005. The role of GIS: Coping with space (and time) in air pollution exposure assessment. Journal of Toxicology and Environmental Health-Part a-Current Issues, 68(13-14): 1243-1261.

Christakos, G., Bogaert, P. and Serre, M., 2002. Temporal GIS: Advanced Field-Based Applications. Springer Verlag, Berlin, Heidelberg, New York, 217 pp.

Dietzel, C., Herold, M., Hemphill, J.J. and Clarke, K.C., 2005. Spatio-temporal dynamics in California's central valley: Empirical links to urban theory. International Journal of Geographical Information Science, 19(2): 175-195.

ElAwad, Y., Chen, Z., G., H., Tao, C.V. and Abdalla, R., 2005. An Integrated Approach for Flood Risk Assessment for the Red River in Southern Manitoba, Annual Conference of the Canadian Society for Civil Engineering, Toronto.

Emergency Management Ontario (EMO), 2003. Emergency Plans Act. Government of Ontario.

Erharuyi, N. and Fairbairn, D., 2003. Mobile geographic information handling technologies to support disaster management. Geography, 88: 312-318.

Farley, J., 1999. Disaster Management Scenarios, Univ. of Arkansas, OGC Discussion Paper.

Giardino, M., Giordan, D. and Ambrogio, S., 2004. GIS technologies for data collection, management and visualization of large slope instabilities: two applications in the Western Italian Alps. Natural Hazards and Earth System Sciences, 4(2): 197-211.

Glover, J., 1995. The Need for Open GIS - Part 1: The Integration Challenge. Mapping Awareness, 9(8).

Goodchild, M.E., 2003. Geographic information science and systems for environmental management. Annual Review of Environment and Resources, 28: 493-519.

Goodchild, M.F., Parks, B.O. and Steyaert, L.T., 1993. Environmental Modeling with GIS. Oxford University Press, New York.

Gupta, P.K. and Singh, A.P., 2005. Disaster management for Nandira watershed district Angul (Orissa) India, using temporal Remote Sensing data and GIS. Environmental Monitoring and Assessment, 104(1-3): 425-436.

Langran, G. and Christman, N.R., 1988. A framework for temporal geographic information. Cartographica, 25(3): 1-14.

Longley, P.A., Goodchild, M.F., Maguire, D.J. and Rhind, D.W., 2005. Geographic Information Systems and Science. Wiley, 517 pp.

Mennis, J.L. and Fountain, A.G., 2001. A spatio-temporal GIS database for monitoring alpine glacier change. Photogrammetric Engineering and Remote Sensing, 67(8): 967-975.

OGC, 2002. OpenGIS Reference Model.

Peuquet, D., 2001. Making Space for Time: Issues in Space-Time Data Representation. Geoinformatica, 5(1): 11-32.

Peuquet, D.J., 2002. Representation of Space and Time. The Guilford Press, New York, London, 380 pp.

Spery, L., Claramunt, C. and Libourel, T., 2001. A spatio-temporal model for the manipulation of lineage metadata. Geoinformatica, 5(1): 51-70.

Tobler, W., 1959. Automation and Cartography. Geographical Review, 49: 526-534.

Visser, U., Stuckenschmidt, H. and Schlieder, C., 2002. Interoperability in GIS – Enabling Technologies, 5th AGILE Conference on Geographic Information Science, Palma (Balearic Islands, Spain).

Waugh, W.L., 1995. Geographic Information-Systems - the Case of Disaster Management. Social Science Computer Review, 13(4): 422-431.

Increasing public and environmental safety through integrated monitoring and analysis of structural and ground deformations

Adam Chrzanowski, Anna Szostak-Chrzanowski, Jason Bond and Rick Wilkins

Canadian Centre for Geodetic Engineering, University of New Brunswick
Fredericton, New Brunswick, E3B 5A3 Canada
adamc@unb.ca

1. Introduction

Any engineered or natural structure, when subjected to loading, undergoes deformation and/or rigid body movements. Once the deformation, its velocity and/or acceleration exceed critical values, the structure fails. The critical values are determined using failure criteria that are based upon either empirical formulae or principles of continuum mechanics. By providing continuous and properly designed deformation monitoring schemes, one may provide information about the new state of the deformation. This information can then be used to provide advance warning of imminent structural failure.

Unfortunately, it is not uncommon for large civil structures, dams, bridges, open pit mines, slopes, and underground excavations to be monitored using outdated and inefficient systems and/or inadequately trained operators. Worse, some operations do not use any deformation monitoring system at all. Large dams are a notable example. There are over 45,000 large dams (taller than 15 m) in the world that should be monitored. A study performed during 1990-92 (Chrzanowski et al. 1993) revealed that most owners of dams, even in the most advanced and industrialized countries do not have any monitoring standards and specifications.

Recent catastrophic disasters such as the collapse of highway overpasses in Canada; failures of levees and bridges in New Orleans; collapses of roofs of large civil structures in Germany, Poland, and Russia; land slides in California, Pakistan, and the Philippines; and rock failures and losses of lives in deep coal mines in China have dramatically increased the demand for the development of new monitoring systems and their broader applications. Automation, multi-sensor integration, continuous data collection, integrated analysis and physical interpretation, and enhanced accuracy and reliability are key issues in the development of such systems.

The Canadian Centre for Geodetic Engineering (CCGE) at the University of New Brunswick is dedicated to the development of new monitoring systems and new methods for integrated analysis, modeling, and prediction of structural and ground deformations using an interdisciplinary approach. Some of the significant recent developments at CCGE include: the development of the ALERT software suite for fully automated and continuous monitoring of deformations with multi-sensor systems; enhancement and full automation of GPS monitoring techniques; augmentation of monitoring schemes with terrestrial emitters of GPS-like signals (i.e., 'pseudolites'); the use of monitoring results in the verification of the deterministic models of deformation; and a new approach to the design of multi-sensor monitoring surveys based on deterministic modeling of deformations using the finite element method. This paper gives a review of these recent developments.

1. Design of Monitoring Schemes

2.1 Design Criteria

The design of a monitoring scheme should be based on a good understanding of the physical process which leads to deformation. The investigated deformable object should be treated as a mechanical system, which undergoes deformation according to the laws of continuum mechanics (Szostak-Chrzanowski et al. 2006). This requires the causative factors (loads) of the process and the characteristics of the object under investigation to be included in the analysis leading to the design. This is achieved by using deterministic modeling of the load-deformation relationship. Thus the design process requires an interdisciplinary cooperation between specialists in various fields of geoscience and engineering, including structural, rock mechanics, and geodetic engineering, depending on the type of the investigated object.

If the monitoring system is to be used as a failure warning system, it must be fully automated to handle continuous or very frequent data collection (depending on the expected rate of deformations). It must be able to perform data processing, and visualization in near-real time, and must have sufficient accuracy and capability to trigger the alarm. In order to minimize triggering false alarms, the system must be capable of distinguishing between the actual deformation signal and noise caused by errors of observations. False alarms are expensive and lead to a wrong evaluation of the physical state of the object which may have large economic and sociological impacts.

Design of the monitoring scheme requires decisions to be made regarding the type, location, density, and accuracy of monitoring sensors. The location of the sensors or the observed targets must include points where maximum or critical deformations are expected (Chrzanowski 1993). Concerning the required accuracy, most of the deformable objects (e.g. bridges, dams, nuclear power stations, open pit mines) require sub-cm or even mm level accuracy.

2.2 Choice of Monitoring Sensors

The sensors used in monitoring measurements are generally grouped into geodetic techniques (terrestrial and space) and geotechnical/structural instruments (e.g., tiltmeters, extensometers, strainmeters). Among the available geodetic and geotechnical/structural technologies, there are very few, if any, sensors that can fully satisfy the above monitoring criteria as a stand alone system. Therefore, in most cases, various techniques must be combined into an integrated monitoring system. Among geodetic techniques, the best for fully automated and continuous monitoring are GPS and robotic total stations (RTS) with automatic target recognition (e.g. Leica TCA 1800). If needed, GPS can be augmented with pseudolites and/or other satellite positioning systems. Other, comparatively new, geodetic techniques include laser scanners, interferometric synthetic aperture radar (InSAR), and digital photogrammetry. They have, however, many limitations and restrictions, which still require further research and enhancements. For example, the satellite born InSAR, provides repeated radar images only every 24-35 days depending on the satellite system. InSAR also suffers from other limitations (Chen et al. 2000). The recently developed ground based InSAR technology (e.g. Pieraccini et al. 2006) promises significant improvement in continuous monitoring of steep slopes and embankments.

All of the discussed geodetic technologies are vulnerable to the effects of changes of atmospheric conditions (changes in the density of air due to the changes in temperature, humidity, and barometric pressure) causing:

- in the case of optical direction measurements, changeable refraction along the lines of sight,
- in the case of electromagnetic distance measurements, errors due to the varying velocity of propagation of electromagnetic waves, and
- in the case of GPS, residual tropospheric delay biases when there are large elevation differences between the receivers (Bond et al. 2005).

Geodetic methods supply information on the absolute and relative displacements (changes in coordinates) from which displacement and strain fields for the monitored object may be derived. Thus, geodetic surveys supply global information on the behavior of the investigated object. In some cases, however, the use of geodetic techniques may be uneconomical and may have inadequate accuracy.

There is a multitude of geotechnical instruments equipped with electromechanical transducers (Dunnicliff 1988) that may easily be adapted for continuous monitoring and telemetric data acquisition. Usually, the geotechnical instruments are embedded in the investigated object for the duration of the monitoring project. These instruments supply only very localized information on a selected component of the deformation (e.g. only local tilt or local extension in one direction when using a tiltmeter or an extensometer, respectively).

Geotechnical instruments require thorough calibration for the effects of environmental temperature, drift of the readout, and conversion constant. Once embedded within the structure, however, the geotechnical/structural instruments cannot be rechecked or recalibrated. Because of this, it is not uncommon that geotechnical instruments provide unreliable data or even fail during the life of the structure. Since geodetic measurements allow for redundancy and the possibility of statistical evaluation of the quality of the data, they generally provide more reliable results. Geodetic and geotechnical measurements compliment each other and, ideally, should be used together creating an *integrated monitoring scheme.* In addition, when the investigated object is located within the influence of seismic activity, the local monitoring system must be integrated with a regional system. A good example illustrating these concepts is given in (Duffy et al. 2001).

To illustrate the use of deterministic modeling in designing an integrated monitoring scheme, an example of a 75 m high, Concrete Face Rockfill Dam (CFRD) resting on a 60 m thick till is given. Fig. 1 shows expected horizontal and vertical displacements caused by filling the reservoir (Szostak-Chrzanowski and Massiera 2006).

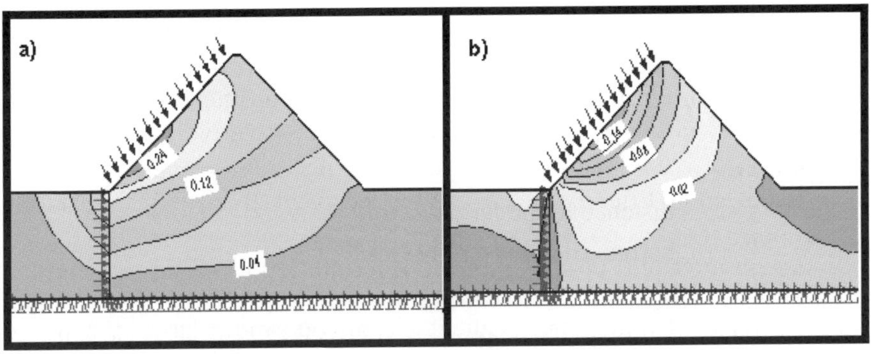

Fig. 1 Predicted displacements [m] after filling the reservoir a) horizontal displacements b) vertical displacements

As one can see from the modelled displacements, the largest displacements are expected to occur at the upstream face of the dam, which is covered by a concrete slab. It is the most crucial area for monitoring the deformation. Since the upstream face is under water, the monitoring scheme should be designed to have geotechnical instruments such as fibre-optic strainmeters and tiltmeters embedded in the concrete slab. The rest of the dam could be monitored by geodetic methods using, for example, robotic total stations and GPS. Besides the deformation sensors, various physical geotechnical sensors must also be used, for example piezometers, seepage gages, and others. Final details of the design including the density of the instrumentation, accuracy requirements and frequency of observations should be discussed between geotechnical and geodetic engineers. For example, by modeling the expected deformation at various water level stages in the reservoir, one can obtain information on the rates (velocity) of deformations. This information will aid in determining the required frequency of repeat surveys.

2.3 Challenges of Geodetic Monitoring Systems

2.3.1 Effects of Atmospheric Refraction

Mitigating the effects of atmospheric refraction on direction measurements is the oldest, unresolved problem of geodetic surveys. The pointing error, e, caused by refraction is a function of the gradient of temperature dT/dl occurring across the line of sight. It can be derived from the basic theory of refraction that the approximate relationship between e and temperature gradient can be expressed as in Eq. (1):

$$e = \frac{3.9Ps^2 10^{-5}}{T^2}\frac{dT}{dL} \tag{1}$$

Where:

s	distance to the target in [m]
P	barometric pressure in [mb]
T	absolute temperature [°K]
$\frac{dT}{dL}$	temperature gradient [°C/m] perpendicular to the line of sight

For example, in atmospheric conditions of 1013 mb and 20°C, over a 1000 m sight length, a uniform change in the temperature gradient from nighttime to daytime of only 0.1°C/m would cause more than a 4 cm change in the determined position of the target. Intensive tests with robotic total stations at two large open pit mines in Chile and in Western Canada (Chrzanowski and Wilkins 2006) and at a large earth dam in California (Duffy et al. 2001), indicate that the temperature gradients within two metres of sun exposed surfaces may change by 2^0C/m or more from nighttime to daytime. Diurnal pointing changes of up to 200 mm were recorded over a distance of 1500 m with the line of sight passing across an open pit mine far away from the pit walls. It must be noted that the effects of refraction are much more severe when large temperature gradients occur near the RTS rather than if they occur closer to the target. This is important criteria to consider when designing the location of the monitoring instruments.

Since the effect of refraction increases proportionally to the square of the distance, it follows that the location of the observing instruments should be as close as possible to the targets being monitored. The diurnal, cyclic effects of refraction can be significantly minimized by daily averaging of the monitoring results. This approach can only be used if it satisfies the update frequency. Alternatively, one should try to model and predict the cyclic effects of refraction as a function of the time of day based on previous observation data. Otherwise, epoch-to-epoch results will show large erroneous displacements, which could trigger false alarms.

Changes in atmospheric conditions cause much smaller errors in distance measurements than in the direction observations. For example, a 1°C change in air temperature causes approximately a 1 ppm change in the distance. This can be further reduced by introducing meteorological corrections to the observed distances. Thus, the design of a geodetic monitoring scheme should rely more heavily on distance observations than direction measurements.

2.3.2 Instability of Reference Points

In deformation surveys, the definition of the datum is adversely affected by the use of reference points that are erroneously assumed stable. This in turn gives a biased displacement pattern that can easily lead to a misinterpretation of what is really happening to the deformable object.

A methodology utilizing an iterative weighted similarity transformation (IWST) of displacements for the identification of unstable reference points was developed at CCGE (Chen et al. 1990) in the early 1980s. The methodology is based on using a similarity transformation of displacement components d_i with the condition that $\sum|d_i| = \min$. The weights of individual displacement components are inversely proportional to the absolute value of the component itself. The transformation is an iterative process that is repeated until subsequent iterations reach a preselected convergence criterion. The IWST method has been successfully applied in all types of engineering projects where reference point stability has been a concern.

3. ALERT Software for Fully Automated Multi-Sensor Monitoring Systems

Initially, the ALERT software suite (Wilkins et al. 2003) developed at CCGE was designed for automated monitoring of deformations using only RTS and meteorological sensors (for correcting observed distances). Very recently, GPS has been added to ALERT to work either as a stand alone fully automated GPS monitoring system or to work together with RTSs as a hybrid RTS/GPS system. In the latter case, GPS is used to control and give positional corrections to the RTSs, which may be setup within the deformation zone. GPS allows the network of RTSs to be connected to stable reference points since it does not require intervisibility between stations. Fig. 2 shows typical observation shelters of the ALERT system. Fig. 2a shows one of eight RTS stations installed at the Diamond Valley Lake Project in Southern California (Duffy et al. 2001) to monitor three large earth dams. Fig. 2b shows a RTS/GPS shelter in a large open pit mine in Western Canada.

Fig. 2 a) RTS shelter with solar power panels, b) Typical RTS/GPS ALERT shelter

The ALERT software suite is composed of a series of modules that automate surveying tasks, handle database management, and provide graphical user interfaces. An initial setup is required to catalog the elements of the monitoring networks (e.g., survey point names, RTS locations, communication parameters, etc.) and create ALERT projects. Once projects are defined, one can create observation, data transfer, and processing schedules for each monitored site to obtain displacement results. Processed data is automatically made available in a near-real time fashion. Using computer network connections, oversight of current data collection activities and modifications to observation schedules can be done remotely.

The system takes advantage of the core functionality of the Microsoft systems (e.g. NT 4.0, Windows 2000, and Windows XP). There is full support for remote operation via LAN and Internet connections and provider-independent database access. In addition, the software's observation and processing tasks are automated according to any desired schedule and the system is able to recover from power outages with no user intervention.

An alarm system has been incorporated into the ALERT software. An alarm definition is created by attaching to it one or more user defined criteria with a list of action items. The criteria can be defined for displacements, velocity, or acceleration. The action items attached to the alarm can be triggered either by individual points or groups of points, when their movement reaches or extends beyond the predefined criteria values associated with the alarm. The action items for each alarm can report to a log file, email a list of recipients, change point status, and/or trigger external alarm hardware (e.g. light, siren, etc.) via a serial port. Fig. 3 gives a more illustrative view of this process in terms of a flow chart.

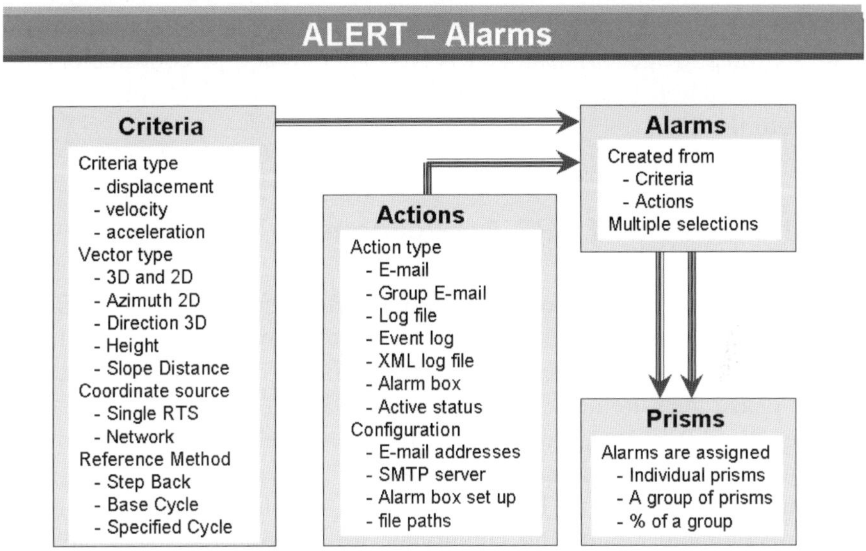

Fig. 3 Flow for creating an alarm

ALERT is completely autonomous with several self-recovery features that are critical for the automated monitoring projects. The computers that run ALERT are configured to automatically reboot if power is lost, allowing a backup service to complete any interrupted data collection tasks

A very unique feature of the ALERT software is its automated handling of multiple-RTS networking. Observations between RTSs, common control points, and multiple shots to object prisms can all be combined automatically to obtain one overall network solution. This takes advantage of the redundant information and gives a more accurate and reliable result. Due to the configuration defects in this type of RTS network, the processing of this data requires a special least squares algorithm that adjusts observation differences with respect to a user defined reference epoch. The results of the network adjustment are further processed using iterative weighted similarity transformation of displacements to identify unstable reference points (Chen et al. 1990) and remove their effect.

The result of data processing is a series of time-tagged coordinate values that are stored in the project database. Plotting utilities allow rapid visualization of displacement and velocity trends as well as vector plots of displacements and velocities with their confidence regions. To increase the accuracy of the results, observation cycles may be grouped into mean values over a selected period of time to minimize, for example, the cyclic effects of atmospheric refraction The database is in a readily accessible format, therefore, the end user can easily extract coordinate values using standard Structured Query Language (SQL) queries and build plotting and analysis tools to meet specialized needs. The storage of coordinate solutions in a relational database makes it very easy to selectively examine subsets of the data.

4. Enhancement of GPS Applications

Implementing GPS for deformation monitoring poses difficult challenges. Displacements encountered in deformation monitoring are frequently at the sub-cm level. Since the practical resolution of an undifferenced GPS carrier-phase measurement is approximately 2 mm (1% of the L1 carrier wavelength of 0.190 m), monitoring millimeter level displacements in near real-time pushes the limits of the system.

Achieving reliable, mm level precision in ‘real-time’ using GPS is not easy in favorable monitoring conditions, let alone in the harsh environments frequently encountered in deformation monitoring projects. Some of the challenges faced in designing a system that meets deformation monitoring needs include (Bond et al. 2007a):

- Satellite Visibility: In deformation monitoring environments where there are obstructions hindering satellite visibility (e.g., dams, open pit mines, buildings), dilution of precision values rise due to the degradation in satellite geometry. The system must be able to cope with periods of the day during which there are too few satellites visible to provide a high enough quality solution to meet project requirements.
- Residual Tropospheric Delay: In deformation monitoring environments where there are significant changes in elevation (e.g., open pit mines, volcanoes), residual tropospheric delay can cause significant positioning biases, especially in height. The differential troposphere causes a 3 to 5 mm relative height error for every millimeter difference in zenith delay between stations (Beutler et al. 1988). Residual tropospheric delay must be accounted for if the desired precision is to be achieved.
- Multipath: In deformation monitoring environments where multipath sources are abundant (e.g., building structures, vehicles) multipath can contaminate the position solutions. Practically every observation site is

affected to some degree by multipath. Multipath biases can reach up to ¼ of the carrier wavelength (≈ 4.8 cm for the original L1 carrier-phase measurement) (Leick 1994).

- Providing On-time Information: Deformation monitoring poses a unique GPS scenario; the points of interest are neither quite static nor kinematic because there is motion but it is usually very small. In providing GPS position updates, it cannot be assumed that the antenna's position at an epoch agrees with that of a prior epoch. One way to handle this is to model the motion as static and to add process noise.

4.1 PPMS Software

Recent efforts at the CCGE at the University of New Brunswick (UNB) to develop GPS software for deformation monitoring in harsh environment conditions have resulted in the emergence of the Precise Position Monitoring System (PPMS) (Bond et al. 2007a). PPMS utilizes a delayed-state Kalman filter to process GPS triple-differenced (TD: differencing consecutive double-differenced observations) carrier phases. Test results have indicated that the software is capable of detecting mm level displacements without having to solve for ambiguity terms. The ability to provide high precision solutions that are independent of ambiguities makes PPMS desirable for deformation monitoring since it is less susceptible to false alarms caused by cycle slips than traditional double-differenced (DD: differencing between receivers followed by differencing between satellites or vice versa) processing methods. The trade-off in using the TD approach is a longer convergence time than for DD methods. This is generally not a concern, however, for deformation monitoring applications where long term structural behaviour is of interest.

The TD approach can be considered an extension of the observation difference, least squares approach used for processing deformation monitoring data (Wilkins et al. 2003). The attractiveness of the TD observation is that it is a time difference of DD observations and consequently any biases common to both observations will be highly correlated and therefore significantly reduced. This strategy has some important benefits:

- The user no longer needs to solve for the ambiguity term, which allows the system to be more robust;
- For observation intervals less than a few seconds, the correlation between atmospheric parameters between consecutive epochs will be large and therefore biases originating from them will be significantly reduced. This is useful for mitigating residual tropospheric delay biases over large height differences;

- For observation intervals less than a few seconds, the correlation in the low frequency component of multipath terms between consecutive epochs will be large and therefore biases originating from them will be significantly reduced. The high frequency component still remains.

The effectiveness of the software is illustrated by comparing solutions obtained using traditional, DD processing techniques employed by commercial software with those obtained using PPMS. Fig. 4 presents the up component solutions (generally the poorest precision) of a GPS baseline observed in an open pit environment. The height difference between master and rover stations is 361 m. It can be seen that a peak-to-peak spread of just over 4 cm exists. Fig. 5 presents the east, north and up components of the same baseline processed using PPMS. There is a height change of a few mm that is detected after hour 50, which is not so easily identified in Fig. 4. The peak-to-peak spread of the up component in Fig. 5 is in the order of 1 cm after the change in height.

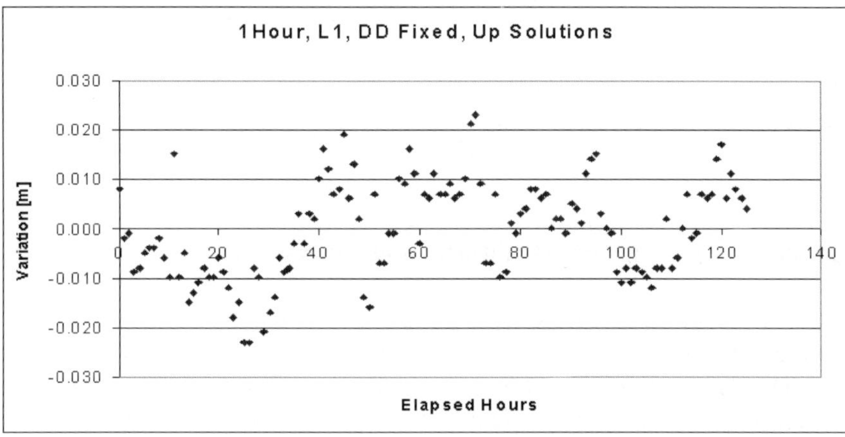

Fig. 4 Commercial, 1 Hour, DD Fixed, 'Up' Component Solutions

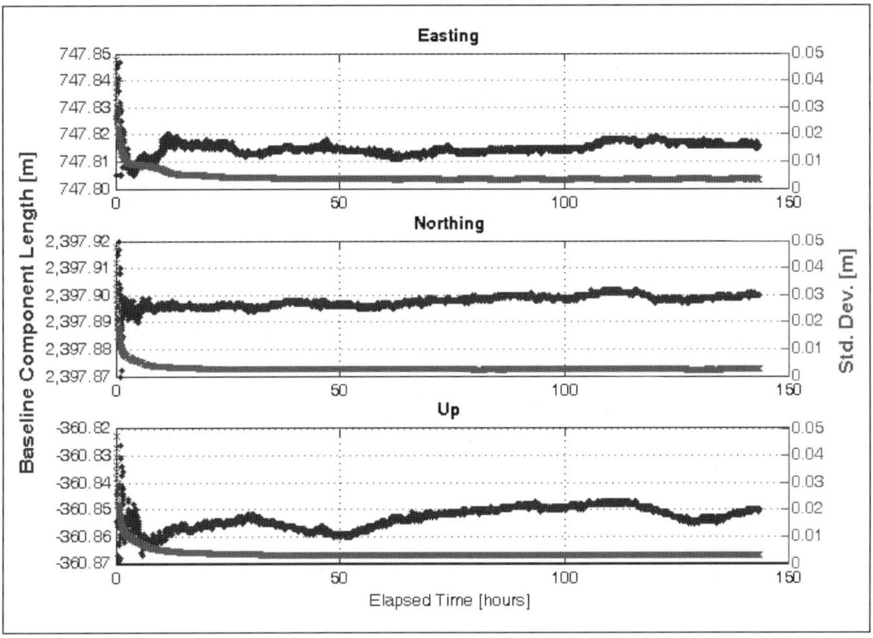

Fig. 5 Continuous, TD, east, north, up solutions using PPMS

4.2 Augmentation of GPS with Pseudolites

GPS data obtained from a large open pit mine indicated that there were several periods during the day for which fewer than 4 satellites were visible. This leads to degradation in the precision of the solutions as well as a reduction in the frequency of updates. Consequently, the potential of augmenting GPS with pseudolites (PLs) has been investigated to improve the frequency and precision of solutions (Bond et al. 2007b).

Being a ground-based transmitter, PL error sources must be handled differently than GPS signal error sources. PPMS was modified to address nuances in PL data processing which include cycle slip detection, PL location determination and PL observation modelling. Fig. 6 illustrates the results of a baseline that was observed in a harsh environment (having a 40 degree elevation cut-off) while using a PL. A slow, 15 mm displacement occurs in the horizontal plane beginning at hour 14 and ending at hour 20. It takes approximately 10 hours for the solution to converge. It can be seen that the combined GPS+PL system allows at least some of the displacement to be detected whereas the results from standalone GPS are inconclusive.

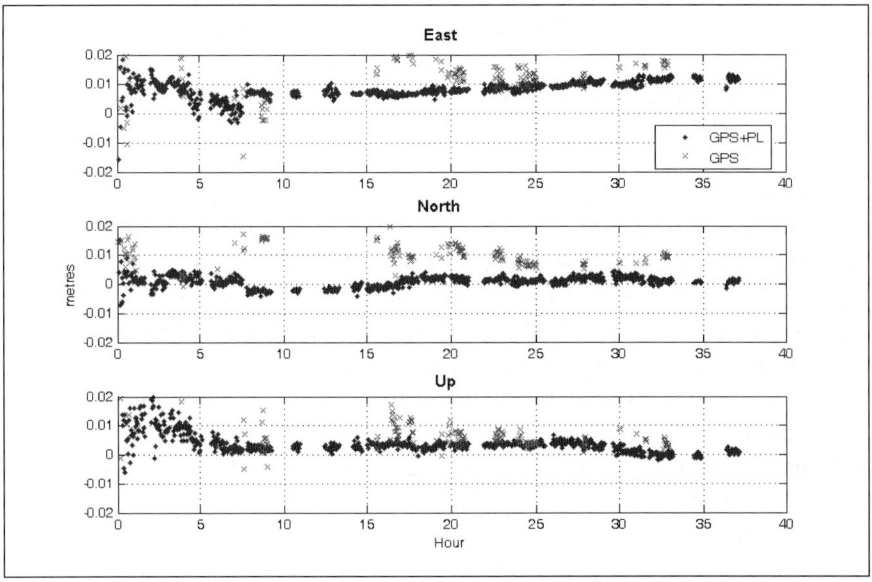

Fig. 6 GPS+PL results in a harsh environment

5. Integrated Analysis and Physical Interpretation of Deformations

Analysis of deformations of any type of deformable body includes geometrical analysis and physical interpretation. Geometrical analysis describes the change in shape and dimensions of the monitored object, as well as its rigid body movements (translations and rotations). The ultimate goal of a geometrical analysis is to determine the displacement and strain fields in the space and time domains for the whole deformable object. The *Generalized Method of Geometrical Deformation Analysis* (Chen 1983; Chrzanowski et al. 1983) allows for a simultaneous analysis of any type of observations (geodetic and geotechnical) even if scattered in space and time. The displacement field is obtained by iterative least squares fitting of an appropriate displacement function to the measured deformation quantities. Examples are given in (Chrzanowski 1993).

Physical interpretation is based on establishing the relationship between causative factors (loads) and deformations. This can be determined either by:

- A statistical method, which analyses the correlation between the observed deformations and loads; or by
- A deterministic method, which utilises the loads, properties of the material, and physical laws governing the stress-strain relationship.

By comparing the geometrical model of deformations derived from the observed deformation quantities with the deformations obtained from the deterministic model, one can determine the actual deformation mechanism (Chrzanowski et al. 1994) and/or verify the designed geomechanical parameters (e.g. Szostak-Chrzanowski et al. 2002). Integrated analysis may also explain the causes of deformation in the case of abnormal behaviour of the investigated object. Thus, the role of monitoring surveys is much broader than serving only as a warning system.

The ultimate goal of deterministic modeling of deformations is to develop a prediction model. The model is developed for the given geometry, loading conditions, boundary conditions, and specific behaviour and properties of the material. Once a prediction model is developed, it may be used for the design of a monitoring scheme.

Deterministic analysis of deformation is based upon continuum mechanics, in which solving differential equations of equilibrium of forces is the main problem. In many cases closed form solutions of the equations may be difficult or impossible to obtain. Consequently, numerical methods, such as the finite element method (FEM) are used. The basic concept of the finite element method (displacement approach) is that the continuum of the deformable body is replaced by an assembly of individual small elements of finite dimensions, which are connected together only at the nodal points of the elements (Zienkiewicz and Taylor, 1989). The elements may be of any shape, but usually three or four nodal elements are chosen for two-dimensional analysis. For each element, one can establish the relationship between the nodal loads and displacement or strain field variables.

The displacements are calculated using the equilibrium equation:

$$Kd = r - f^{b} + f^{\sigma_0} + f^{\varepsilon_0} \qquad (2)$$

Where:

d	is the vector of nodal displacements
K	is the total stiffness matrix
r	is the vector of external forces concentrated at nodal points
f^{b}	is the loading vector of body forces
f^{σ_0}	is the vector from initial stresses

f^{ε_0} is the loading vector from initial strains

The global matrices and vectors are calculated through a superimposition of local (at each element or at each node of the FEM mesh) matrices K_e and vectors r_e, $f_e^{b}, f_e^{\sigma_0}, f_e^{\varepsilon_0}$

The stiffness matrix for an element in two dimensions is given as:

- $K_e = \iint B_e^T D B_e t dx dy$ (3)

Where:

B_e is the matrix relating strains in the element to its nodal displacements

D is the constitutive matrix of the material which in the case of linear-elastic material contains Young modulus and Poisson ratio

t is the unit thickness of the elements

Since the total stiffness matrix K is singular, boundary conditions must be applied in order to solve Eq. (2) for the displacements.

The most critical problem in modeling and predicting deformations, particularly in rock or soil material, is to obtain in-situ characteristics of the materials. The process of collecting samples to determine in-situ characteristics is very difficult and very costly. Often the data is incomplete. Additionally, this process has the following limitations:

it is impossible that the selected samples will represent the true characteristics in all locations;

the samples may be disturbed during the collection process; and

laboratory loading conditions may differ from natural conditions.

The difficulty in determining material characteristics is the main cause of uncertainty in deterministic modeling of deformations.

Results of properly designed monitoring schemes may be used to enhance the deterministic model (e.g. by correcting the material parameters of the observed object). This can be achieved using forward or back analysis (Szostak-Chrzanowski et al. 1994). In turn, the enhanced deterministic model may be used in improving the monitoring scheme.

Recent research at CCGE has demonstrated how to successfully incorporate deterministic modeling and monitoring in the analysis of engineered and natural structures. In particular, research was implemented in ground subsidence studies caused by mining activity (Chrzanowski and Szostak-Chrzanowski 2004) and in modeling deformations of large earth and rock filled dams (Szostak-Chrzanowski et al. 2005). In these projects, a 'large-scale' approach has been used. This approach is characterized by an introduction of a concept of equivalent (averaged) material properties. The investigated object is treated as being homogeneous or is treated as being built of blocks in case of discontinuities of the material. The approach to modeling rock deformations is supported by a method known as the S-C method (Szostak-Chrzanowski et al. 2005). The method was developed to model the behaviour of brittle and evaporate rock material. In the case of brittle rock, it is modeled as a non-tensional material. Evaporates (e.g. salt rock), which have characteristics of viscous material, are modeled as a non-Newtonian liquid. The behaviour of the soil material, for instance in modeling behaviour of earth-filled dams, may be determined by using a non-linear hyperbolic model of the stress-strain relation developed by Kondner (1963).

6. Conclusions

Significant progress has been made at the CCGE in the development of fully automated monitoring systems and in the deterministic design and analysis of deformation surveys. The effects of changeable atmospheric conditions on geodetic measurements and the effects of improper calibration and poor reliability of in-situ geotechnical/structural instrumentation still remain as the main problems of current monitoring systems. Further research must be devoted to the development of integrated monitoring systems in which the two types of measurements complement each other which will increase reliability. Geodetic engineers should become acquainted with principles of continuum mechanics. They should utilize deterministic modeling of deformations in order to make sound decisions regarding the design and analysis of monitoring surveys.

Acknowledgements

The research reviewed in this paper has been supported by the Natural Sciences Research Council of Canada, Atlantic Canada Opportunities Agency, and the Canadian Wireless Telecommunications Association.

References

Beutler G, Bauersima I, Gurtner W, Rothacher M, Schildknecht T, Gieger A (1988) Atmospheric refraction and other important biases in GPS carrier phase observations. Atmospheric Effects on Geodetic Space Measurements, Monograph 12, School of Surveying, University of New South Wales, pp 15-43

Bond J, Chrzanowski A, Kim D (2007a) Bringing GPS into Harsh Environments for Deformation Monitoring. GPS Solutions (in press)

Bond J, Chrzanowski A, Kim D (2007b) Augmenting GPS with Pseudolites for Deformation Monitoring in Harsh Environments. Proceedings of the ION National Technical Meeting, January 22-24, San Diego, CA, Available at http://www.ion.org/ and on CD-ROM.

Bond J, Chrzanowski A, Wilkins F (2005) Using GPS for Augmenting Deformation Monitoring Systems in Open Pit Mines- Problems and Solutions. Geomatica, 59(1):73-82

Chen YQ (1983) Analysis of deformation surveys – A generalised method. Department of Geodesy and Geomatics Engineering, Technical Report No. 94, University of New Brunswick, Canada

Chen YQ, Chrzanowski A, Secord JM (1990) A strategy for the analysis of the stability of reference points in deformation surveys. CISM Journal, 44(2):141-149

Chen YQ, Zhang G, Ding X, Li Z (2000) Monitoring Earth Surface Deformations with InSAR Technology: Principle and Some critical Issues. Journal of Geospatial Engineering 2(1):3-21

Chrzanowski, A (1993) Modern Surveying Techniques for Mining and Civil Engineering, Chapter 33 in: Comprehensive Rock Engineering (Pergamon Press, Oxford, New York), 3:773-808

Chrzanowski A, Chen YQ, Secord JM (1983) On the strain analysis of tectonic movements using fault crossing geodetic surveys. Tectonophysics 97:297-315

Chrzanowski A, Chen YQ, Szostak-Chrzanowski A, Ogundare J (1994) Separability of Combined Deterministic and Geometrical Models of Deformation. Proceedings of the 20th FIG Congress, Melbourne, Australia, March 5-12, paper No. 652.1

Chrzanowski A, Frodge SL, Avella S (1993) The Worldwide Status of Monitoring and Analysis of Dam Deformations. Proceedings of the 7th Int. Symposium on Deformation Measurements and 6th Canadian Symposium on Mining Surveying, Banff, Alberta, May 2-6, pp 77-88

Chrzanowski A, Szostak-Chrzanowski A (1995) Identification of Dam Deformation Mechanism. Proceedings of the International Conference of the Malaysian Water Association on Dam Engineering (ed. J.S.Y. Tan), Kuala Lumpur, August 1-2, pp 179-187

Chrzanowski A, Szostak-Chrzanowski A (2004) Physical Interpretation of Ground Subsidence Surveys – A Case Study. Journal of Geospatial Engineering, pp 21-29

Chrzanowski A, Szostak-Chrzanowski A, Massiéra M, Whittaker C (2002) Monitoring and Numerical Modelling of Deformations of Large Dams – a Case Study. Journal of Technical Sciences, Publisher University of Warmia and Mazury, Olsztyn, pp 47-60

Chrzanowski A, Wilkins R (2006) Accuracy Evaluation of Geodetic Monitoring of Deformations in Large Open Pit Mines. Proceedings of the 3rd IAG Symposium on Geodesy for Geotechnical and Structural Engineering and 12th FIG Symposium on Deformation Measurements, eds. H. Kahmen and A. Chrzanowski, Baden, Austria, May 21-24, CD-ROM

Duffy M, Hill C, Whitaker C, Chrzanowski A, Lutes J, Bastin G (2001) An automated and integrated monitoring program for Diamond Valley Lake in California. Proceedings of the 10th Int. Symp. on Deformation Measurements , Orange, CA, March 19-22, pp K1-K8 Available at: http://rincon.gps.caltech.edu/FIG10sym/

Dunnicliff, J (1988) *Geotechnical Instrumentation for Monitoring Field performance*, John Wiley & Sons, New York, 577 p.

Kondner RL (1963) Hyperbolic stress-strain response: cohesive soils, Journal of the Soil Mechanics and Foundation Division, ASCE, 89(1963), (SM1):115-143

Leick A (1994) GPS Satellite Surveying, 2nd ed. John Wiley & Sons, Inc., New York.

Pieraccini M, Luzi G, Mecatti D, Noferini L, Macaluso G, Atzeni C (2006) Ground –Based Radar Interferometry for Monitoring Unstable Slopes. Proceedings of the 3rd IAG / 12th FIG Symposium on Deformation Measurements, eds. H. Kahmen and A. Chrzanowski, Baden, May 22-24, CD-ROM

Szostak-Chrzanowski A, Chrzanowski A (1991) Modeling and Prediction of Ground Subsidence using an Iterative Finite Element Method. Proceedings of the 4th International Symposium on Land Subsidence, Ed. A. I. Johnson, Houston, Texas, May 12-17, 200, pp 173-180

Szostak-Chrzanowski A, Chrzanowski A, and Massiéra M (2005) Use of Geodetic Monitoring Measurements in Solving Geomechanical Problems in Engineering and Geosciences. Engineering Geology 70(1-2), Application of Geodetic Techniques in Engineering Geology, ed: S. Stiros and A.Chrzanowski, June 3, pp 3-12

Szostak-Chrzanowski A, Massiéra M (2006) Relation between Monitoring and Design Aspects of Large Earth Dams. Proceedings of the 3rd IAG Symposium on Geodesy for Geotechnical and Structural Engineering and 12th FIG Symposium on Deformation Measurements, ed. H. Kahmen and A. Chrzanowski, Baden, Austria, May 21-24 , CD-ROM

Szostak-Chrzanowski A, Proszynski W, and Gambin W (2006) Continuum Mechanics as a Support for Deformation Monitoring, Analysis and Interpretation. Proceedings of the 3rd IAG Symposium on Geodesy for Geotechnical and Structural Engineering and 12th FIG Symposium on Deformation Measurements, ed. H. Kahmen and A. Chrzanowski, Baden, Austria, May 21-24, CD-ROM

Wilkins R, Chrzanowski A, Bastin G (2003) ALERT - A fully automated real time monitoring system. Proceedings of the 11th International (FIG) Symposium on Deformation Measurements, (ed, Stathis Stiros, Patras University, Greece), Santorini, Greece, May 25-28, pp. 209-216.

Zienkiewicz OC, Taylor RL (1989) The Finite Element Method. 4th ed., McGraw Hill, London.

Using GPS for Monitoring Landslides in Kala Reservoir of the Yalong River Area

Zhihong Xue[1], Guangyun Li[2], Zongchun Li[2], Xiaoping Wu[2] and Jiandong Wei[2]

[1]Department of Surveying and Instrumentation
[2]Department of Geodesy and Navigation
Zhengzhou Institute of Surveying and Mapping, 66 Longhai Road, Zhengzhou, Henan 450052, China
guangyun@public2.zz.ha.cn

Abstract

Landslide is a kind of serious geologic disaster. It's very important to monitor the deformation of the landslide area that threatens the lives or constructions. There are some successful applications in deformation monitoring using GPS. In certain circumstance, the continuous observation and real-time 3D coordinates with high precision can be carried out automatically using GPS. However, most of the landslide areas lie in canyons where the signals from GPS satellites are sheltered seriously. So in this case, to get precise and reliable results in deformation monitoring using GPS is a problem which needs being discussed and solved. In this paper, the landslides monitoring project in the Kala Reservoir area using GPS in China is introduced firstly. The design of deformation monitoring GPS network and the data processing method are discussed subsequently. The results of the deformation monitoring indicate that the requirements of the project are accomplished.

1 Introduction

Kala Reservoir of the Yalong River locates in Muli County of Sichuan Province, China. It lies in a mountainous area and fits for building of hydropower station since abundant of water resources are available. Geological information indicates that five huge landslides seated along the riversides for about 10 km, see Fig. 1. The stability of these landslides plays the key role in the selection of the dam's location. So the deformation of the landslide areas must be monitored in pre-study period of the hydropower station project so as to help choosing the optimal location of the dam.

Conventional geodetic techniques such as total station and leveling were adopted widely to monitor the deformation of landslide in the past and there were many successful experiences. However, in this mountainous area, it is not the good choice because of the heavy field works and the traffic conditions. So GPS is determined to monitor the deformation of the landslide area in this project.

The GPS has been used for deformation monitoring in many cases, especially for outer deformation monitoring of dams [3][4]. Although 3D deformation information with high precision can be derived automatically from permanent surveying GPS system, there are many problems in this project. Such as wide landslide area, signals sheltering due to the mountains along both sides of the river, pool electric power supply and traffic need be considered. Moreover, only 4 Trimble 5700 GPS receivers are available in the pre-study period. How to accomplish the requirements of the project needs further research.

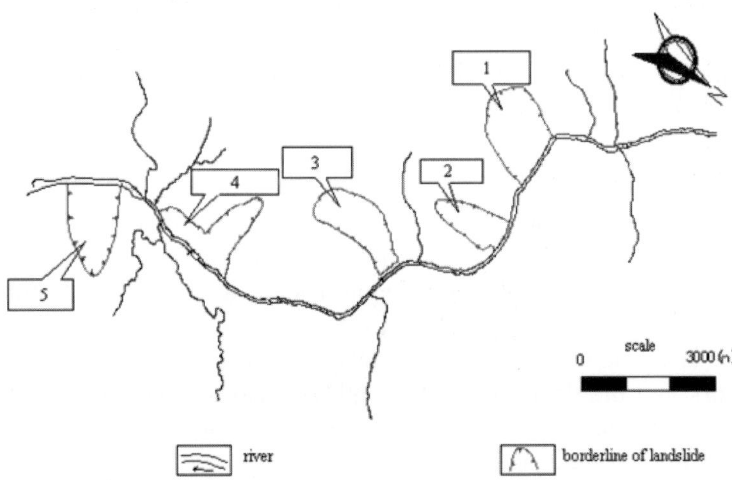

Fig.1 The distribution of landslides

2. Design of the Scheme

According to the specifications of deformation measurement and the condition of this project, the determined accuracy of coordinates is ±3mm. The GPS networks are divided into two orders i.e. reference network and monitoring network. The reference network covers the whole area of landslides and nearby each landslide there exists three reference points. The control points of reference network are made of concrete cylinder pillars with forced centering equipment and antenna orientation mark. In order to reduce the errors related to instruments, the same GPS receiver in the different measurement period occupies the same point. The orientation of antenna is pointed to the orientation mark and the height of antenna is measured precisely.

2.1 Design of the reference network

Reference network consists six reference points that are all constructed in bedrocks outside the deformation area. The distance from point II06 to II01 is about 10.5km. See Fig.2. There are 12 independent baselines in the net. Considering the traffic and accommodation of surveyor, GPS data with two 5 hours session is collected. Sampling interval is set to 10 seconds and elevation cut-off angle is 15° [5]. In order to improve the precision of baseline solution, the high precision GPS single positioning was conducted at II03, which is in the center of the network. The data was recorded for 24 hours continuously and its coordinates in WGS-84 were determined from 4 nearby IGS stations.

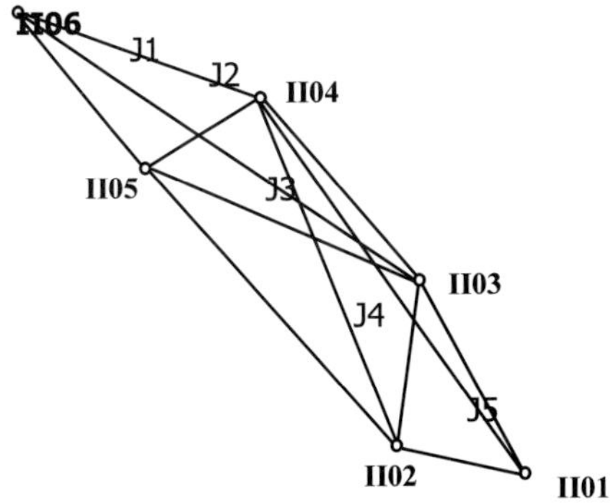

Fig.2 The reference network

2.2 Design of the monitoring network

There are total of 36 monitoring points in 5 landslides. The monitoring points in the same landslide make a monitoring network. So, there are total of 5 monitoring networks, which named J1 to J5, see Fig.1 and Fig.2. The point's distribution of J3 monitoring network is illustrated in Fig.3. Each monitoring network consists 3 reference points. For simplifying data processing and analysis, the observation and processing of each monitoring network is independently. Two GPS receivers are fixed on two reference points, and the other two receivers are occupied on each monitoring point and the 3rd reference point in turn to observe 2 hours. The sampling interval is set to 10 seconds and the elevation cut-off angle is 15°. The field observation time of each monitoring network is one day. For lack of the superfluous observation in monitoring network, the prediction of satellite ephemeris is necessary.

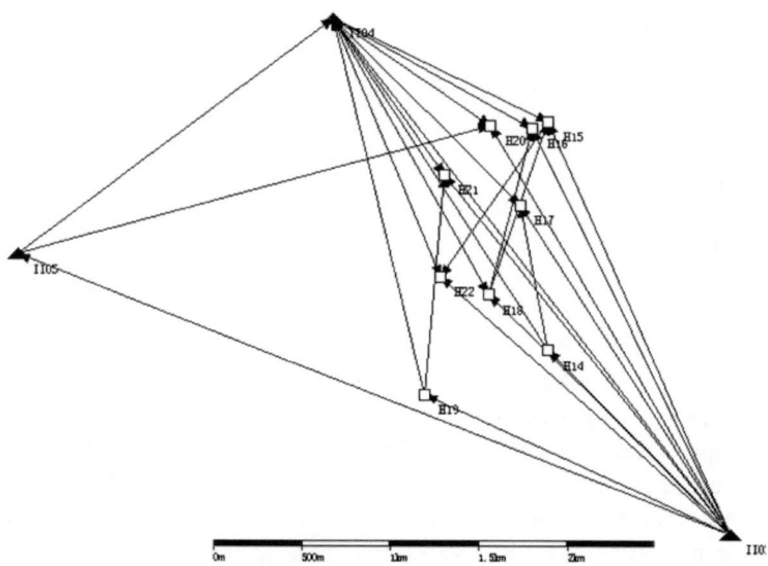

Fig. 3 J3 monitoring network

2.3 Accuracy of given data

The quality of GPS baseline solution is affected by the precision of coordinates of control point, which can be calculated approximately by

$$\frac{\Delta b}{b} \approx \frac{\Delta s}{r} \tag{1}$$

where Δs is the error in the direction of baseline of given coordinate, r is the average distance from GPS satellites to the earth, which is approximately 2×10^7 m, $\frac{\Delta b}{b}$ is the relative error of baseline.

According to the designed accuracy of this project and the approximate distance of baseline, if the error of baseline due to coordinate of given point is expected to less than 0.2mm, then the accuracy of coordinate of given point should be better than 0.4m.

2.4 Ephemeris to use

The position error of GPS due to satellite orbit can be given approximately by

$$\Delta x \approx \frac{1}{5} \cdot \frac{b}{r} \cdot \Delta X \tag{2}$$

where b is the distance of baseline, and r is the average distance from satellite to the earth, ΔX is the satellite orbit error, Δx is the error of baseline calculation.

In this project, the error of baseline due to satellite orbit is expected to less than 0.2 mm. The orbit error should be less than 2 m for 10 km baseline and less than 10m for 2km baseline. At present, IGS can provide precise ephemeris with orbit accuracy of about 5 cm, and without SA policy the accuracy of the broadcast ephemeris is approximately 5m [6]. So the precise ephemeris should be adopted when processing reference network while the broadcast ephemeris can be used in processing of monitoring network.

3. Data processing

3.1 Connection of IGS station and II03

The GPS data of point II03 was collected continuously with two 24 hours session separately in day of 212 and 214 in 2006. The coordinates were adjustment with the data of 4 nearby IGS stations. See Fig. 4. IGS precise ephemeris and Bernese software was adopted.

To verify the accuracy of solution, adjustment was conducted by fixing different IGS station and the results were compared. Selecting different IGS station that is set to fix, the maximum coordinate difference of point II03 in x, y and z direction is 6.5mm, 29mm and 10.5mm separately. So, the coordinate accuracy of II03 can be considered as higher than 2cm.

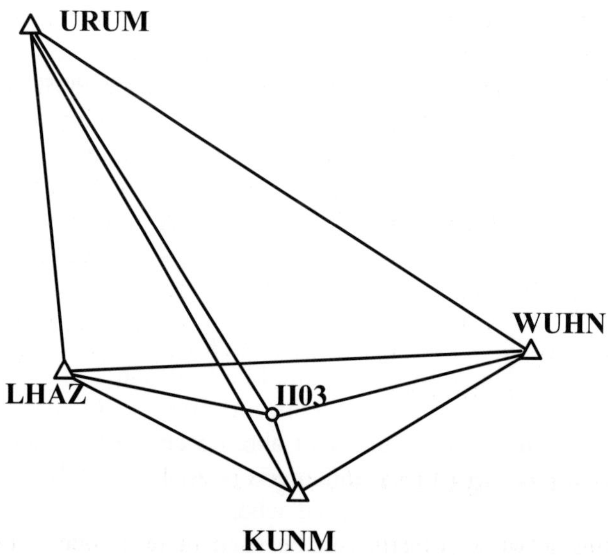

Fig. 4 Point II03 and IGS stations

3.2 Data processing of reference network

In reference network adjustment, the geocentric coordinate of II03 is set to fix. After verifying of simultaneous and independent observation loop, the free network adjustment was performed. The coordinates and RMS error of reference points in WGS84 are showed in table 1. The central meridian of Gauss-Krueger projection is selected to the approximate longitude of II03.

Table 1: Coordinates and accuracy of reference network

Point	North (m)	RMS (mm)	East (m)	RMS (mm)	Height(m)	RMS (mm)
II01	32832.7346	1.4	11571.9365	1.3	2404.6290	2.1
II02	33266.7077	1.0	9754.6430	1.0	2464.9201	1.5
II03	35856.2170	0.0	10085.9250	0.0	2614.1480	0.0
II04	38751.6940	1.0	7833.7605	0.9	2471.3274	1.4
II05	37423.7671	1.1	6032.5999	0.9	2955.2981	1.4
II06	40222.7303	1.6	4115.6822	1.5	2368.7253	2.1

Table 1 indicates that the accuracy in horizontal and height is about 2mm, which meets the specifications of the reference network.

The stability of reference network is the key to insure the reasonable deformation information, so the regular testing of the stability of the points in reference network is essential. If the testing is passed, the result of the first campaign of the reference network will be adopted as the datum of deformation calculation. If it is found that some points in the network have deformed, the deformation must be eliminate to ensure that the acquired deformation is reasonable. To test the stability of reference network between two campaigns, which have been finished yet, the classical method of deformation analysis is used. The results indicate that there is no remarkable deformation between two epochs.

3.3 Data processing of monitoring network

In monitoring network adjustment, the coordinate of one reference point must be fixed as the position datum of the network before baseline solution. Then all baseline observations are adjusted using free network adjustment in WGS84 coordinate system. The accuracy of monitoring network J3 is showed in Table 2. Since monitoring points are located in landslide, the view field is narrow and the loss of signal is happened frequently, the quality of observation is affected seriously. From the result in table 2, the accuracy of monitoring network is within the requirement of ±3mm.

Table 2: RMS of J3 using adjustment of free network

Point	RMS of North (mm)	RMS of East (mm)	RMS of Height (mm)
H14	0.9	0.9	1.8
H15	1.1	0.9	1.8
H16	0.9	0.9	1.4
H17	1.0	1.0	1.9
H18	0.8	0.8	1.3
H19	1.0	0.9	1.3
H20	0.9	0.8	1.3
H21	1.0	0.9	1.3
H22	1.0	1.0	1.6
II03	0.5	0.5	0.7
II04	0.5	0.4	0.7
II05	1.0	0.9	1.4

Fixing the position datum in network adjustment is not enough for deformation monitoring data processing using GPS [1]. The coordinate results of networks obtained in different periods could have a tiny variation in azimuth and scale due to some reasons such as the satellites orbit error, atmospheric refraction, measurement error etc. So, the deformation information got from the adjustment by fixing the position datum will contain the influence of the datum change in azimuth and scale. To get the correct deformation value, the variation in azimuth and scale must be eliminated obviously.

Taking the J3 monitoring network for example, II03, II04, II05 were selected as common points and the coordinate transformation parameters were computed between two sets of coordinates corresponding to reference network and J3 monitoring network using free network adjustment. The seven parameters are showed in table 3.

Table 3. Transformation parameters between reference network and J3 monitoring network

X_0	-1.4550(m)	θ_X	0.7425535323(s)
Y_0	0.9077(m)	θ_Y	-0.3085309295(s)
Z_0	4.7255(m)	θ_Z	0.0605765202(s)
Scale	0.461896(ppm)	M_0	1.2(mm)

The result in Table 3 confirms that there exist small changes between the two sets of coordinates. The coordinate translation is the main factor. However, rotation and scale parameter could not be neglected. For the general land survey, the translation factor is normally taken into account. But for the deformation monitoring with high accuracy requirement, the datum change in azimuth and scale should also be considered.

In practice, after the free network adjustment, the coordinates of monitoring network must be transformed to the datum of the first epoch reference network. Then the reasonable deformation information can be obtained.

Five campaigns of monitoring observation have been accomplished in Kala Reservoir area to the present. The coordinate changes of monitoring points in J3 network are showed in Fig. 5. It indicates that after the second period campaign most of the points in this landslide have uprising tendency and then down slowly. Geological analysis confirms that the reason is the explosion for geological exploration in nearby area. Eliminating this affect, the tendency of the monitoring points in J3 landslide is falling to the northeast in general, which means that this landslide is sliding toward the direction of Yalong River slowly, especially for the point No.15 which located in the front of the landslide. The deformation result obtained by GPS is verified by the result of geological methods, which is got by the strain sensor and inclinometer measurements in the same area.

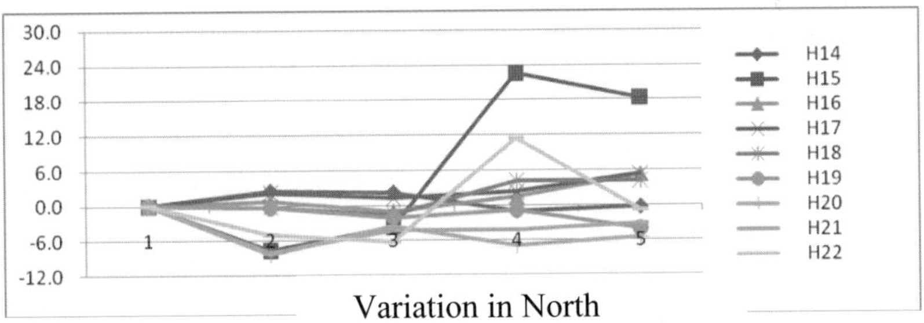

Variation in North

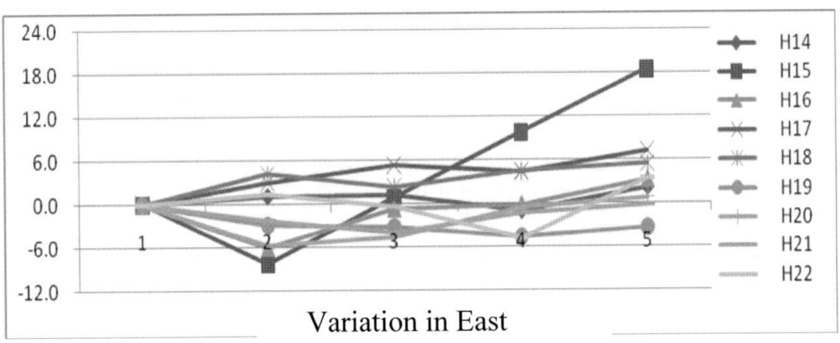

Variation in East

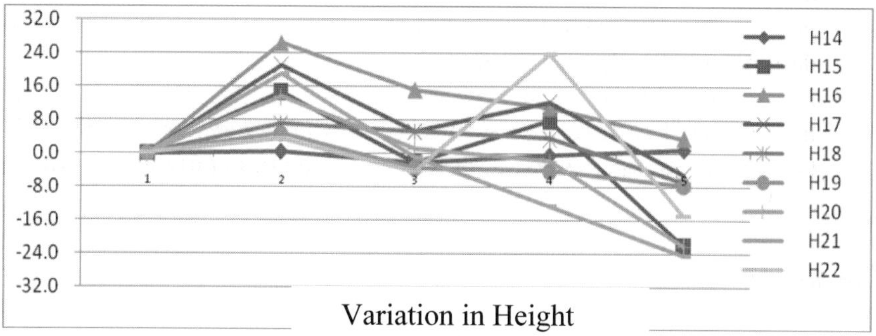

Fig. 5 Deformation trend of monitoring points (mm)

4. Conclusions

The requirements of landslide monitoring are reached by the optimal scheme design and precise data processing in a wide range area using GPS static survey technique. From this project, some experiences are concluded as followings.

1. For landslides extended to 10km, the accuracy specification of landslide monitoring can be met using GPS by reasonable scheme design and strict field operation even in the area of canyons.
2. Long baseline should be avoided in GPS network design. For the baseline of about 7 km, the accuracy of ± 2 mm could be reached by 5 hours continuous occupation. It is a good practice to divide the whole landslide area into a number of smaller parts. The observation and processing of those smaller networks can be carried out independently, which can avoid the interaction between those nets and simplifying the field observation and data processing.
3. The errors caused by the difference between instruments can be eliminated by occupying same point using same instrument, fixing antenna to the same direction and measuring antenna height with high precision. If possible, antenna joint staff with fixed length is recommended to use, which can remove the error of antenna height.
4. Ephemeris prediction is important before observation, especially for monitoring network that has no sufficient redundant observation or only a few instruments are available and satellite signals are sheltered seriously. The time span must be avoided when GDOP is poor.
5. To get the reliable deformation value from serial campaigns, the same coordinate datum should be adopted in data processing. Generally speaking, the stable reference points in engineering deformation

monitoring network could be found, and the coordinates of these points got in the first period of reference network can be used as the datum for subsequent adjustment.

6. In the areas where the sky view is narrow and the signals from satellites are sheltered seriously just like in this project, integrated GPS-GLONASS dual-system receiver is expected to get higher accuracy.

References

Huang QH (2006) The application of GPS in engineering deformation monitoring, Western China Exploration Engineering (in Chinese), 7: 231-233

Industrial Standard GB/T 18314-2001of the People's Republic of China, Specifications for Global Positioning System (GPS) Surveys (in Chinese).

Li CH, Zhang LY, Cai MF (2005) Datum design and data processing of landslip monitoring using GPS, Colored Metal, 3: 33-35

Li ZH, Xu SQ (2002) Recent advancements of GPS, Journal of Geomatics, 3: 32-35

Wang J, Zhang HH, Xu RL (2001) Data processing method in GPS deformation monitoring, Jiangsu Surveying and Mapping (in Chinese), 4: 12-14

Yang RG, Ou JK, Wen BD (2006) GPS broadcast ephemeris error and its effect on positioning, Journal of Geomatics, 1: 1-3

Contributors

Rifaat Abdalla, PhD Student, GeoICT Lab, Department of Earth and Space Science and Engineering, York University, 4700 Keele Street, Toronto, Ontario M3J 1P3, Canada, rifaat.abdalla@gmail.com

Mohamed Bakillah, PhD Student, Department of Geomatics Sciences, Laval University, 1342 Pavillon Louis-Jacques-Casault, Quebec City, Quebec G1K 7P4, Canada, mohamed.bakillah.1@ulaval.ca

John Barlow, PhD, Post Doctoral Research Fellow, University of Saskatchewan, Saskatoon, Saskatchewan S7N 5A5, Canada, john.barlow@usask.ca

Yvan Bédard, PhD, Professor, Department of Geomatics Sciences, Laval University, 1342 Pavillon Louis-Jacques-Casault, Quebec City, Quebec G1K 7P4, Canada, yvan.bedard@scg.ulaval.ca

Jason Bond, PhD Student, Department of Geodesy and Geomatics Engineering, University of New Brunswick, P.O. Box 4400, Fredericton, New Brunswick E3B 5A3, Canada.

Jean Brodeur, Centre for Research in Geomatics, Laval University, 1342 Pavillon Louis-Jacques-Casault, Quebec City, Quebec G1K 7P4, Canada.

Yves A. Buehler, PhD Student, NPOC Remote Sensing Laboratories, Department of Geography, University of Zurich, Winterthurerstrasse 190, 8057 Zurich, Switzerland, yves.buehler@geo.uzh.ch

Enrico Cadau, The San Marco Project Research Centre, University of Rome, Via Salaria, 851, 00138 Rome, Italy

Rosaline R. Canessa, PhD, Assistant Professor, Department of Geography, University of Victoria, PO Box 3050 STN CSC, BC V8W 3P5, Canada, rosaline@uvic.ca

Daniele Cerra, European Union Satellite Centre (EUSC), Apdo de Correos, 511. E28850 Torrejon de Ardoz, Madrid, E-28850, Spain.

Zheng Chang, PhD Student, Department of Civil Engineering, Ryerson University, 350 Victoria Street, Toronto, Ontario M5B 2K3, Canada, zchang@ryerson.ca

Michael A. Chapman, PhD, Professor, Department of Civil Engineering, Ryerson University, 350 Victoria Street, Toronto, Ontario M5B 2K3, Canada, mchapman@ryerson.ca

Dongmei Chen, PhD, Assistant Professor, Department of Geography, Queen's University, Kingston Ontario K7L 3N6, Canada, chendm@post.queensu.ca

Zhi Chen, PhD, Assistant Professor, Department of Building, Civil, and Environmental Engineering, Faculty of Engineering and Computer Science, Concordia University, Montreal, Quebec H3G 1M8, Canada, zhichen@alcor.concordia.ca

Quiming Cheng, PhD, Professor, Department of Earth and Space Science and Engineering, Department of Geography, York University, Toronto, Ontario M3J 1P3, Canada, qiuming@yorku.ca

Adam Chrzanowski, PhD, Professor Emeritus, Department of Geodesy and Geomatics Engineering, University of New Brunswick, P.O. Box 4400, Fredericton, New Brunswick E3B 5A3, Canada, adamc@unb.ca

Antonio De la Cruz, European Union Satellite Centre (EUSC), Apdo de Correos, 511. E28850 Torrejon de Ardoz, Madrid, E-28850, Spain.

Mahmoud Reza Delavar, PhD, Assistant Professor and Head, Department of Surveying and Geomatics Engineering, Engineering Faculty, University of Tehran, North Kargar, 11365/4563, Iran, mdelavar@ut.ac.ir

Xiaoli Ding, PhD, Professor, Department of Land Surveying & Geo-Informatics, The Hong Kong Polytechnic University, Hung Hom, Kowloon, Hong Kong, China, lsxlding@polyu.edu.hk

Andrew Farrar, Department of Geography, Queen's University, Kingston Ontario K7L 3N6, Canada

Andrew U. Frank, PhD, Professor and Head, Institute of Geoinformation and Cartography, Technical University of Vienna, Gusshausstr. 27-29 E127, A-1040 Vienna, Austria, frank@geoinfo.tuwien.ac.at

Steven E. Franklin, PhD, Professor and Vice-President (Research), University of Saskatchewan, Saskatoon, Saskatchewan S7N 5A5, Canada, steven.franklin@usask.ca

M.J. Garcia, European Union Satellite Centre (EUSC), Apdo de Correos, 511. E28850 Torrejon de Ardoz, Madrid, E-28850, Spain.

Jiming Guo, PhD, Professor, School of Geodesy and Geomatics, Wuhan University, Wuhan, Hubei 430079, China

Armin Grün, PhD, Professor, Institute for Geodesy and Photogrammetry, Swiss Federal Institute of Technology (ETH) Zürich, CH-8093 Zürich, Switzerland, agruen@geod.baug.ethz.ch

L. Hajibabai, Center of Excellence in Geomatics Engineering and Disaster Management, Department of Surveying and Geomatics Engineering, Engineering Faculty, University of Tehran, North Kargar, 11365/4563, Iran, lbabai@geomatics.ut.ac.ir

Leila Hashemi Beni, PhD Student, Department of Geomatics Sciences, Laval University, 1342 Pavillon Louis-Jacques-Casault, Quebec City, Quebec G1K 7P4, Canada, Leila.hashemi.1@ulaval.ca

Quanyi Huang, PhD, Professor, Center of Public Safety Research, Tsinghua University, Beijing 100084, China

Gracia Joyanes, European Union Satellite Centre (EUSC), Apdo de Correos, 511. E28850 Torrejon de Ardoz, Madrid, E-28850, Spain.

Zhizhong Kang, PhD, Postdoctoral Researcher, Section of Optical and Laser Remote Sensing, Department of Earth Observation and Space Systems, Faculty of Aerospace Engineering, Delft University of Technology, Kluyverweg 1, 2629 HS Delft, The Netherlands, z.kang@tudelft.nl

C. Peter Keller, PhD, Professor and Dean of the Faculty of Social Sciences, University of Victoria, PO Box 3050 STN CSC, BC V8W 3P5, Canada, pkeller@uvic.ca

Tobias W. Kellenberger, PhD, Senior Scientist, NPOC Remote Sensing Laboratories, Department of Geography, University of Zurich, Winterthurerstrasse 190, 8057 Zurich, Switzerland, tobias.kellenberger@geo.uzh.ch

Giovanni Laneve, The San Marco Project Research Centre, University of Rome, Via Salaria, 851, 00138 Rome, Italy

G. D. Lewis, Instructor, School of Agriculture & Environment, Assiniboine Community College, 1430 Victoria Avenue East Brandon, Manitoba R7A 2A9, Canada, lewisG@Assiniboine.net

Guangyun Li, Professor and Chair, Department of Geodesy and Navigation, Zhengzhou Institute of Surveying and Mapping, 66 Longhai Road, Zhengzhou, Henan 450052, China, guangyun@public2.zz.ha.cn

Hui Li, Department of Geodesy and Geomatics Engineering, University of New Brunswick, P.O. Box 4400, Fredericton, New Brunswick E3B 5A3, Canada

Jonathan Li, PhD, Associate Professor, Department of Geography, University of Waterloo, 200 University Avenue West, Waterloo, Ontario N2L 3G1, Canada, junli@uwaterloo.ca

Songnian Li, PhD, Associate Professor, Department of Civil Engineering, Ryerson University, 350 Victoria Street, Toronto, Ontario M5B 2K3, Canada, snli@ryerson.ca

Yu Li, PhD Student, Department of Geography, University of Waterloo, 200 University Avenue West, Waterloo, Ontario N2L 3G1, Canada, y62li@fes.uwaterloo.ca

Zongchun Li, Department of Geodesy and Navigation, Zhengzhou Institute of Surveying and Mapping, 66 Longhai Road, Zhengzhou, Henan 450052, China,

Alan M. MacEachren, PhD, Professor and Director, The GeoVISTA Center, Department of Geography, Pennsylvania State University, 302 Walker Building, University Park, PA 16802, USA, maceachren@psu.edu

M. R. Malek, PhD, Assistant Professor, Center of Excellence in Geomatics Engineering and Disaster Management, Department of Surveying and Geomatics Engineering, Engineering Faculty, University of Tehran, North Kargar, 11365/4563, Iran, malek@ncc.neda.net.ir

Cindy A. Marven , PhD Student, Spatial Sciences Laboratories, Department of Geography, University of Victoria, PO Box 3050 STN CSC, BC V8W 3P5, Canada, cmarven@shaw.ca

Marcin Mieleczwky, European Union Satellite Centre (EUSC), Apdo de Correos, 511. E28850 Torrejon de Ardoz, Madrid, E-28850, Spain.

Mir Abolfazl Mostafavi, PhD, Assistant Professor, Department of Geomatics Sciences, Laval University, 1342 Pavillon Louis-Jacques-Casault, Quebec City, Quebec G1K 7P4, Canada, Mir-Abolfazl.Mostafavi@scg.ulaval.ca

Ronald Pelot, PhD, Associate Professor, Department of Industrial Engineering, Dalhousie University, PO Box 1000, Halifax, Nova Scotia B3J 2X4, Canada, ronald.pelot@dal.ca

Jianbing Peng, PhD, Professor and Dean, School of Geological and Geomatics Engineering, Chang'an University, 126 Yanta Road, Xian, Shaanxi 710054, China

Scott Pezanowski, Research Analyst , The GeoVISTA Center, Department of Geography, Pennsylvania State University, 302 Walker Building, University Park, PA 16802, USA, spezanowski@psu.edu

Jacynthe Pouliot, PhD, Professor, Department of Geomatics Sciences, Laval University, 1342 Pavillon Louis-Jacques-Casault, Quebec City, Quebec G1K 7P4, Canada, Jacynthe.Pouliot@scg.ulaval.ca

Mehdi Rezaeian, PhD Student, Institute for Geodesy and Photogrammetry, Swiss Federal Institute of Technology (ETH) Zürich, CH-8093 Zürich, Switzerland, mehdi.rezaeian@geod.baug.ethz.ch

Giancarlo Santilli, The San Marco Project Research Centre, University of Rome, Via Salaria, 851, 00138 Rome, Italy

Jamal Shahrabi, PhD, Assistant Professor, Faculty of Industrial Engineering, Amirkabir University of Technology, Tehran, Iran, jamalshahrabi@aut.ac.ir

Sharmin Sultana, MASc Student, Department of Building, Civil, and Environmental Engineering, Faculty of Engineering and Computer Science, Concordia University, Montreal, Quebec H3G 1M8, Canada, s_sulta@encs.concordia.ca

Anna Szostak-Chrzanowski, PhD, Adjunct Professor and Senior Research Associate, Department of Geodesy and Geomatics Engineering, University of New Brunswick, P.O. Box 4400, Fredericton, New Brunswick E3B 5A3, Canada, amc@unb.ca

C. Vincent Tao, PhD, GeoICT Lab, Department of Earth and Space Science and Engineering, York University, 4700 Keele Street, Toronto, Ontario M3J 1P3, Canada, tao@yorku.ca

Brian Tomaszewski, PhD Student, The GeoVISTA Center, Department of Geography, The Pennsylvania State University, 302 Walker Building, University Park, PA 16802 USA, bmt139@psu.edu

Albrecht Weiser, Institute for Spatial Information and Surveying Technology, Department for Geoinformatics and Surveying, FH Mainz, University of Applied Sciences, Holzstrasse 36, 55116 Mainz Germany.

Renqiang Wen, PhD Student, School of Geodesy and Geomatics, Wuhan University, Wuhan, Hubei 430079, China

Rick Wilkins, PhD Student, Department of Geodesy and Geomatics Engineering, University of New Brunswick, P.O. Box 4400, Fredericton, New Brunswick E3B 5A3, Canada, rwilkins@unb.ca

Xiaoping Wu, Professor, Department of Geodesy and Navigation, Zhengzhou Institute of Surveying and Mapping, 66 Longhai Road, Zhengzhou, Henan 450052, China, geowxp@gmail.com

Yan Wu, PhD Student, Department of Industrial Engineering, Dalhousie University, PO Box 1000, Halifax, Nova Scotia B3J 2X4, Canada, ywu3@dal.ca

Pingping Xie, Department of Geodesy and Geomatics Engineering, University of New Brunswick, P.O. Box 4400, Fredericton, New Brunswick E3B 5A3, Canada

Yubin Xin, PhD Student, Department of Earth and Space Science and Engineering, York University, Toronto, Ontario M3J 1P3, Canada, xin@yorku.ca

Wei Xu, PhD Student, Section GIS Technology, OTB Research Institute for Housing, Urban and Mobility Studies, Delft University of Technology, The Netherlands, wei.xu@tudelft.nl

Zhihong Xue, Lecturer, Department of Surveying Engineering and Instrumentation, Zhengzhou Institute of Surveying and Mapping, 66 Longhai Road, Zhengzhou, Henan 450052, China

Jianqing Zhang, Professor, School of Remote Sensing and Information Engineering, Wuhan University, Wuhan, Hubei 430079, China, jqzhang@supresoft.com.cn

Qin Zhang, PhD, Professor and Associate Dean, School of Geological and Geomatics Engineering, Chang'an University, 126 Yanta Road, Xian, Shaanxi 710054, China, zhangqinle@263.net.cn

Yun Zhang, PhD, Associate Professor, Department of Geodesy and Geomatics Engineering, University of New Brunswick, P.O. Box 4400, Fredericton, New Brunswick E3B 5A3, Canada, yunzhang@unb.ca

Zuxun Zhang, Professor, School of Remote Sensing and Information Engineering, Wuhan University, Wuhan, Hubei 430079, China, zxzhang@supresoft.com.cn

Chaoying Zhao, PhD Student, School of Geological and Geomatics Engineering, Chang'an University, 126 Yanta Road, Xian, Shaanxi 710054, China, zhaochaoying@163.com

Qiansheng Zhao, PhD Student, School of Geodesy and Geomatics, Wuhan University, Wuhan, Hubei 430079, China, zh_qsh@hotmail.com

Shaobo Zhong, PhD Student, Center of Public Safety Research, Tsinghua University, Beijing 100084, China

Alexander Zipf, PhD, Professor, Institute for Spatial Information and Surveying Technology, Department for Geoinformatics and Surveying, FH Mainz, University of Applied Sciences, Holzstrasse 36, 55116 Mainz Germany, zipf@geoinform.fh-mainz.de

Sisi Zlatanova, PhD, Assistant Professor, Section GIS Technology, OTB Research Institute for Housing, Urban and Mobility Studies, Delft University of Technology, The Netherlands, s.zlatanova@tudelft.nl

Printing: Krips bv, Meppel
Binding: Stürtz, Würzburg